SYSTEM MODELLING AND OPTIMIZATION

IFIP - The International Federation for Information Processing

IFIP was founded in 1960 under the auspices of UNESCO, following the First World Computer Congress held in Paris the previous year. An umbrella organization for societies working in information processing, IFIP's aim is two-fold: to support information processing within its member countries and to encourage technology transfer to developing nations. As its mission statement clearly states,

IFIP's mission is to be the leading, truly international, apolitical organization which encourages and assists in the development, exploitation and application of information technology for the benefit of all people.

IFIP is a non-profitmaking organization, run almost solely by 2500 volunteers. It operates through a number of technical committees, which organize events and publications. IFIP's events range from an international congress to local seminars, but the most important are:

- The IFIP World Computer Congress, held every second year;
- open conferences;
- working conferences.

The flagship event is the IFIP World Computer Congress, at which both invited and contributed papers are presented. Contributed papers are rigorously refereed and the rejection rate is high.

As with the Congress, participation in the open conferences is open to all and papers may be invited or submitted. Again, submitted papers are stringently refereed.

The working conferences are structured differently. They are usually run by a working group and attendance is small and by invitation only. Their purpose is to create an atmosphere conducive to innovation and development. Refereeing is less rigorous and papers are subjected to extensive group discussion.

Publications arising from IFIP events vary. The papers presented at the IFIP World Computer Congress and at open conferences are published as conference proceedings, while the results of the working conferences are often published as collections of selected and edited papers.

Any national society whose primary activity is in information may apply to become a full member of IFIP, although full membership is restricted to one society per country. Full members are entitled to vote at the annual General Assembly, National societies preferring a less committed involvement may apply for associate or corresponding membership. Associate members enjoy the same benefits as full members, but without voting rights. Corresponding members are not represented in IFIP bodies. Affiliated membership is open to non-national societies, and individual and honorary membership schemes are also offered.

SYSTEM MODELLING AND OPTIMIZATION

Methods, Theory and Applications

19th IFIP TC7 Conference on
System Modelling and Optimization
July 12–16, 1999, Cambridge, UK

Edited by

M.J.D. Powell
Department of Applied Mathematics and Theoretical Physics
University of Cambridge, UK

S. Scholtes
Judge Institute of Management Studies
University of Cambridge, UK

KLUWER ACADEMIC PUBLISHERS
BOSTON / DORDRECHT / LONDON

Distributors for North, Central and South America:
Kluwer Academic Publishers
101 Philip Drive
Assinippi Park
Norwell, Massachusetts 02061 USA
Telephone (781) 871-6600
Fax (781) 871-6528
E-Mail <kluwer@wkap.com>

Distributors for all other countries:
Kluwer Academic Publishers Group
Distribution Centre
Post Office Box 322
3300 AH Dordrecht, THE NETHERLANDS
Telephone 31 78 6392 392
Fax 31 78 6546 474
E-Mail <services@wkap.nl>

Electronic Services <http://www.wkap.nl>

Library of Congress Cataloging-in-Publication Data

IFIP TC7 Conference on System Modelling and Optimization (19th : 1999 : Cambridge, England)
System modelling and optimization : methods, theory, and applications : 19th IFIP TC7 Conference on System Modelling and Optimization, July 12-16, 1999, Cambridge, UK / edited by M.J.D. Powell, S. Scholtes
p. cm. — (IFIP ; 46)
Includes bibliographical references.
ISBN 0-7923-7881-4 (alk. paper)
1. Control theory—Congresses. 2. Mathematical optimization—Congresses. 3. Automatic control—Congresses. I. Powell, M.J.D. (Michael James David), 1936– II. Scholtes, Stefan. III. Title. IV. International Federation for Information Processing (Series) ; 46.

QA402.3 .I4536 1999
629.8'312—dc21 00-033105

Printed on acid-free paper.

Printed in the United States of America.

Contents

Foreword

IFIP TC7 Conferences on System Modelling and Optimization are held every two years. Their subjects cover a wide range of methods, theory and applications. The present volume contains most of the invited papers and a few of the submitted ones that were presented at the 19th Conference, held in Cambridge, England from July 12th to 16th, 1999.

The meeting was attended by about 210 participants from 37 countries. 10 invited and 143 submitted talks were presented. These proceedings, however, include only 14 of the papers, in order that each article can be long enough to provide a coherent and substantial contribution to knowledge. Therefore all of the work that was submitted for possible publication was judged by unusually high standards. The editors are very grateful to the referees who helped in this way. A list of the other 139 papers that do not appear now is also given. Most of them were not offered to the proceedings, because they addressed work in progress or will be published elsewhere.

The official co-sponsors of the conference were IFIP (International Federation for Information Processing) and DAMTP (Department of Applied Mathematics and Theoretical Physics, University of Cambridge, UK). Their support was crucial and most welcome. Further, the following companies and institutions made generous donations: Barrodale Computing Services (Canada), British Computer Society (UK), Cambridge University Press (UK), DOT Products (USA), General Motors (USA), IBM Unternehmensberatung (Germany), Kluwer Academic Publishers (The Netherlands), London Mathematical Society (UK), Nomura International (UK), Numerical Algorithms Group (UK), Pareto Partners (UK) and Terrasciences (USA). Their sponsorship was highly important to the academic excellence of the programme, to the participation from a wide range of countries, to the enjoyment of the social activities and to the publication of these proceedings.

We also ask many individuals to accept our thanks. The registration desk helpers come to mind immediately, namely Caroline Powell, Catherine (and Georgia) Powell, Alice Powell and Ortelia Bejancu. We are grateful for the addresses by Alec Broers (Vice Chancellor of Cambridge University), David Hartley (President of the British Computer Society) and Peter Kall (Chairman of IFIP TC7) at the Opening Session and also for the speech by Josef Stoer at the conference dinner. Further, the invited speakers gave some brilliant talks and the authors of the contributed papers enhanced the quality of the occasion. We received very valuable assistance from the University of Cambridge through staff on the New Museums Site, in DAMTP and in the Judge Institute of Management Studies. We are indebted to the University Centre and to Pembroke College, especially Paula Hunt and Ken Smith, for the success of the social events. Finally, every word that you read in these proceedings has been produced in its present form by Hans-Martin Gutmann. It is a pleasure to acknowledge all of these contributions.

M.J.D. Powell Cambridge
S. Scholtes

TOPOLOGY DESIGN OF STRUCTURES, MATERIALS AND MECHANISMS – STATUS AND PERSPECTIVES

Martin P. Bendsøe
Department of Mathematics
Technical University of Denmark
DK-2800 Lyngby, Denmark
M.P.Bendsoe@mat.dtu.dk

1. INTRODUCTION

The field of variable-topology shape design in structural optimization has its origins in theoretical studies of existence of solutions in variational problems, in particular shape optimization problems, and in studies in theoretical material science of variational bounds on material properties. Progress in computational methods and the ever increasing computer power has oriented the area towards applications, with significant developments being achieved over the last decade, leading to a fairly widespread use of the methodology in industry.

In this short paper we outline some of the basic ideas and methods of existing methods, but it is not our purpose to cover all work and approaches in this field. Instead we refer to existing literature containing rather comprehensive surveys, see e.g., [7, 8, 31]. Moreover, note that reference is mostly made to recent papers that include bibliographies useful for on overview of the area. Thus the presentation does not try to present a complete historical perspective.

The area of computational variable-topology shape design of continuum structures is presently dominated by methods which employ a material distribution approach for a fixed reference domain, in the spirit of the so-called 'homogenization method' for topology design ([3, 9]). That is, the geometric representation of a structure is similar to a grey-scale rendering of an image, in discrete form corresponding to a raster representation of the geometry on a fixed reference domain. The physics of the problem is also represented by boundary conditions and forcing

M.J.D. Powell and S. Scholtes (Eds.), *System Modelling and Optimization: Methods, Theory and Applications.*

terms defined on this fixed reference domain, much in analogy to fictitious domain methods for FEM analysis.

One can normally distinguish between three versions of raster based geometry models for continuum topology optimization. The basic problem is an unrestricted '0-1' integer design problem (generalized shape optimization), that is, a design specifies unambiguously whether there is solid material or void at every point in a candidate design region. Otherwise, there are no restrictions on the shape. Unfortunately, in general, this class of problems is ill-posed in the continuum setting (cf., [11, 20]). Well-posed problems can be obtained by either extending the space of admissible solutions to obtain their relaxed versions, usually by incorporating microstructure (see, e.g., [3]), or by restricting the space of admissible solutions. The latter can be accomplished by enforcing an upper bound on the perimeter of the structure (see [28], and references therein), by imposing constraints on the slopes of the parameters defining geometry (see [29], and references therein), by the introduction of a filtering function limiting the minimum scale (see [10, 36] for an overview), or one can introduce a ground structure with a fixed number of design degrees of freedom (like a fixed mesh for design). Here the first three restriction methods are well-posed in the setting of a continuum description of the design problem, while in the latter case existence relies entirely on the finite dimension.

Relaxation usually yields a set of continuously variable design fields to be optimized over a fixed domain, so the algorithmic problems associated with the discrete 0-1 format of the basic problem statement are circumvented. The continuum relaxation approach can be very involved theoretically (see for example [1, 13]) and much work is still needed in this area.

The restriction approach leads to 'classical designs' and there is no ambiguity as to the physical modelling (local material response is determined solely by the presence or absence of the given solid material). However, the major challenge is the solution of a large-scale integer programming problem. Due to the high cost of function calls for these problems, solving the 0-1 formulation directly, for example by genetic algorithms or simulated annealing, is, in general, not viable for very large-scale problems, and this is an area that should have more focus in the coming years (see also below). Another – and the most commonly used – approach is to replace the integer variables with continuous variables and then introduce some form of penalty that steers the solution to discrete 0-1 values. A key part of these methods is the introduction of interpolation functions (often interpreted as material densities) that express various physical quantities (e.g., material stiffness, cost, etc.) as a function of the continuous variables. Moreover, geometric properties also require suitable interpretation. Although there is a strong resemblance to the relaxed formulations, it is important to recognize that the continuous format is then merely part of a computational strategy which does not alter the ultimate goal of solving an integer problem, i.e., to obtain a black-and-white design.

2. BASIC PROBLEM STATEMENT

In continuum topology design we seek the optimal distribution of material in a fixed reference domain Ω in $\mathbf{R}^2$ or $\mathbf{R}^3$, with the term 'optimal' being defined through choice of objective and constraint functions. The objective and constraint functions involve some kind of physical modelling that provides a measure of efficiency within the framework of a given area of applications, here structural mechanics. Here we thus consider a mechanical element as a body occupying a domain Ω^m which is part of Ω on which applied loads and boundary conditions are defined (this reference domain is often called the ground-structure). Referring to the reference domain Ω we can define an example problem as a minimization of force times displacement, over admissible designs and displacement fields satisfying equilibrium (the minimum compliance problem):

$$
\begin{aligned}
&\min_{C,u} \int_\Omega f u \, d\Omega \\
&\text{subject to:} \\
&\qquad \int_\Omega C_{ijkl}(x)\epsilon_{ij}(u)(x)\epsilon_{kl}(v)(x)\, d\Omega \;=\; \int_\Omega f v \, d\Omega \\
&\qquad\qquad\qquad\qquad\qquad\qquad\qquad\qquad \text{for all } v \in U, \\
&\qquad\qquad\qquad\qquad C_{ijkl} \;\in\; E_{\text{ad}}
\end{aligned}
\tag{1}
$$

Here the equilibrium equation is written in its weak, variational form, with U denoting the space of kinematically admissible displacement fields, u the equilibrium displacement, f the forces, and $\epsilon(u)$ linearized strains. The rigidity tensor C_{ijkl} is the design variable of our problem and various definitions of the set of admissible rigidity tensors is what distinguishes various settings for the design problem. This type of problem is what is often labelled as a problem of 'control in the coefficients' (c.f., [22]), here with the controls entering in the high order part of the governing differential operator.

A classical variant of problem (1) is the so-called variable thickness problem, where the set of admissible designs is defined through a thick-

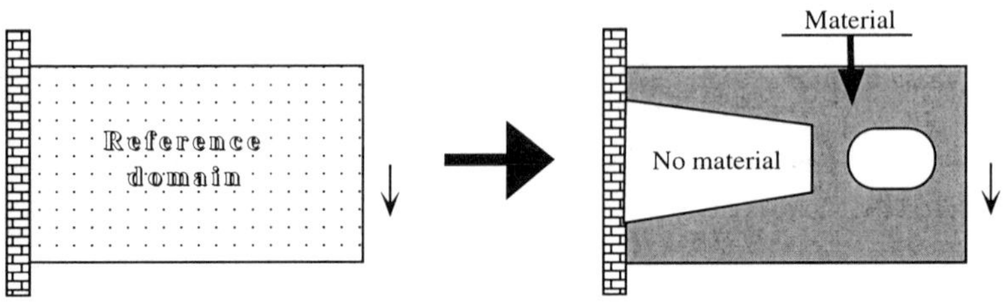

Figure 1 The generalized shape design problem of finding the optimal material distribution.

ness function h:

$$C_{ijkl}(x) = h(x)C^0_{ijkl}, \qquad h \in L^\infty(\Omega),$$
$$\int_\Omega h \, d\Omega = V, \qquad 0 \le h_{min} \le h \le h_{max}, \tag{2}$$

for given material properties C^0_{ijkl} and given volume V. This problem is rather unique, as one here has existence of solution in the 'naive' setting, i.e., in the formulation which immediately springs to mind when modelling the problem (see [27] and references therein).

For a topology design setting, defined through designs given as domains Ω^m of material points, the admissible designs are defined by a point-wise volume fraction of a given material, and this density can only attain the values zero or one (a black-and-white design), c.f., Figure 1:

$$\begin{aligned} &C_{ijkl}(x) = \Theta(x)C^0_{ijkl}, \\ &\Theta(x) = 1, \ x \in \Omega^m; \quad \Theta(x) = 0, x \in \Omega \setminus \Omega^m \\ &\mathrm{Vol}(\Omega^m) = \int_\Omega \Theta(x) \, d\Omega = V \end{aligned} \tag{3}$$

Here existence of solutions usually require further consideration as to the modelling of the problem. Loosely speaking, materials with a structural hierarchy allow for stiffer structures, as seen in nature in bone, wood, etc. and used in composite structures (see Figure 2). This can lead to a lack of existence of solutions, as such composites can be constructed as limiting sequences of designs defined by (3); however, composites are not covered by (3) (if C^0_{ijkl} is isotropic, the material in (3) is at any point isotropic, but a composite will usually be anisotropic). One says that the set designs is not closed under G-convergence (or H-convergence) (cf., e.g., [20, 23]).

One technique to obtain a well-posed problem, as mentioned in the introduction, is to introduce a constraint $G(\Omega^m) \le \gamma$ on for example

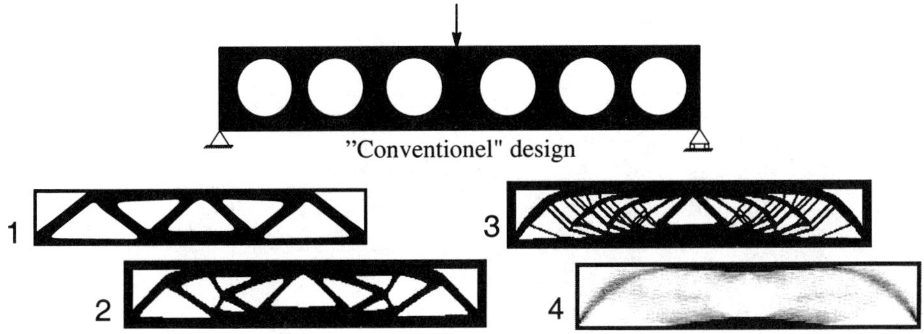

Figure 2 A 'traditional' design of a beam with holes, as seen in aircraft etc., compared with elements of a minimizing sequence of designs with finer and finer scales, with the limit (4) consisting of composites in large areas. Note that even design (1), which is an optimal topology within a limit on geometric complexity, is 40% stiffer than the design with round holes, for the same amount of material.

the perimeter of the domain, thus limiting the geometric complexity of the domain Ω^m so that a structural hierarchy cannot be generated (existence still requires a formal proof, as seen in [4, 10]). If the geometric constraint is not imposed in (1), the continuum problem is ill-posed, and an alternative for obtaining existence of solutions is to extend the problem to its relaxed form using composites formed by mixing void and the given material. This changes the nature of the problem from one of seeking black-and-white designs to a situation where 'grey-scale' structures may also appear.

Problem (1) with design set (3) (regularized through a geometric constraint) is a large scale discrete optimization problem and it has in this form only recently been handled with success computationally ([6]). The most popular method for solving (1), and an approach that has been extremely successful for many applications, is to consider formulations in terms of continuous variables, with the goal of using derivative based mathematical programming algorithms. This means that one changes the model for the material properties to a situation where the volume fraction is allowed to take on any value between zero and one. It may also involve finding an appropriate method for limiting geometric complexity, for example, exchanging the total variation of a density for the perimeter of a domain. This means that one pursues the optimal black-and-white design through an iterative computational procedure which at intermediate steps operate with 'grey' designs. This can be rather confusing, as the computational procedures for the geometrically constrained black-

and-white design problem and for the relaxed formulation, cunningly so, can be very similar.

3. A FRAMEWORK FOR COMPUTATIONS

Computational means are central for solving problems of the type (1). Typically a finite dimensional version of the problem is generated first, normally with finite element approximations of displacement fields and design variables. This leads to a non-linear, non-convex, large scale optimization problem of the generic form

$$\begin{aligned} \inf \quad & p^T u \\ K(D)u &= p \\ \sum D_i &= V, \quad 0 < D_{min} \le D_i \le D_{max}, \end{aligned} \tag{4}$$

where u here is the vector of nodal displacements and D the design variables, entering the equilibrium equation through the stiffness matrix K. This problem could be solved directly via large-scale general purpose mathematical programming techniques or by use of specially developed algorithms that utilizes the problem structure, as seen in 4. Such techniques treat both displacements and design variables as independent variables and the equilibrium equation is solved by the optimization code (an example of this can be found in [41]).

Such an approach is not very common, as one traditionally rewrites the problem as a problem in the design variables only, as

$$\begin{aligned} \inf \quad & F(D) \\ \sum D_i &= V \\ 0 < D_{min} \le D_i &\le D_{max} \ , \end{aligned} \tag{5}$$

where the equilibrium equation is considered as a function-call:

$$\begin{aligned} & F(D) = p^T u, \\ & \text{where } u \text{ solves } K(D)u = p. \end{aligned} \tag{6}$$

If function gradients are required by the optimization algorithm, sensitivity analysis becomes a central theme, i.e., one should devise efficient means to compute the derivatives of functions involving the displacements. This usually involves the so-called adjoint system, as also known from control theory. Here, provided the stiffness matrix is positive definite, we have here the simple expression (the computation of the derivative of the stiffness matrix with respect to design is simple for topology design):

$$\frac{\partial F}{\partial D_i} = -u^T \frac{\partial K}{\partial D_i} u, \\ \text{where } u = (K(D))^{-1} p. \tag{7}$$

However, in many cases this analysis is not so simple. Sensitivity analysis is actually, in its own right, an important area of computer aided design, as it provides information on changes in performance with respect to design variables, see for example papers in [18].

In many situations it is convenient to operate on (5), as one here treats the equilibrium analysis as a function call, as represented by, e.g., commercial software packages for finite element analysis. Moreover, in many design problems, such as shape design via a CAD definition of boundaries, the number of design variables is much smaller than the number of degrees of freedom of the displacement, and an optimization code used for (5) will thus treat a moderately sized problem (with 'expensive' function calls). However, for topology design, even problem (5) is large-scale; but in many cases one can formulate relevant problems involving only a moderate number of constraints, which are not box-constraints (as in the model problem used here). Thus dual methods have become popular, and especially techniques based on separable, convex approximations (CONLIN, see [16], and MMA, see [39]) have been found to be particularly effective.

We close this brief section on computational issues by noting that the problem at hand is a *two field* problem which may develop checkerboards in computations (as for a Stokes flow problem). Constraints limiting geometric complexity usually removes such effects (for a fine enough mesh), but in a relaxed setting (with existence of solutions, but no geometric constraints) special care must be taken to avoid such 'polluted' results. See [36] for an overview and further discussion.

4. HOMOGENIZATION MODELS WITH ANISOTROPY

The initial work on numerical methods for topology design of continuum structures was based on using composite materials as the basis for describing varying material properties in space ([9]). This approach was inspired by theoretical studies on generalized shape design in conduction and torsion problems and by numerical and theoretical work related to plate design (see, e.g., [11, 17, 23]). Initially, composites consisting of square or rectangular holes in periodically repeated square cells were used for planar problems. Later so-called ranked laminates (layers) have become popular, both because analytical expressions of their effective properties can be given and because investigations proved the optimality of such composites, in the sense of bounds on effective properties (for an overview, see for example [2]). Also, with layered materials existence of solutions to the minimum compliance problem for both single and multiple load cases is obtained, without any need for additional constraints on the design space (e.g., without constraints on the geometric complexity), and thus interpolation and relaxation are both provided for. However, the relaxed models seldom result in black-and-white designs in themselves. For all the models mentioned here, homogenization techniques for computing effective moduli of materials play a central role. Hence the use of the phrase 'the homogenization method for topology design' for procedures involving this type of modelling.

The homogenization method for topology design involves working with orthotropic or anisotropic materials. This adds to the requirements of the finite element analysis code, but the main additional complications is the extra design variables required to describe the structure. Thus, a so-called rank-2 microstructure with two orthogonal layers of material (at two scales and in dimension two) require three distributed variables, as the material properties at each point of the structure will depend on two size-variables characterizing the layer thicknesses and on one variable characterizing the angle of rotation of the material axes (the axes of the layers).

5. THE SIMP MODEL

Complementing the use of the homogenization method, where anisotropic composites are a priori accepted as part of the design space, a popular method to model material properties which are *isotropic* at intermediate densities is the so-called penalized, proportional fictitious material SIMP-model (SIMP: Solid Isotropic Material with Penalization), see for example [8] for an overview. In this model a continuous variable ρ is introduced, with $0 \leq \rho \leq 1$, resembling a density of material by the fact that the volume of the structure is evaluated as

$$V = \int_{\Omega} \rho(x)\; d\Omega \tag{8}$$

The relation between this density and the material tensor C_{ijkl} in the equilibrium analysis is written as

$$C_{ijkl}(\rho) = \rho^p C^0_{ijkl}, \quad p > 1, \tag{9}$$

where the given material has stiffness given by C^0_{ijkl}. The interpolation 9 satisfies that $C_{ijkl}(0) = 0$ and $C_{ijkl}(1) = C^0_{ijkl}$, meaning that if a

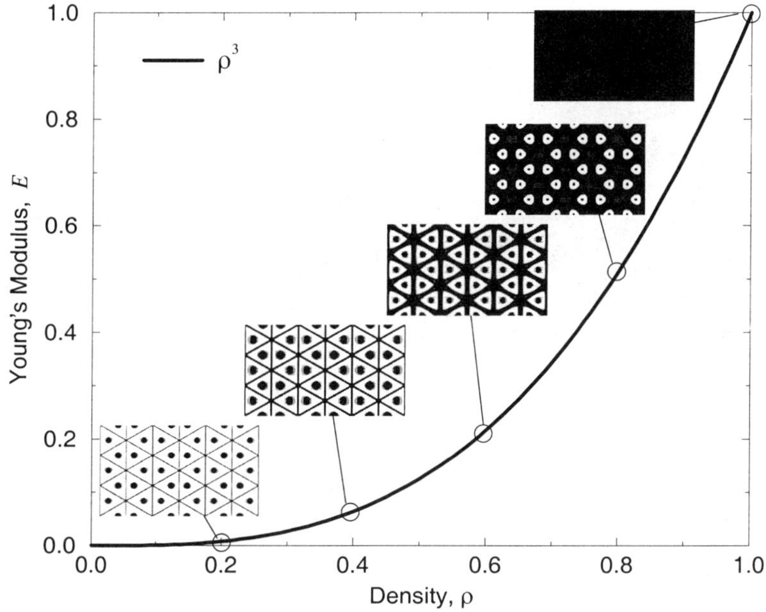

Figure 3 Microstructures of material and void realizing the material properties of the SIMP model with $p = 3$ (Poisson's ratio is 0.33). As stiffer material microstructures can be constructed from the given densities, non-structural areas are seen at the cell centers. See [8].

final design has density zero and one in all points, this design is a black-and-white design for which the performance has been evaluated with a correct physical model. For problems where the volume constraint is active, experience shows that optimization does actually result in such designs if one chooses the power p sufficiently big (in order to obtain true '0-1' designs, $p \geq 3$ is usually required). The reason is that for such a choice one imposes a penalization on intermediate densities (volume is proportional to ρ but stiffness is less than proportional).

For the SIMP interpolation 9 it is not immediately apparent that areas of grey can be interpreted in physical terms. However, it turns out that under fairly simple conditions on p, any stiffness used in the SIMP model can be realized as the stiffness of a composite made of void and an amount of the base material corresponding to the relevant density, see Figure 3. Thus using the term 'density' for the interpolation function ρ is quite natural. The geometries in Figure 3 represent periodic composites with repetitive cells, obtained through the methodology of inverse homogenization (material design) described in [33]. This method employs topology design (using SIMP!) for the cell of the microstructure

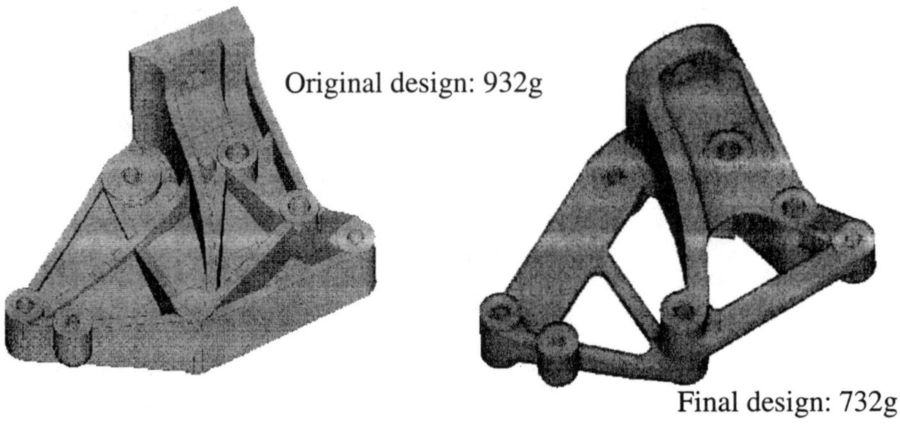

Figure 4 Design of a bracket for an automobile application. Original design and design achieved by using topology design in the initial design phase. By courtesy of Altair Engineering.

to obtain specified material properties of the composite, making the dog bite its tail.

6. APPLICATIONS

Computational means for topology design have now been developed into reliable tools for considering such problems involving stiffness (including multiple load problems) and vibration criteria (as a reinforcement problem, see, e.g., [12]). The efficiency of topology design as a tool in the *initial phase* of a design process and the appearance of commercial software has lead to a broad acceptance of the method in industry, , see, e.g., [5, 32, 40] (see also Section 8). Research has also lead to considering extended classes of problems, encompassing more involved settings for the physical model, for the optimization formulation, as well as for the design parametrization. Thus non-linear behaviour (large displacements, non-linear material behaviour), local constraints (i.e., constraints imposed on all points in the domain) as well as multi-material problems are covered in current research. The areas of applications are also being extended, research moving into such fields as design of materials, see Figure 5, and design of multiple physics devices (like MEMS, Micro-Electro-Mechanical-Systems), see Figure 6. For example literature we refer to [14, 19, 24, 25, 34, 37].

It is important in this context to note that direct links between specific classes of composites and proofs of existence for many of these extended

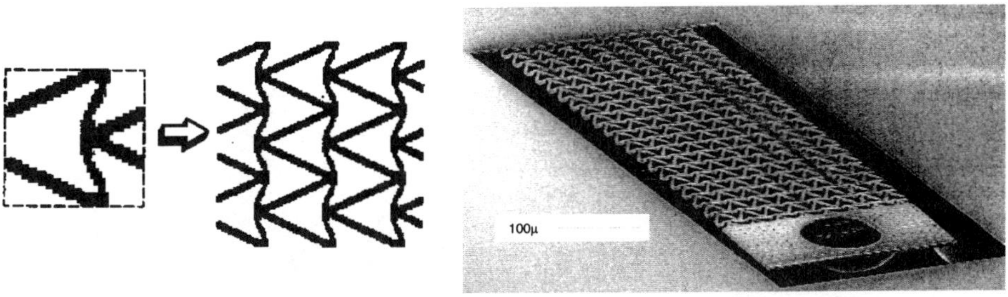

Figure 5 Design of a periodic composite with *negative* Poisson ratio. Topology optimization is applied to the unit cell of periodicity, and the effective parameters of the composite are computed via homogenization. The right-hand picture shows a micro-device representing the composite. By courtesy of Ole Sigmund, Ulrik Darling Larsen and Siebe Boustra, see also [21, 33].

problems have yet to be discovered, i.e., it is not known for these problems what composites will make relaxation intrinsic in such an approach. The emphasis in the field is currently on modelling and on the development of computational means, mostly relying on restriction methods as an underlying implicit guarantee of well- posedness. Thus, in the context of theory, the pace of implementations for new problem types has far overtaken the full mathematical understanding of the interplay between methods for obtaining 'classical' solutions and methods for deriving relaxed problem settings, and here lies an immense challenge for the future. For a mathematician there is thus an abundance of (hard) problems to study.

For exemplification of modelling consideration we will here briefly outline some aspects of treating multiple physiscs problems. The phrase 'multiple physics' is here used to cover topology design where several physical phenomena are involved in the problem statement, thus covering situations where for example elastic, thermal and electromagnetic analyses are involved. When modelling such situations, the basic concept of the density distribution method for topology design provides a general framework for computations, but here the initial obstacle is the need for interpolation of not only stiffness but also other physical properties. If only linear models are considered, one possibility is to use the theory and computational framework of homogenization of composite media to compute effective elastic, thermal and electromagnetic properties of a given type of composites and use such relationships between intermediate density and material properties in the design problem. An example of this approach for thermo-elastic problems can be found in [30]. However, the less complex design description of the SIMP approach has lead also

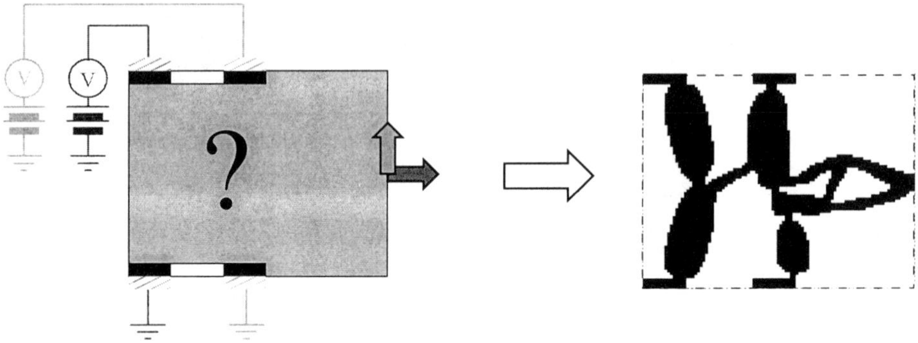

Figure 6 Design of a thermo-electro-mechanical actuator. The mid right-hand point should move right or up when voltage is applied at the corresponding poles. The movement is through elastic deformation due to the temperature change which occurs when a current passes through the structure. By courtesy of O. Sigmund, [35]

to the development of such interpolation schemes for multiple physics problems. As an example of this, in reference [37], microstructures with extreme thermal expansion are designed by combining a material interpolation of elastic properties with a similar interpolation of the thermal expansion coefficients.

7. DIFFICULTIES – FUTURE WORK

The models described here all refer to an approach to topology design where material is distributed in a fixed domain. A pivotal aspect of this idea in computational implementations is the use of a fixed FEM mesh for the domain. This is not an inherent requirement, but is useful for computational efficiency. If a topology of material and void is the goal of the design process, this will imply that low density areas are also included in the analysis for each feasible design. For stress constraints this leads to the difficulty of the so-called 'stress singularity phenomenon', where low density regions may have high stress but are structurally insignificant for the final design, and where regularization (in the sense of mathematical programming) is needed for numerically solving such problems ([14]). A similarly elusive problem arising from this basic design representation appears in situations involving stability and vibration criteria. The relevant data to consider in such situations are the eigenvalues of the structurally relevant parts of the structure, i.e., the buckling loads and the vibration frequencies of the 'black' part of a black- and-white design. In a true black-and-white design this are the non-zero eigenvalues, but at intermediate steps of an iterative optimization method implemented with interpolation schemes it can become unclear what are the relevant values to consider. Examples of this are localized modes which appear in low density regions and which should be filtered out in order that the optimization deals with the structurally interesting modes. Moreover, one should cater for certain aspects which complicates modelling, for example that no structure what-so-ever has the highest eigenfrequency of all structures with only structural mass. See references [24, 26] for further discussion.

For future work it would be beneficial to reconsider the way geometry representation is currently handled. This should include studying alternatives to the current close linking of geometry and physics where the same finite element grid is used to approximate the geometry and the mechanical response fields. Here grid-less methods may prove useful. Typically, in fixed domain material distribution methods, the grid is a uniform, rectangular partition of space and the design variables are assumed to be constant within each element. Thus, a raster model is imposed on the geometry. This approach has a number of very beneficial features in terms of computational efficiency, but it also has some intrinsic less than desirable features (non-smooth boundary representation etc.) which are typically not too severe if they are recognized and addressed in the implementation as well as in the interpretation of the computer generated designs. However, as outlined above for the stress constrained and vibration constrained problems, the fixed grid representation for geometry and analysis implies that it can be rather tricky to handle the correct formulation of objectives and constraints as well as interpolating physical data for intermediate densities, a feature which gets exaggerated when considering non- linear material behaviour. The latter could be circumvented by further success in handling the 0-1 problem directly, an area of utmost importance. For the problems arising due to the fictitious domain type approach to analysing on a fixed domain a new approach in the area of topology design may be the only option. In view of this, it seems that future research could also benefit from investigations into alternative approaches to topology design for black-and-white design, with an emphasis on the geometric modelling aspect. Here the concept of the bubble method ([15]) should no doubt receive further attention, especially in the light of recent work on the concept of a general topological derivative, see [38]. Work in this direction could also be helpful in an effort to rejuvenate the field of optimal boundary shape design, an important area, unfortunately currently lingering an idle life.

8. USING THE WEB

Much useful information on topology design can be found on the Internet. A free topology optimization test-site (make your own design!) can be found at `http://www.topopt.dtu.dk/`. Here one can also find tutorials on the subject, a free 99 line MATLAB topology optimization code, as well as interesting links to academic and commercial sites.

Acknowledgments

The author would like to thank Ole Sigmund for permission to use several illustrations shown in this presentation. Kind assistance by Jens Gravesen when preparing this paper is also gratefully acknowledged.

References

[1] G. Allaire, E. Bonnetier, G. Francfort and F. Jouve (1997), Shape optimization by the homogenization method, *Numerische Mathematik*, 76, pp. 27–68.

[2] G. Allaire and R. Kohn (1993a), Optimal bounds on the effective behaviour of a mixture of two well-ordered elastic materials, *Quaterly of Appled Mathematics*, 51, pp. 643–674.

[3] G. Allaire and R. Kohn (1993b), Optimal design for minimum weight and compliance in plane stress using extremal microstructures, *European Journal of Mechanics A*, 12, pp. 839–878.

[4] L. Ambrosio and G. Buttazzo (1993), An optimal design problem with perimeter penalization, *Calculus and Variation*, 1, pp. 55–69.

[5] A. Back-Pedersen (1999), Taking advantage of using both topology and shape optimization for practical design, in *Proc. NAFEMS World Congress '99, Newport, Rhode Island*, NAFEMS ltd, pp. 889–899.

[6] M. Beckers (1999), Topology optimization using a dual method with discrete variables, *European Journal of Mechanics A*, 17, pp. 14–24.

[7] M. Bendsøe (1995), *Optimization of Structural Topology, Shape, and Material*, Springer Verlag, Heidelberg.

[8] M. Bendsøe and O. Sigmund (1999), Material interpolation schemes in topology optimization, *Archives of Applied Mechanics*, 69, pp. 635–654.

[9] M. Bendsøe and N. Kikuchi (1988), Generating optimal topologies in structural design using a homogenization method, *Computer Methods in Applied Mechanics and Engineering*, 71, pp. 197–224.

[10] B. Bourdin (1999), Filters in topology optimization, Danish Center for Applied Mathematics and Mechanics, DCAMM Report No. 628.

[11] G. Cheng and N. Olhoff (1981)], An investigation concerning optimal design of solid elastic plates, *International Journal of Solids and Structures*, 17, pp. 305–323.

[12] A. Díaz and N. Kikuchi (1992), Solutions to shape and topology eigenvalue optimization problems using a homogenization method, *International Journal for Numerical Methods in Engineering*, 35, pp. 1487–1502.

[13] A. Diaz and R. Lipton (1997), Optimal material layout for 3D elastic structures, *Structural Optimization*, 13, pp. 60–64.

[14] P. Duysinx and M. Bendsøe (1998), Topology optimization of continuum structures with local stress constraints, *International Journal for Numerical Methods in Engineering*, 43, pp. 1453–1478.

[15] H. Eschenauer, V. Kobelev and A. Schumacher (1994), Bubble method of topology and shape optimization of structures, *Structural Optimization*, 8, pp. 42–51.

[16] C. Fleury (1989), CONLIN: an efficient dual optimizer based on convex approximation concepts, *Structural Optimization*, 1, pp. 81–89.

[17] J. Goodman, R. Kohn and L. Reyna (1986), Numerical study of a relaxed variational problem from optimal design, *Computer Methods in Applied Mechanics and Engineering*, 57, pp. 107–127.

[18] M. Kamat (1995), *Structural Optimization – Status and Promise*, AIAA, Washington D.C.

[19] N. Kikuchi, S. Nishiwaki, J. Fonseca and E. Silva (1998), Design optimization method for compliant mechanisms and material microstructure, *Computer Methods in Applied Mechanics and Engineering*, 151, pp. 401–417.

[20] R. Kohn and G. Strang (1986), Optimal design and relaxation of variational problems, *Communications in Pure and Applied Mathematics*, 39, pp. 113–137, pp. 139–182, pp. 353–377.

[21] U. Larsen, O. Sigmund and S. Bouwstra (1997), Design and fabrication of compliant mechanisms and material structures with negative poissons ratio, *Journal of MicroElectroMechanical Systems*, 6, pp. 99–106.

[22] J.-L. Lions (1981), *Some Methods in the Mathematical Analysis of Systems and Their Control*, Gordon and Breach Science Publishers, Inc., New York.

[23] K. Lurie, A. Cherkaev and A. Fedorov (1982), Regularization of optimal design problems for bars and plates, I, II, III, *Journal of Optimization Theory and Applications*, 37, pp. 499–543, 42, pp. 247–282.

[24] M. Neves, H. Rodrigues and J. Guedes (1995), Generalized topology design of structures with a buckling load criterion, *Structural Optimization*, 10, pp. 71–78.

[25] J. Ou and N. Kikuchi (1996), Optimal design of controlled structures, *Structural Optimization*, 11, pp. 19–28.

[26] N.L. Pedersen (1999), Maximization of eigenvalues using topology optimization, Danish Center for Applied Mathematics and Mechanics, DCAMM Report No. 620.

[27] J. Petersson (1999a), A finite element analysis of optimal variable thickness sheets, *SIAM Journal of Numerical Analysis*, 36, pp. 1759–1778.

[28] J. Petersson (1999b), Some convergence results in perimeter-controlled topology optimization, *Computer Methods in Applied Mechanics and Engineering*, 171, pp. 123–140.

[29] J. Petersson and O. Sigmund (1998), Slope constrained topology optimization, *International Journal for Numerical Methods in Engineering*, 41, pp. 1417–1434.

[30] H. Rodrigues and P. Fernandes (1995), A material based model for topology optimization of thermoelastic structures, *International Journal for Numerical Methods in Engineering*, 38, pp. 1951–1965.

[31] G. Rozvany (1997), *Topology Optimization in Structural Mechanics*, Springer Verlag, Heidelberg.

[32] U. Schramm (1999), The use of structural optimization in automotive design - state of the art and vision, in Bloembaum, C., editor, *Proc. Third World Congress of Structural and Multidisciplinary Optimization*, University of New York at Buffalo, pp. 200–202.

[33] O. Sigmund (1995), Tailoring materials with prescribed elastic properties, *Mechanics of Materials*, 20, pp. 351–368.

[34] O. Sigmund (1997), On the design of compliant mechanisms using topology optimization, *Mechanics of Structures and Machines*, 25, pp. 495–526.

[35] O. Sigmund (1998), Topology optimization in multiphysics problems, in *Proc. 7th AIAA/USAF/NASA/ISSMO Symposium on Multidisciplinary Analysis and Optimization St. Louis MI Sept. 2-4 '98*, AIAA, pp. 1492–1500.

[36] O. Sigmund and J. Petersson (1998), Numerical instabilities in topology optimization, *Structural Optimization*, 16, pp. 68–75.

[37] O. Sigmund and S. Torquato (1997), Design of materials with extreme thermal expansion using a three-phase topology optimization method, *Journal of the Mechanics and Physics of Solids*, 45, pp. 1037–1067.

[38] J. Sokolowski and A. Zochowski (1999), On the topological derivative in shape optimization, *SIAM Journal on Control and Optimization*, 37, pp. 1251–1272.

[39] K. Svanberg (1987), The method of moving asymptotes – a new method for structural optimization, *International Journal for Numerical Methods in Engineering*, 24, pp. 359–373.

[40] R. Yang and A. Chahande (1995), Automotive applications of topology optimization, *Structural Optimization*, 9, pp. 245–249.

[41] J. Zowe, M. Kocvara and M. Bendsøe (1997), Free material optimization via mathematical programming, *Mathematical Programming B*, 79, pp. 445–466.

MIP: THEORY AND PRACTICE – CLOSING THE GAP

Robert E. Bixby
ILOG CPLEX Division
889 Alder Avenue
Incline Village, NV 89451, USA
and
Department of Computational and Applied Mathematics
Rice University
Houston, TX 77005-1892, USA
bibxy@caam.rice.edu

Mary Fenelon
ILOG CPLEX Division
889 Alder Avenue
Incline Village, NV 89451, USA

Zonghao Gu
As above

Ed Rothberg
As above

Roland Wunderling
As above

1. INTRODUCTION

For many years the principal solution technique used in the practice of mixed-integer programming has remained largely unchanged: Linear programming based branch-and-bound, introduced by Land and Doig (1960). This, in spite of the fact that there has been significant progress

M.J.D. Powell and S. Scholtes (Eds.), *System Modelling and Optimization: Methods, Theory and Applications.*

in the theory of integer programming and in the closely related field of combinatorial optimization. Many of the ideas developed there have received extensive computational testing, but, until recently, relatively little of that work has made it into the commercial codes used by practitioners. That situation has now changed. Several such codes, among them LINGO[1], OSL[2], and XPRESS-MP[3], as well as the CPLEX[4] code studied in this paper, now include cutting-plane capabilities as well as other ideas from the backlog of accumulated theory. As suggested by the title of this paper, the gap between theory and practice is indeed closing.

In order to fix ideas, we begin with a formal definition. A *mixed-integer program* (MIP) is an optimization problem of the form

$$\begin{array}{ll} \text{minimize} & c^T x \\ \text{subject to} & Ax = b \\ & l \leq x \leq u \\ \multicolumn{2}{l}{\textbf{some or all } x_j \textbf{ integral,}} \end{array}$$

where A is an $m \times n$ matrix, called the *constraint matrix*, x is a vector of *variables*, c is the *objective function*, and l and u are vectors of bounds. Thus, a MIP is a linear program (LP) plus an integrality restriction on some or all of the variables. This last restriction is what makes MIPs difficult (NP-hard, in the technical sense); it takes a well understood, convex problem and makes it non-convex. It also makes the mixed-integer modeling paradigm a powerful tool in representing real-world business applications.

The power of the mixed-integer modeling paradigm was recognized almost immediately, dating back to the 50s and 60s, and numerous attempts were made to apply it. Unfortunately, while the modeling paradigm was strong, the available software and computers for solving the models were not. The result was disillusionment, some of which persists to this day. Many potential practitioners still believe that mixed-integer programming is nice to talk about, but has limited practical applicability. An important message of this paper is that this situation has changed, and changed dramatically just in the last year. It is now possible to solve many difficult, interesting, and practical mixed-integer models using off-the-shelf software.

The following is an outline of the contents of the paper. We begin with a discussion of advances in methods for solving linear programming

[1]LINGO is a trademark of Lindo Systems, Inc.
[2]OSL is a trademark of IBM Corporation
[3]XPRESS-MP is a trademark Dash Associates Ltd.
[4]CPLEX is a trademark of ILOG, Inc.

problems. First we give a snapshot overview of developments in the period from the mid-80s to 1998, and then we look at 1999. One reason to begin with linear programming, rather than directly with mixed-integer programming, is that linear programming is an enabling technology for solving MIPs. Given this first reason, the real motivation for including a discussion of linear programming here is that 1999 has seen some remarkable and unexpected improvements in the classical simplex method.

The discussion of linear programming will be followed by the mixed-integer programming part of the paper. The presentation emphasizes features. Specific topics to be discussed will include node presolve, heuristics for finding feasible solutions, and cutting planes. These will be followed by extensive computational results.

The discussion of mixed-integer programming features can be viewed as having two main parts. The first discusses features that attempt to decrease the "upper bound" (e.g., heuristics to find better integral solutions). The second discusses features that attempt to increase the lower bound (e.g., cutting planes). When the upper and lower bounds become equal, the computation is finished.

An important guiding principle of our mixed-integer algorithmic developments is that solving MIPs often requires a "barrage" of different, but cooperating ideas. In other words, we try to take advantage of structures that are common to many real-world MIPs, hoping that some or all will contribute to a better solution for a particular model. To do so, it is essential to develop good defaults, and implement the individual ideas in such a way that they help when they can, and otherwise hurt as little as possible. This approach is perhaps different from that of most theoretical investigations, where the goal is typically to demonstrate the efficacy of a particular new idea, usually in isolation.

Finally, we consider several examples. Two of these examples will provide a counterbalance to the idea that good defaults are sufficient to handle all models. While we would like to run mixed-integer programming codes much as we run linear-programming codes, as black boxes, there will always be instances that demand some sort of tuning or reformulation.

We close this section with one general remark. For many of the computational results presented in this paper, we will use geometric means as a method to summarize results. On occasion, when doing so, we will simply use the word "mean." This usage will always refer to the geometric mean, and not the more common arithmetic mean. Arithmetic means can be quite misleading when applied to a set of ratios, as would often be the case in this paper.

2. LINEAR PROGRAMMING

2.1. PROGRESS SOLVING LPS: MID-80S TO 1998

No attempt is made here to discuss linear-programming improvements in this period in detail. We will present just one table, followed by a brief discussion. A detailed discussion is a topic unto itself.

As the following table illustrates, over the past ten years there has been steady progress in our ability to solve linear programming problems[5].

Model: PDS-30, Patient Distribution System
49944 rows, 177628 columns, 393657 nonzeros

Version	Time (seconds)
CPLEX 1.0 (1988)	57840
CPLEX 3.0 (1994)	4555
CPLEX 5.0 (1996)	3835

The model PDS-30 is one of a class of models introduced in Carolan, et al., (1990), and is well-known within the linear-programming community. For reasons that are hopefully apparent given the CPLEX 1.0 data in the table, the larger instances in this class (e.g., PDS-30) were considered very difficult when first introduced. The runtimes in the table were produced using a modern workstation, a 296 MHz Sun UltraSparc. Considering the improvement in machine speeds between 1990 and the present, probably exceeding a multiple of 1,000, describing this model as being very difficult in 1990 is an understatement.

Some remarks are in order before we move on to the developments of the last year. First, while it is not illustrated in this table, the first release of CPLEX was already a significant improvement over at least one of the standard portable codes available at that time, XMP developed by Marsten (1981). Thus, the fifteen-fold improvement for the one problem in the table in the period 1988 to 1998 can be viewed as an underestimate. Second, and much more important in our view, the most significant development of the last decade is not really reflected in this

[5]In CPLEX 1.0 only a primal simplex algorithm was available. In subsequent versions, primal and dual simplex algorithms, and a barrier algorithm were available. We used the dual simplex algorithm when solving PDS-30 with CPLEX 3.0 and 5.0.

table. This development was the leap forward in robustness of linear programming codes. They have not only become more robust in terms of solve times, but also much more robust at handling numerical difficulties and problems related to degeneracy. In short, linear programming has become more-and-more a tool that practitioners can simply use, embedding it as a black box in other applications without having to worry whether it will do its job.

2.2. PROGRESS SOLVING LPS: 1999

We began the work described in this section by making a simple observation: LPs have become larger. This is the same sort of observation that was made ten years ago at the start of the developments highlighted in the previous section. Here it led us to focus specifically on models with at least 10,000 constraints. It also led us to focus on the simplex method, since it was the simplex method that seemed to be underperforming on these large models. It didn't take long to discover where the bottleneck lay: The solution of the two (sometimes three) linear systems that are necessary at each simplex iteration. These linear systems are commonly called *BTRAN* and *FTRAN* (see Chvatál (1983)).

It is not strictly necessary to know what the *FTRAN* and *BTRAN* systems refer to here. The basic idea is quite simple. Imagine we are to solve a large linear system $Lx = a$, where L is a triangular matrix, a is extremely sparse, and x turns out to be very sparse as well. Both vectors often have fewer than 100 nonzeros among them, in spite of the fact that L is of order 10,000 or more (corresponding to a linear programming problem with 10,000 or more constraints). Clearly, when a and x contain this few nonzeros, it is unlikely that the cause was cancellation during the solve; more likely is that the number of nonzeros touched in L, in order to compute x, was very small as well. Thus, the key to reducing the cost of the solve is to do it in an amount of time linear in this number of nonzeros. As it turns out, though this fact was apparently not being exploited in linear programming codes, the existence of such an algorithm has long been known in the sparse linear algebra community. It is equivalent to a certain, natural reachability problem in a graph. See Gilbert and Peierls (1988).

When the above bottleneck was removed, it then made possible further improvements to the simplex method itself. This is where the *real* progress occurred. Two examples:

- **The dual simplex algorithm:** It can be shown that variables with two finite bounds often do not need to be binding in the ratio test. Exploiting that fact leads to what one might call a "long-step" dual simplex algorithm.

- **Fast pricing update:** If the solutions of the key linear systems at each iteration are sparse, then it is reasonable to expect that only a small number of reduced costs will change, and hence that appropriate update schemes can be introduced to accelerate the choices of entering and leaving variables in the primal and dual simplex algorithms, respectively.

For PDS-30, the resulting improvement is significant indeed:

Model: PDS-30, Patient Distribution System
49944 rows, 177628 columns, 393657 nonzeros

Version	Time (seconds)
CPLEX 1.0 (1988)	57840
CPLEX 3.0 (1994)	4555
CPLEX 5.0 (1996)	3835
CPLEX 6.5 (1999)	165

Of course, PDS-30 is just one problem, used here as an illustration. Much more extensive tests were done to evaluate the effects of the changes introduced with CPLEX 6.5. In addition to the changes outlined above for the simplex method, there were also important, if not quite as dramatic, improvements in the barrier implementations. These barrier improvements can be summarized as due to two things: (a) Better ordering algorithms for the computation of the Cholesky factorization, see Rothberg and Hendrickson (1998), and (b) better exploitation of the available level-two cache in modern computing architectures, see Rothberg and Gupta (1991).

In what follows, an extensive set of test results are given to evaluate the performance improvements in CPLEX 6.5. The results are broken into two parts: Small models and large models. Before plunging into the details, it is perhaps worthwhile to point out the philosophy of the way the improvements were implemented. The overall target was "robustness and scalability" in the algorithms. At least as important as making the algorithms better on larger models was that performance did not degrade, and hopefully improved, on the broad middle-range of models that dominate in practice. Indeed, while the improvements on large models were exciting and were the impetus behind this work, the real effort was expended in making sure that these improvements didn't get in the way when they didn't help. The same theme was mentioned earlier for mixed-integer programming.

2.3. PERFORMANCE ON SMALL LPS

Using a 400 MHz Pentium II running a Linux operating system, CPLEX 5.0 was run on all models in the CPLEX library of linear programming problems, a library that has been collected over a period now exceeding ten years. For each of the primal and dual simplex algorithms, we collected all the models that had solve times of less than 100 seconds using CPLEX 5.0. Performance on these models was then compared to CPLEX 6.5. The following table summarizes the results, where, as discussed below, a ratio bigger than 1.0 means that 6.5 was faster than 5.0:

Performance Improvements: Small LPs
CPLEX 5.0 to 6.5

Segment	Primal Ratio	Number	Dual Ratio	Number
0 to 1 secs	1.02	145	1.02	139
1 to 10 secs	1.36	99	1.42	101
10 to 100 secs	1.59	87	1.88	102

Thus, for example, there were 99 models where the solve time with 5.0 using primal simplex was at least 1 second and no more than 10 seconds (and 101 such models for dual simplex). For primal simplex, taking the solve time for each model with 5.0 and dividing it by the solve time for 6.5 resulted in 99 ratios of solve times. Computing the geometric mean of these ratios gave a value of 1.36. Similarly, for the dual the computed mean was 1.42. Thus, based upon this last number, one might say that for models in the 1 to 10 seconds segment, the 6.5 dual was 42% faster than the 5.0 dual. These results were a pleasant surprise. It was only for larger models that we were certain there would be substantial improvements.

Below, for completeness, we also list some size statistics for the above groups, using both the geometric mean and the median. These are problem sizes after *presolve* was applied, where presolve refers to a set of problem reduction routines applied prior to calling the optimization routines. Some of the original, unpresolve model sizes are quite substantial and would be misleading in the current context. See Brearley, et al.

(1975), and Anderson and Anderson (1995) for a discussion of presolve reductions.

Problem Sizes

Primal Simplex

	Means			Medians		
	Rows	Cols	Nonzeros	Rows	Cols	Nonzeros
0 to 1 secs	307	556	3379	337	491	2667
1 to 10 secs	1396	3316	17046	1357	2814	15114
10 to 100 secs	2774	9060	53962	3016	6912	43298

Dual Simplex

	Means			Medians		
	Rows	Cols	Nonzeros	Rows	Cols	Nonzeros
0 to 1 secs	287	488	2919	323	515	2961
1 to 10 secs	1424	2973	15927	1377	2814	14295
10 to 100 secs	2824	7597	45860	3266	6831	41402

As one final statistic, we give the geometric means of the solve times using CPLEX 6.5 for each of the above groups, using primal and dual:

CPLEX 6.5: Mean Solution Times (seconds)

Segment	Primal Simplex	Dual Simplex
0 to 1 secs	0.2	0.1
1 to 10 secs	2.4	2.3
10 to 100 secs	19.5	17.4

2.4. PERFORMANCE ON LARGE LPS

We also went through the entire CPLEX library of LPs, previously mentioned, and collected all instances which, after application of CPLEX 5.0 presolve, had at least 10,000 rows. From these models an attempt was made to remove those that appeared to be just minor variations on other models in the collection. In the same vein, there were several instances, such as the PDS models, where a whole family of models with increasing sizes were found. In these cases, the largest instance from the family was included in the test-set and the others deleted.

All runs were done on PCs with 400 MHz Pentium II processors running a Linux operating system. Models were included in the final performance numbers only if, in presolved form, they were solvable within one-half Gigabyte of physical memory. This limitation was dictated by memory availability on our test machines. A time limit of 500,000 seconds (about 6 days) was also imposed for each run. A limit this large may seem excessive, but it was deemed necessary for the tests since the expectation was that several models would solve in several thousands of seconds with CPLEX 6.5 and would be a large multiple slower with CPLEX 5.0. Comparisons would not have been possible otherwise. Note that in the final analysis of the data, where ratios are used to compare the various algorithms, models that exceeded the memory limit were not included. However, those that reached the time limit were included, in such cases using 500,000 seconds as the run time. As a result the comparisons we made underestimated the actual improvements.

Table 1 in the appendix gives size statistics for the models generated, ordered by the number of constraints in the model. Generic names (LP01 through LP90, ordered by increasing numbers of constraints) have been used since many of these models are proprietary. The mean number of constraints was about 50,000, with three models having over 1,000,000 constraints. The following two tables summarize comparative performance as problem size grows. There are distinct tables for barrier and simplex (plus best) since the sets of models not meeting the memory restriction were different.

Performance Improvements: Large LPs
CPLEX 5.0 to 6.5

	Mean Ratios		
Problems	Primal Simplex	Dual Simplex	Best
Biggest 10	8.5	22.3	18.0
Biggest 20	7.9	18.8	12.2
Biggest 30	7.4	20.2	11.3
Biggest 40	6.4	14.0	8.0
Biggest 50	5.5	11.9	6.7
Biggest 60	5.2	10.1	6.2
Biggest 70	5.1	9.1	5.6
Biggest 80	4.5	8.2	5.2
All	4.4	8.0	5.2

Performance Improvements: Large LPs – Barrier
CPLEX 5.0 to 6.5

Problem	Mean Ratios
Biggest 10	11.9
Biggest 20	5.5
Biggest 30	4.1
Biggest 40	3.7
Biggest 50	3.6
Biggest 60	3.7
Biggest 70	3.6
All	3.6

The first table above refers to simplex results and the best of primal and dual simplex and barrier, where barrier includes crossover to a basic solution. The total number of models in the *All* category for simplex was 86; four models failed the memory-limit test. 75 models are in the *All* category for barrier; fifteen failed the memory-limit test.

For each model passing the memory test and for each algorithm, two runtimes were produced, one for CPLEX 5.0 and one for CPLEX 6.5. Given these sets of numbers, ratios were computed of the 5.0 time divided by the corresponding 6.5 time. Thus, a ratio bigger than 1.0 meant that 6.5 was faster.

To understand how the summary numbers in the tables were constructed, consider the row labeled Biggest 30 in the first table, for the primal and dual simplex algorithms and "best." To get the numbers in this row, we computed the geometric means of the time ratios for models LP58 to LP90 (excluding LP85, LP89, and LP90, which failed the memory test) doing so for each of the three algorithms. The results for the primal simplex algorithm indicate a speedup of 7.4 on average, for CPLEX 6.5 versus CPLEX 5.0; for the dual the speedup was 20.2 on average; and for the best of primal, dual, and barrier, the speedup was 11.3.

In summary, the overall improvements are very large indeed. The magnitude of these improvements was unexpected.

One thing the first of the above tables does indicate quite clearly is that the dual simplex algorithm experienced a larger improvement than the other algorithms. This observation leads to the question of how the various algorithms compare to each other. Which is best? Here is a summary:

CPLEX 6.5: Algorithm Comparison

Algorithms	Instances	Mean	Wins for Dual
Primal/Dual	86	2.6	56
Barrier/Dual	78	1.2	41

Thus, dividing the primal solve time for each model by the dual solve time and computing the geometric means of the resulting ratios gives the result that the dual was a factor of 2.6 faster overall. 86 models were included in the test. In 56 of the instances dual was the winner. In 30 instances primal won. For the barrier versus dual comparison, it was much closer, with dual winning by only a small margin, but winning nevertheless, with a mean ratio of 1.2. Dual was the faster algorithm in 41 instances, while barrier won in 37 instances.

Missing from the table, because of the focus on the dual, is the comparison between barrier and primal. In that comparison barrier won 46 times and primal 32 times, and the mean ratio was 1.8, with barrier the winner. Finally, doing a comparison among all algorithms, using all 90 models (see Remark 3, below), we obtain the interesting result that primal won 18 times, dual 33 times, and barrier 39 times.

Remarks:

1 Among the 86 instances in which CPLEX 6.5 primal and dual were compared, primal and dual reached the 500,000 second time limit on one common model. This model contributed a 1.0 to the mean ratio. There were six additional instances in which the primal reached the time limit, and no additional instances for the dual.

2 There were four models too large to be solved with any of the algorithms within the one-half Gigabyte limit: LP13 (because of the density of the LU and Cholesky factors), LP85, LP89, and LP90. In all four cases, limited tests were run on machines with more available physical memory. In each of these cases, barrier was clearly the superior algorithm. One of the models, LP89, has yet to be solved with a simplex algorithm.

In addition to the four models just listed, there were eight models—LP26, LP63, LP71, LP78, LP79, LP80, LP81, and LP86—that could not be run with CPLEX 6.5 barrier within the one-half Gigabyte memory limit, but could be run with both primal and dual. Partial barrier tests were run with these models on larger-memory machines. In each case the simplex method dominated the performance of the barrier algorithm. In six of the cases, dual was the superior algorithm, in one primal, and in one case primal and dual produced similar performance.

3 All of the numbers presented here can be viewed as biased against the barrier algorithm in the following two senses. First, the floating-point performance on the machines we used, X86 PC's, is markedly inferior to that on most UNIX workstations. Floating-point performance is key to the performance of the barrier algorithm. If these tests had been run on machines with better floating-point performance, barrier likely would have "won" the comparison with dual. Second, the barrier algorithm can be run in parallel on shared-memory machines, and produces good speedups over a wide range of model characteristics. No such parallelism appears to be available for the simplex algorithms. If this difference in parallel performance had been exploited, even to the extent of using two processors, again barrier would have won.

What can one say in general about the best way to solve large models? Which algorithm is best? If this question had been asked in 1998, our response would have been that barrier was clearly best for large models. If that question were asked now, our response would be that there is no clear, best algorithm. Each of primal, dual, and barrier is superior in a significant number of important instances.

3. MIXED-INTEGER PROGRAMMING

This is a discussion focused on features. We will consider the following topics:

- Heuristics
- Node Presolve
- Cutting Planes

As mentioned earlier, a guiding principle of our MIP developments was to apply a "barrage" of different techniques to each model. By applying every technique to every model, we benefit if any of the techniques are effective, and we free the users from having to determine which techniques are appropriate for their specific models. An unanticipated benefit from this approach was that the techniques often combine to produce results that would not have been possible with any one technique. The obvious downside is that we pay the cost of every technique, even when the technique is not effective for that model. Much of the work of implementing the techniques we will now discuss went into creating aggressive strategies for determining that a technique is not helping and turning it off automatically.

The standard technique for solving mixed-integer programming problems is a version of divide-and-conquer known as linear-programming based branch-and-bound, or, what is now a more correct name, *branch-and-cut*. This algorithm begins by solving the linear-programming relaxation, obtained by simply deleting the integrality restrictions. If the solution x^* of this LP satisfies all the integrality restrictions, we are done; otherwise, some integrality restriction is violated. Picking an integral variable x_j that is currently fractional with value x_j^*, we *branch*, creating two separate "child" problems from the single "parent" problem, one of which has the added restriction $x_j \leq \lfloor x_j^* \rfloor$ and the other of which has the added restriction $x_j \geq \lceil x_j^* \rceil$. At any point, if a cutting plane is identified that cuts off the solution to the current LP, that constraint is added to the LP. The procedure is repeated.

Two important quantities that are generated during the branching process are an objective function upper bound and an objective function lower bound. Upper bounds are obtained by finding feasible integral solutions. Lower bounds are obtained by taking the smallest optimal objective value for a linear-programming relaxation among all current active branch-and-cut *nodes*. In terms of these two bounds, we can think of node presolve and heuristics as contributing to the upper bound, and both node presolve and cutting planes as contributing to the lower bound.

3.1. NODE PRESOLVE

It is now standard to apply problem simplification routines to linear programming problems prior to solving. For integer programming, such "root" reductions seem to be even more important. We begin by applying a restricted form of the reductions for linear programming, those that are valid for integer programs. We then apply several additional reductions, the main two being "bound strengthening" and "coefficient reduction." See Hoffman and Padberg (1991) and Savelsbergh (1994) for discussions of mixed-integer "root" presolve.

The above is a description of what we do before the branching process is started. What do we do within the tree? In the integer-programming literature there are several proposals that perform rather extensive sets of presolve operations at the nodes. However, our presolve needs to work for general-purpose models, and has to have the property that it is not too expensive in the event that it does not produce positive results for a particular model. We have thus selected a very restricted kind of node presolve, one that does not make any changes that affect the constraint matrix: We implemented a fast, incremental form of bound strengthening. The following is an illustration of how bound-strengthening works.

Example: Resource allocation.
The problem is to decide how to split up a minute of available time among various possible jobs. Here is a constraint:

$$40x_1 + 30x_2 + 60x_3 + 60x_4 + 30x_5 + 20x_6 + 60x_7 + 40x_8 = 60$$

On the right-hand side, 60 is the total number of seconds of time available. There are eight possible choices for individual jobs. The variables, each of which must take on the value 0 or 1, determine which of the jobs are selected. Imagine that this constraint is part of a larger formulation. Down in the branch-and-cut tree, it might happen that the variable x_2 is fixed to 1 at some node (e.g., due to previous branching on that variable). The right-hand side of the above constraint may then be updated, reducing it by 30 units. If we then compute upper bounds on each of the remaining variables and round, we deduce $x_1 = x_3 = x_4 = x_7 = x_8 = 0$. These fixings are the result of one pass of bound strengthening. A second pass allows us to conclude $x_5 = 1$, and a third pass $x_6 = 0$.

As noted earlier, node presolve attacks both the lower and upper bounds simultaneously. By deriving tighter bounds on integer variables, it often increases the objective value of the associated relaxation and thus improves the lower bound. By excluding fractional values, it also increases the likelihood that the solution of the linear-programming relaxation at a node is integer feasible, thus potentially improving the upper bound. As discussed in the next section, node presolve is also used as part of the node heuristics.

Returning to the general case, it was important first to make the code incremental so that it could benefit, during branch-and-cut, when the processing of a node was followed immediately by the processing of one of its children. Also important were good choices for defaults. The choices we made were the following:

- The number of repeated applications was limited; instances exist where, unrestricted, long sequences of reductions can occur.

- Apply only to non-(0,±1) matrices; otherwise, no rounding occurs. If there is no rounding, all bounds that are deduced are implied by the LP. We are interested in mixed-integer reductions.

- Node presolve is applied for the first 100 nodes processed, then optionally discontinued depending upon its effectiveness during those initial 100 nodes. Every 100 nodes thereafter, node presolve is applied again, and optionally reactivated.

3.2. NODE HEURISTICS

The idea of a node heuristic is simple. Instead of waiting for branching to force integrality, we consider isolating the MIP at a particular node and applying local operations within that node to determine an integral solution. Typically these operations make use of the x vector generated as the solution of the linear-programming relaxation at that node and then perform some sort of "dive," fixing an increasingly large number of variables until either a new, best integral solution is found, a new *incumbent*, or the fixings that are made result in infeasibility or an objective value worse than the current incumbent.

What are the reasons heuristics may help? First, having a good integral solution as early as possible helps the overall branch-and-cut procedure. It helps in reducing the number of nodes that are processed, and it speeds the processing of individual nodes by providing a tight objective-cutoff for the dual simplex algorithm (the method of choice for reoptimizing at the nodes). Second, in many real-world problems, high quality integral solutions are of much more importance than proofs of optimality.

We used the following ingredients in our implementations:

- List fixing with different orders: All our heuristics involve diving, employing a sequence of fixings. These fixings can, for example, be done with basic variables first or non-basics first, or in some combination. Each alternative gives a different sequence.

- Periodic linear solves: We optionally solve LPs during the dive. These solves are expensive, relative to other steps, and so we limit the number of solves to five.

- Reduced-cost fixing: When an LP is solved, new reduced-cost information is generated, and that can be used to determine new reduced-cost fixings.

- Quick and dirty node presolve: Here we leverage the existence of the node presolve by using a restricted version to deduce implied fixings from the preceding fixings.

With these ingredients, five different heuristics were implements. Each of the five is applied at the root by default. The "most successful" is applied periodically at subsequent nodes.

3.3. CUTTING PLANES

This is the area in which the bulk of the theoretical work has been done. CPLEX 6.5 includes the implementation of six different kinds of cutting plane routines, each with its own defaults determining when and how often it is applied.

The kinds of cuts that are applied are listed below together with a limited set of references.

- **Knapsack Covers:** Crowder, Johnson, and Padberg (1983); Weismantel (1997).
- **GUB Covers:** Gu, Nemhauser, and Savelsbergh (1998).
- **Flow Covers:** Padberg, Van Roy, and Wolsey (1985); Gu, Nemhauser, and Savelsbergh (1999).
- **Cliques:** Johnson and Padberg (1983); Atamturk, Nemhauser, and Savelsbergh (1998).
- **Implied Bounds:** Hoffman and Padberg (1991)
- **Gomory Mixed-Integer Cuts:** Gomory (1960).

Knapsack covers were the first cuts to find extensive use in general purpose solvers, and have been successfully used in commercial codes for several years. GUB Covers are a mild extension of knapsack covers that exploit the existence of GUB constraints ($\sum_j x_j \leq 1$) intersecting a given knapsack constraint. Flow covers can be viewed as closely related to knapsacks. This class of constraints appears to be very special-purpose, but is really quite general. The separation step, that of actually finding a flow cover violated by a given x vector, uses the same approach as for knapsack covers. The lifting step, which attempts to strengthen the initially found cut by increasing the dimension of its intersection with the underlying convex hull of integral feasible solutions, is particularly important for this class, but also quite complex. Cliques are touched upon briefly in Example 2. Implied bound cuts are discussed below. Gomory cuts are the classic mixed-integer cuts introduced by Gomory in 1960, and recently reinvestigated by Balas, et al. (1996). As we shall see, the power of these cuts, long neglected, is significant.

Knapsack Extensions. Knapsack covers have been recognized in CPLEX since version 3.0. The lifting was improved significantly in version 5.0 using ideas suggested by Martin and Weismantel (1995). In version 6.5 the applicability of the existing routines was extended in the following ways:

- **Equality constraints:** Equality constraints that would be candidates for knapsack separation, if they were inequalities, are replaced by pairs of opposing inequalities.
- **Continuous variables:** Where possible, continuous variables are replaced by appropriate bounds, depending upon the sign of the corresponding constraint coefficient and the sense of the constraint.
- **Surrogate knapsacks:** Given a collection of constraints of the form

  $$\sum_{j=1}^{n} x_j \geq b, \quad x_j \leq a_j y_j \ (j = 1, \ldots, n), \quad y_j \in \{0,1\} \ (j = 1, \ldots, n).$$

  We replace each x_j by the expression $a_j y_j$.

Implied Bounds. It is standard wisdom in integer programming that one should disaggregate variable upper bound constraints on sums of variables. These are constraints of the form:

$$x_1 + \ldots + x_n \leq (u_1 + \ldots + u_n)y, \ y \in \{0,1\}.$$

where u_j is a valid upper bound on $x_j \geq 0$ $(j = 1, \ldots, n)$. This single constraint is equivalent, given the integrality of y, to the following collection of "disaggregated" constraints:

$$x_j \leq u_j y \ (j = 1, \ldots, n)$$

The reason the second, disaggregated formulation is preferred is that, while equivalent given integrality, its linear-programming relaxation is stronger. However, given the ability to automatically disaggregate the first constraint, these "implied bound" constraints can be stored in a pool and added to the LP only as needed. Where n is large this latter approach will typically produce a much smaller, but equally effective LP.

Gomory Mixed-Integer Cuts. Gomory mixed-integer cuts were among the first introduced but for years have had the unfortunate reputation that they were not effective in practice. That reputation seems to be based upon two phenomena. First, Gomory cuts are often "dense," adding a significant number of nonzeros to the constraint matrix. The linear-programming solvers of the day just couldn't handle the resulting increased density. Second, in the early tests, cuts were applied in a way that today seems obviously bad, but was quite natural at the time. Gomory's algorithm, not simply the cuts he introduced, was being viewed as a potential complete solution to integer programming, just as the simplex method was a "complete" solution for linear programming. Thus, instead of adding groups of cuts, where a group consists of as many "good" violated cuts as could be found, cuts were added one at a time, and branching was ignored. The result was that convergence was either very slow, or simply did not occur.

Times have changed. Linear-programming solvers are better and we know cuts should be added in groups; moreover, we don't expect cuts to solve the entire problem. We now realize how strong an ally intelligent branching can be. With these thoughts in mind, Gomory cuts become a very natural choice. They are the most general cuts that we have (one can always find a violated Gomory mixed-integer cut), they are easy cuts to implement, and they have the interesting, well-known property that they combine two important ideas: Rounding and disjunction. In effect, through disjunction they capture some of the effect of branching without increasing the number of active nodes.

There is a nice geometry corresponding to Gomory mixed-integer cuts, as well as a simple, straightforward algebraic derivation. Given the importance of these cuts, we sketch both.

First, the geometry. Consider a simple mixed inequality $x + y \geq 3.5$, where $x \geq 0$ and y is integral (not necessarily nonnegative). The feasible region for the linear-programming relaxation has exactly one fractional extreme point, $(0, 3.5)$. Removing this point is easy. We round, $y \leq \lfloor 3.5 \rfloor$ and $y \geq \lceil 3.5 \rceil$, and intersect the feasible region with the resulting pair of inequalities. The result is a pair of disjoint polyhedra, in effect, a *disjunction*. This disjunction can be removed by taking the convex hull of the two polyhedra. Equivalently, we can add the cutting plane $2x + y \geq 4$ to the original defining inequality. This cut is exactly the associated Gomory mixed-integer cut, perhaps more properly viewed as a *mixed-integer rounding* cut in this case. See Wolsey (1998) for a further discussion of these issues.

Note that it is sometimes observed that Gomory cuts are weak relative to some of the combinatorially-derived cuts, those that can be shown to be facet defining. However, at least in this case, the Gomory cut is as strong as it can be. It defines the integer hull.

Now for the algebra. Let $y, x_j \in \mathbf{Z}_+$, and consider the equation

$$y + \sum_j a_{ij} x_j = d = \lfloor d \rfloor + f, \quad f > 0.$$

Think of this equation as a row of an optimal simplex tableau. Now rounding, and introducing the notation $a_{ij} = \lfloor a_{ij} \rfloor + f_j$, we may define

$$t = y + \sum (\lfloor a_{ij} \rfloor x_j : f_j \leq f) + \sum (\lceil a_{ij} \rceil x_j : f_j > f) \in \mathbf{Z}_+.$$

Subtracting yields

$$\sum (f_j x_j : f_j \leq f) + \sum ((f_j - 1) x_j : f_j > f) = d - t.$$

Now applying a disjunction, effectively branching on t, we have

$$\begin{aligned} t \leq \lfloor d \rfloor \quad &\Longrightarrow \quad \textstyle\sum (f_j x_j : f_j \leq f) \geq f, \text{ and} \\ t \geq \lceil d \rceil \quad &\Longrightarrow \quad \textstyle\sum ((1 - f_j) x_j : f_j > f) \geq 1 - f \end{aligned}$$

Dividing by the right-hand side in each case, we obtain a quantity that is always nonnegative and, for the corresponding regions of t values, is at least 1. Hence, the sum is at least 1:

$$\sum \left(\frac{f_j}{f} x_j : f_j \leq f \right) + \sum \left(\frac{1 - f_j}{1 - f} x_j : f_j > f \right) \geq 1$$

This inequality is a Gomory mixed-integer cut. For simplicity, we have described it for a pure integer constraint, but adding continuous variables is easy and really contributes nothing to understanding these inequalities.

We remark that, for a variety of reasons, it has become standard in courses on integer programming to present Chvátal-Gomory integer rounding cuts. These cuts are closely related to the above, but are simpler to describe. They also have very nice, easily described theoretical properties. On the other hand, even for pure integer problems, it is the mixed-integer cuts that are computationally most useful. And, as we are about to see, the mixed-integer cuts *really* do work.

3.4. COMPUTATIONAL RESULTS

We present results for two test-sets of models. The first is MIPLIB 3.0, see Bixby, et al. (1998). This is a public-domain collection of problems that is used by many as the standard test-set for evaluating mixed-integer programming codes. To obtain the models and a complete set of statistics, see `http://www.caam.rice.edu/~bixby/miplib/miplib.html`

We ran the following test, comparing CPLEX 6.0, which contains none of the enhancements described in this section, and CPLEX 6.5. The tests were run on a 500 MHz DEC Alpha 21264 computer with 1 Gigabyte of physical memory. Runs were made with a time limit of 7200 seconds.

The MIPLIB test-set includes 59 models. Of these 59, 22 were solved[6] with both codes in less than ten seconds using default settings. Of the remaining 37, ten hit the time limit with version 6.0 but were solved with version 6.5. The geometric mean of the CPLEX 6.5 solution times for these ten models was 48.5 seconds. Removing these ten leaves 27 models. Eight of these 27 models were solved by neither code. In these eight cases, we compared the gap between the incumbent and the best bound at termination. Version 6.0 produced a gap that was better in one case, by about 0.1%. Taking the ratios of the percentage gaps in all eight cases, dividing the 6.0 gap by the 6.5 gap, and taking the geometric mean yielded 3.3. Thus, the mean gap for version 6.5 was 3.3 times better. Removing the eight models that were solved by neither code left 19 models. These are the ones that were (a) reasonably hard, and (b) solvable by both codes. The geometric mean of the solution-time ratios in this case was 11.2. That is, CPLEX 6.5 was over 11 times faster on average on these models. These results are summarized in the following table:

MIPLIB 3.0 - Defaults
CPLEX 6.0 versus CPLEX 6.5
7200 second time limit

- 22 models solved by both codes in less than 10 seconds
- 10 models solved by CPLEX 6.5 and not CPLEX 6.0
- 8 models solved by neither: CPLEX 6.5 3.3 times better gap
- 19 models solved by both: CPLEX 6.5 11.2 times faster

There is a second test-set of largely proprietary models that we prefer to MIPLIB 3.0 in evaluating performance. This test-set was assembled about two years ago, from the CPLEX model library, in the following way. On some machine (what was then the fastest machine available to us), and using the then current version of CPLEX, we ran each model using defaults. Any model that solved to optimality in less than 100

[6] The default CPLEX optimality tolerance for MIPs is a gap of 0.01%

seconds was excluded from further testing. We then made an extensive set of runs on the remaining models, some runs extending to several days, trying a variety of parameter settings. All the models that could be solved in this way were included in the test-set, with the exception of a few models (less than five) that were solvable, but took over about one-half day to solve. With these exceptions, one might characterize the resulting test-set as the models that appeared to be difficult but solvable, assuming tuning was allowed. Statistics for the 80 models in the test-set are given in Table 2 in the appendix ("GIs" stands for *general integer* variables). For the present paper, we made several kinds of runs. All used a 500 MHz DEC Alpha 21264 system and were run with a time limit of 7200 seconds.

First, we ran CPLEX 6.5 and CPLEX 6.0 with defaults. The result was that 6.5 did not solve three of the models to within default tolerances within the allotted two hours (MIP09, which is the MIPLIB 3.0 model *arki001*, MIP40, and MIP50). CPLEX 6.0 did not solve 31 models. The three models not solved by 6.5 were among these 31. Excluding these three, there was one model where the solution times were identical (and small). Version 6.5 was faster in 66 of the remaining cases, and 6.0 was faster in ten cases. Dividing the 6.0 time by the 6.5 time and taking the geometric mean[7] gave a mean speedup of 22.3.

We next compared CPLEX 6.5 running defaults with tuned CPLEX 6.0 times, using the best parameter settings that are known to us. Version 6.5 was faster in 56 cases, and 6.0 in 22 cases. The mean speedup for version 6.5 using default settings compared with 6.0 using tuned settings was 3.8.

Finally, we performed two kinds of tests to evaluate the effects of some of the mixed-integer programming features that have been discussed in this paper. In the first test we started with defaults, turned off individual features one at a time, and measured, using geometric means of ratios of solve times, the effects of these changes. Our second set of tests was performed, effectively, in the opposite direction. We turned off all six kinds of cutting planes, made a set of test runs, and then turned on the individual cuts one at at time, making comparisons using ratios and geometric means. The results are given below:

---

[7]MIP45 was excluded from these ratios. It terminated prematurely with CPLEX 6.0 because of excessive basis singularities in the simplex method while solving some node LP.

Performance Impact: Relative to Defaults

Cuts		Other	
Implied bounds	0%	Node presolve	9%
Cliques	0%	Node heuristics	9%
GUB covers	0%		
Flow covers	12%		
Covers	16%		
Gomory cuts	35%		

Performance Impact: Individual Cuts

Implied bounds	-1%
Cliques	0%
GUB covers	10%
Flow covers	18%
Covers	58%
Gomory cuts	97%

The big winner here, and perhaps the biggest surprise, was Gomory cuts. They were clearly the most effective cuts in our tests.

4. EXAMPLES

We close with some examples. In the previous sections we have attempted to demonstrate that great progress has been made in building general-purpose mixed-integer solvers, solvers that run well with default settings. This development is critical to the wider use of mixed-integer programming in practice. Most users of mixed-integer programming are not interested in the details of how the codes work. They simply want to be able to run a code and get results. Nevertheless, there still are, and probably always will be, many examples of interesting, important MIPs that are solvable, but not without taking advantage of problem structure in some special way.

Example 1 The first example is from a customer who was primarily interested in finding feasible solutions. His criteria was, stop after finding a feasible integral solution with gap less than 1%. CPLEX 6.0 was incapable of meeting this criteria. Indeed, this model was left running for a period of several days on a fast workstation, a 600 MHz Alpha 21164 computer, and not a single feasible solution was found. Below is a CPLEX 6.5 run for this model using a 500 MHz DEC Alpha 21264 computer:

```
Problem 'unnamed.mps.gz' read.
New value for passes for generating fractional cuts: 0
New value for mixed integer optimality gap tolerance: 0.01
Reduced MIP has 7787 rows, 7260 columns, and 22154 nonzeros.
Clique table members: 533
Root relaxation solution time =    0.37 sec.

        Nodes                                                Cuts/
   Node  Left     Objective  IInf  Best Integer     Best Node    ItCnt     Gap

      0     0  -2.8298e+07   224                 -2.8298e+07     3095
               -2.7769e+07   160                   Cuts:  616     4173
               -2.7720e+07   156                   Cuts:  118     4548
               -2.7703e+07   176                    Cuts:  54     4790
               -2.7689e+07   177                    Cuts:  36     4916
               -2.7685e+07   180                    Cuts:  29     4999
               -2.7685e+07   181                Flowcuts:  6     5035
    100   100  -2.7590e+07    58                 -2.7684e+07     6174
    200   200  -2.7446e+07    12                 -2.7684e+07     6673
*   239   236  -2.7434e+07     0  -2.7434e+07   -2.7684e+07     6843   0.91%

GUB cover cuts applied:  6
Cover cuts applied:  44
Implied bound cuts applied:  66
Flow cuts applied:  295
Fractional cuts applied: 212

Integer optimal solution (0.01/1e-06):  Objective =   -2.7433577522e+07
Current MIP best bound =   -2.7684321743e+07 (gap = 250744, 0.91%)
Solution time =   31.90 sec.  Iterations = 6843  Nodes = 240 (235)
```

So, this *is* an example of a model that now solves well with default settings. One interesting aspect of the solution is that it is a case in which several features combined to produce the result. Clearly cuts were involved, and, although it is not clear from the output, the node presolve was also important. Each of several, separate features helps, but it's the combination that leads to a solution.

Example 2 Our second example illustrates how defaults are sometimes not enough. In CPLEX 6.5, several degrees of *probing* on binary variables are available. These options are not turned on by default. As is well known, even with an efficient implementation of probing, computation times can experience a combinatorial explosion.

Probing occurs in three phases in CPLEX 6.5 when activated at its "highest level." In the first phase, it is applied to individual binary variables, as suggested in Brearley, et al. (1975). Thus, each binary variable is fixed in turn to 0 and then to 1, applying bound strengthening after each such fixing. For an individual variable, the result can include the fixing of the variable being probed (when one of the tested values forces the infeasibility of the whole model), implied bounds on continuous variables—hence, implied bound cuts become stronger when probing is activated—and 2-cliques. The 2-cliques that result from this first phase are collected and merged together with the cliques given by GUB constraints, those that are explicit in the original formulation. The result is an initial clique table. This table is then further expanded in a second phase by applying *lifting* directly to the cuts in this table. See Suhl and Szymanski (1994). This idea was suggested to us by Johnson (1999).

Finally, since, in general, there can be an exponential number of maximal cliques, it is not possible to explicitly store all such cliques. Within the branch-and-cut tree we use the clique table and the current solution to the linear-programming relaxation, as suggested by Atamturk, et al., (1998), to generate further clique cuts.

When we first tried to solve the present example model, it appeared not to be possible with CPLEX 6.0. The optimal objective value of the root linear-programming relaxation was 1.0, and the best-bound value never moved above 2.0, even though several parameter settings were tried and several long runs were made.

In the CPLEX 6.5 run displayed below, probing was set to its highest level. The result was that a large number of clique inequalities were generated at the root. These were crucial, pushing the lower bound at the root to 20.8. At the same time, one of the five heuristics that are applied at the root succeeded in finding a feasible solution of value 21. Since, as the output indicates, the objective function in the model could be proven to take on only integral values, it followed that 21 was optimal, and the run terminated without branching.

```
Problem 'unnamed.lp.gz' read.
New value for probing strategy: 2
Elapsed time   10.22 sec. for 47% of probing
Elapsed time   20.30 sec. for 94% of probing
Probing time =   21.53 sec.
Clique table members: 1068
Root relaxation solution time =  142.18 sec.
Objective is integral.

        Nodes                                               Cuts/
   Node  Left     Objective  IInf  Best Integer     Best Node     ItCnt     Gap
      0     0        1.0000  4766                      1.0000     20704
                    20.8000   439                 Cliques:  500     36839
                    20.8000   203                  Cliques:  17     40402
Heuristic: feasible at 22.0000, still looking
Heuristic complete
*     0+    0       21.0000     0       21.0000       20.8000     40402   0.95%

Clique cuts applied:  349

Integer optimal solution:  Objective =    2.1000000000e+01
Solution time =  792.62 sec.  Iterations = 40402  Nodes = 0
```

Example 3 The *noswot* model is one of the smaller, but more difficult models in the MIPLIB 3.0 test-set. It has only 128 variables, 75 of which are binary, and 25 of which are general integers.

This model is very difficult to solve with the currently available branch-and-cut codes. With CPLEX 6.0 it appeared to be unsolvable, even after days of computation. It *is* now solvable with CPLEX 6.5, running defaults, but the solution time of 22445 seconds (500 MHz DEC Alpha 21264), and the enumerated 26,521,191 branch-and-cut nodes, are not a pleasant sight.

In contrast, this model does suddenly become easy if the following eight constraints are added:

```
c184: X21 - X22 >= 0
c185: X22 - X23 >= 0
c186: X23 - X24 >= 0
c187: 2.08 X11 + 2.98 X21 + 3.47 X31 + 2.24 X41
      + 2.08 X51 + 0.25 W11 + 0.25 W21 + 0.25 W31 + 0.25 W41 + 0.25 W51
      <= 20.25
c188: 2.08 X12 + 2.98 X22 + 3.47 X32 + 2.24 X42
      + 2.08 X52 + 0.25 W12 + 0.25 W22 + 0.25 W32 + 0.25 W42 + 0.25 W52
      <= 20.25
c189: 2.08 X13 + 2.98 X23 + 3.4722 X33 + 2.24 X43
      + 2.08 X53 + 0.25 W13 + 0.25 W23 + 0.25 W33 + 0.25 W43 + 0.25 W53
      <= 20.25
c190: 2.08 X14 + 2.98 X24 + 3.47 X34 + 2.24 X44
      + 2.08 X54 + 0.25 W14 + 0.25 W24 + 0.25 W34 + 0.25 W44 + 0.25 W54
      <= 20.25
c191: 2.08 X15 + 2.98 X25 + 3.47 X35 + 2.24 X45
      + 2.08 X55 + 0.25 W15 + 0.25 W25 + 0.25 W35 + 0.25 W45 + 0.25 W55
      <= 16.25
```

Where do these constraints come from? Some time ago, one of the authors of this paper discovered what looked like a high degree of symmetry among some of the variables in the model: X21, X22, X23, and X24. He tried the following idea. There are 24 different ways of forming triples of constraints from these variables, in the way indicated above by constraint c184–c186, with each of these triples removing the symmetry on the variables. Being uncertain that his symmetry observation was really valid for the entire model, he then simply solved the 24 individual instances, and, in so doing, the entire model.

As some explanation for why this approach, creating 24 related instances, could be effective, consider taking several disjoint copies of the same model and putting them side by side in a single model. Doing so is not a good idea; the models, even if they are slightly different, should be solved individually. However, at least for pure LPs, something reasonable will happen, and the total solution time will grow in some way that is not too-highly nonlinear in the number of disjoint copies that have been combined. Indeed, in the case of a barrier algorithm, the total computation time can be expected to grow something close to linearly in the number of copies. However, doing this kind of replication with an integer program is an entirely different matter. There the number of nodes in the search, and hence the solution time, can be expected to grow like the *product* of the number of nodes in the individual search trees.

Returning to the *noswot* instance, the above result prompted one of our co-workers, Irv Lustig (1999), to "reverse engineer" the original model, and give a representation using the OPL modeling language (see Van Hentenryck (1999)). Another co-worker, Jean-Francois Puget (1999), then studied this representation and noticed that it could be given an interpretation as a resource allocation model on five machines, with scheduling, horizon constraints, and transition times. It was then clear that four of the five "machines" were indeed identical, and hence that constraints c184-c186 were valid. In other words, it was necessary to solve only one of the 24 instances mentioned above. In addition, it was also observed that the transition-time constraints could be strengthened by adding five additional cuts that exploited the fact that there was actually a minimum positive transition cost of 0.25. Essentially the argument was that if a machine performs k different jobs, then it must pay at least $0.25(k-1)$ in transition cost. These last constraints are also due to Puget.

With these added constraints, the model becomes solvable. Here are the results using CPLEX 6.0 and 6.5 on a 400 MHz Pentium II Laptop running a Linux operating system:

CPLEX 6.0:	142 seconds	169090 nodes
CPLEX 6.5:	16 seconds	9807 nodes

So, this is a case where good modeling makes the biggest difference, but having a stronger code is also valuable.

5. SUMMARY

In this paper we have discussed recent advances in linear and mixed-integer programming. The linear-programming improvements were most striking for larger models, but are effective for small and medium-sized models as well. One important consequence of this work is that for large models barrier algorithms are no longer dominant; each of primal and dual simplex, and barrier is now the winning choice in a significant number of cases.

For mixed-integer programming, the improvements were dramatic. These resulted from mining an extensive backlog of theoretical ideas from the scientific literatures for integer programming and combinatorial optimization. Particular attention was given to developing good default implementations of these ideas so that they could be applied in concert, each helping on the problems to which they applied, while causing a minimal degradation in performance when they didn't apply.

Acknowledgments

The authors would like to thank John Gregory and Irv Lustig, both of ILOG, for carefully reading this manuscript and contributing significantly to improving the exposition.

Appendix: Problem Size Statistics

Table 1 Large LP Statistics

Model	Rows	Columns	Nonzeros	Model	Rows	Columns	Nonzeros
LP01	10295	50040	150110	LP36	30190	57000	623730
LP02	13005	77133	361567	LP37	30258	492266	1162517
LP03	14738	33025	151383	LP38	31770	272372	829040
LP04	15014	37372	103866	LP39	33440	56624	161831
LP05	15051	34553	132295	LP40	34994	87510	208179
LP06	15349	35215	162709	LP41	35519	43582	557466
LP07	15455	59942	225514	LP42	35645	34675	208769
LP08	15540	23752	86753	LP43	36400	92878	246006
LP09	16223	28568	88340	LP44	38782	261079	1508199
LP10	16768	39474	203112	LP45	39951	125000	381259
LP11	17681	165188	690273	LP46	41340	64162	370839
LP12	18262	23211	136324	LP47	41344	163569	1928534
LP13	18750	84375	9993717	LP48	41366	78750	2110518
LP14	19103	33490	276895	LP49	43387	107164	189864
LP15	19374	180670	5392558	LP50	43687	164831	722066
LP16	19519	45832	124280	LP51	44150	200077	4966017
LP17	19844	55528	152952	LP52	44211	37199	321663
LP18	19999	85191	170369	LP53	47423	81915	228565
LP19	21019	115761	728432	LP54	48097	150138	1195800
LP20	22513	99785	337746	LP55	48548	163200	617683
LP21	22797	63995	172018	LP56	54447	326504	1807146
LP22	23610	44063	154822	LP57	55020	117910	391081
LP23	23700	23005	169045	LP58	55463	191233	840986
LP24	23712	31680	81245	LP59	60384	100078	485414
LP25	24377	46592	2139096	LP60	63856	144693	717229
LP26	26618	38904	1067713	LP61	66185	157496	418321
LP27	27349	97710	288421	LP62	67745	111891	305125
LP28	27441	15128	96118	LP63	69418	612608	1722112
LP29	27899	26243	261968	LP64	72258	226090	2242086
LP30	28240	55200	161640	LP65	84840	316800	1899600
LP31	28420	164024	505253	LP66	95011	197489	749771
LP32	29002	111722	2632880	LP67	99578	326504	2102273
LP33	29017	20074	2001102	LP68	105127	154699	358171
LP34	29147	9984	1013168	LP69	108393	112955	602948
LP35	29724	98124	196524	LP70	118158	487427	974854

Table 1 (continued) Large LP Statistics

Model	Rows	Columns	Nonzeros	Model	Rows	Columns	Nonzeros
LP71	123964	93288	459680	LP81	269640	1205640	6481640
LP72	125211	159109	457198	LP82	280756	920198	5936426
LP73	129181	467192	1025706	LP83	319256	638512	1231403
LP74	155265	377918	930166	LP84	344297	559428	1909649
LP75	175147	358239	1211488	LP85	363458	146096	11470110
LP76	179080	707556	1570514	LP86	589250	1533590	5327318
LP77	185929	189867	2787708	LP87	716772	1169910	2511088
LP78	186441	23732	397080	LP88	1000000	1685236	3370472
LP79	209760	363092	1061495	LP89	1204750	1229623	4693571
LP80	238969	772273	5795991	LP90	1709857	1903725	4959650

Table 2 Large MIP Statistics

Model	Rows	Columns	Binaries	GIs
MIP01	230	2025	1800	0
MIP02	759	17561	17561	0
MIP03	4089	121871	121870	1
MIP04	4116	41428	41427	1
MIP05	823	8904	8904	0
MIP06	426	7195	7195	0
MIP07	1095	11005	10940	65
MIP08	1838	807	807	0
MIP09	1048	1388	415	123
MIP10	2597	2288	1166	1122
MIP11	123	133	39	32
MIP12	105	117	34	30
MIP13	91	104	30	28
MIP14	8619	5428	1305	2
MIP15	37	526	526	0
MIP16	396	162	146	8
MIP17	631	783	28	0
MIP18	2176	6000	6000	0
MIP19	113	392	391	0
MIP20	236	1282	1277	0
MIP21	827	961	152	0
MIP22	2588	435	435	0
MIP23	15	154	0	153
MIP24	852	1337	19	0
MIP25	80	500	500	0
MIP26	4036	769	190	0
MIP27	41	49	0	30
MIP28	516	47311	47311	0
MIP29	582	55515	55515	0
MIP30	363	1298	1254	0

Table 2 (continued) Large MIP Statistics

Model	Rows	Columns	Binaries	GIs
MIP31	2291	1992	174	12
MIP32	6256	8537	197	0
MIP33	1392	1224	240	168
MIP34	1392	1224	240	168
MIP35	1248	1224	384	336
MIP36	1368	1152	216	168
MIP37	1224	1152	336	336
MIP38	2407	1214	802	0
MIP39	3147	2505	388	1
MIP40	192	845	845	0
MIP41	1799	1008	0	1008
MIP42	43	51	0	39
MIP43	146	578	444	0
MIP44	2094	5592	443	3212
MIP45	684	1564	235	0
MIP46	68	151	150	0
MIP47	13	151	150	0
MIP48	12	151	150	0
MIP49	148	1280	1280	0
MIP50	788	645	140	0
MIP51	212	260	259	0
MIP52	2054	10724	10724	0
MIP53	908	129	31	0
MIP54	4480	10958	96	0
MIP55	291	422	98	0
MIP56	2280	1090	0	1090
MIP57	36	87482	87482	0
MIP58	176	548	548	0
MIP59	755	2756	2756	0
MIP60	45	86	55	0
MIP61	246	240	64	0
MIP62	1192	840	48	0
MIP63	2984	1451	1451	0
MIP64	291	556	300	15
MIP65	249	690	690	0
MIP66	314	5111	41	0
MIP67	20022	17665	17664	0
MIP68	23259	29342	13215	0
MIP69	524	1197	1100	96
MIP70	331	45	45	0
MIP71	146	578	444	0
MIP72	42	17419	17419	0
MIP73	3228	15541	15540	0
MIP74	1359	1959	0	1959
MIP75	234	378	168	0
MIP76	234	378	168	0
MIP77	4277	2417	1364	0
MIP78	845	3345	235	0
MIP79	10108	3836	1862	0
MIP80	27	26306	26306	0

References

[1] E. D. Anderson and K. D. Anderson (1995), Presolving in Linear Programming, *Mathematical Programming*, 71, No. 2, pp. 221–245.

[2] A. Atamturk, G. L. Nemhauser and M. W. P. Savelsbergh (1998), Conflict Graphs in Integer Programming, Report LEC-98-03, Georgia Institute of Technology.

[3] E. Balas, S. Ceria, G. Cornueljols and N. Natraj (1996), Gomory Cuts Revisited, *Operations Research Letters*, 19, pp. 1–10.

[4] R. E. Bixby, S. Ceria, C. M. McZeal and M. W. P. Savelsbergh (1998), An Updated Mixed Integer Programming Library: MIPLIB 3.0, *Optima*, 54, pp. 12–15.

[5] A. L. Brearley, G. Mitra and H. P. Williams (1975), Analysis of Mathematical Programming Problems Prior to Applying the Simplex Algorithm, *Mathematical Programming*, 8, pp. 54–83.

[6] W. J. Carolan, J. E. Hill, J. L. Kennington, S. Niemi and S. J. Wichmann (1990), An Empirical Evaluation of the KORBX Algorithms for Military Airlift Applications, *Operations Research*, 38, No. 2, pp. 240–248.

[7] V. Chvatál (1983). *Linear Programming*, Freeman, New York.

[8] H. P. Crowder, E. L. Johnson and M. W. Padberg (1983), Solving Large-Scale Zero-One Linear Programming Problems, *Operations Research*, 31, pp. 803–834.

[9] J. R. Gilbert and T. Peierls (1988), Sparse Partial Pivoting in Time Proportional to Arithmetic Operations, *SJSSC*, 9, pp. 862–874.

[10] R. E. Gomory (1960), An Algorithm for the Mixed Integer Problem, RM-2597, The Rand Corporation.

[11] Z. Gu, G. L. Nemhauser and M. W. P. Savelsbergh (1998), Lifted Cover Inequalities for 0-1 Integer Programs, *INFORMS Journal on Computing*, 10, pp. 417–426.

[12] Z. Gu, G. L. Nemhauser and M. W. P. Savelsbergh (1999), Lifted Flow Covers for Mixed 0-1 Integer Programs, *Mathematical Programming*, 85, pp. 439–467.

[13] K. Hoffman and M. Padberg (1991), Improving Representations of Zero-one Linear Programs for Branch-and-Cut, *ORSA Journal of Computing*, 3, pp. 121–134.

[14] E. Johnson (1999), Private communication.

[15] E. L. Johnson and M. W. Padberg (1983), Degree-two Inequalities, Clique Facets, and Bipartite Graphs, *Annals of Discrete Mathematics*, 16, pp. 169–188.

[16] A. H. Land and A. G. Doig (1960), An Automatic Method for Solving Discrete Programming Problems, *Econometrica*, 28, pp. 497–520.

[17] I. J. Lustig (1999), Private communication.

[18] R. E. Marsten (1981), XMP: A Structured Library of Subroutines for Experimental Mathematical Programming, *ACM Transactions on Mathematical Software*, 7, pp. 481–497.

[19] A. Martin and R. Weismantel (1995), Private communication.

[20] M. W. Padberg, T. J. Van Roy and L. A. Wolsey (1985), Valid Linear Inequalities for Fixed Charged Problems, *Operations Research*, 33, pp. 842–861.

[21] Jean-Francois Puget (1999), Private communication.

[22] E. Rothberg and A. Gupta (1991), Efficient Sparse Matrix Factorization on High Performance Workstations—Exploiting the Memory Hierarchy, *ACM Transactions on Mathematical Software*, 17, No. 3, pp. 313–334.

[23] E. Rothberg and B. Hendrickson (1998), Sparse Matrix Ordering Methods for Interior Point Linear Programming, *INFORMS Journal on Computing*, 10, No. 1, pp. 107–113.

[24] M. W. P. Savelsbergh (1994), Preprocessing and Probing for Mixed Integer Programming Problems, *ORSA Journal on Computing*, 6, pp. 445–454.

[25] U. H. Suhl and R. Szymanski (1994), Supernode Processing of Mixed-Integer Models, *Computational Optimization and Applications*, 3, pp. 317–331.

[26] P. Van Hentenryck (1999), *The OPL Optimization Programming Language*, MIT Press.

[27] R. Weismantel (1997), On the 0/1 Knapsack Polytope, *Mathematical Programming*, 77, pp. 49–68.

[28] L. A. Wolsey (1998), *Integer Programming*, Wiley.

ON THE ROLE OF NATURAL LEVEL FUNCTIONS TO ACHIEVE GLOBAL CONVERGENCE FOR DAMPED NEWTON METHODS

H. Georg Bock
Interdisciplinary Center for Scientific Computing (IWR),
University of Heidelberg,
Im Neuenheimer Feld 368, D-69120 Heidelberg, Germany.
bock@iwr.uni-heidelberg.de

Ekaterina Kostina
As above
ekaterina.kostina@iwr.uni-heidelberg.de

Johannes P. Schlöder
As above
j.schloeder@iwr.uni-heidelberg.de

Abstract The paper discusses a new view on globalization techniques for Newton's method. In particular, strategies based on "natural level functions" are considered and their properties are investigated. A "restrictive monotonicity test" is introduced and theoretically motivated. Numerical results for a highly nonlinear optimal control problem from aerospace engineering and a parameter estimation for a chemical process are presented.

1. INTRODUCTION

It is well-known that stepsize strategies based on suitable merit functions can globalize the convergence of the damped Newton method. Experience shows, however, that the standard choices for merit functions may enforce very small stepsizes when the problems are only mildly ill-conditioned, even in the domain of full-step local convergence, thus

M.J.D. Powell and S. Scholtes (Eds.), *System Modelling and Optimization: Methods, Theory and Applications.*

making the method very inefficient. So called "natural level functions", originally introduced by Deuflhard, can avoid this effect, but up to now lack a rigorous convergence theory. The present paper presents a new view on successful globalization strategies based on these merit functions. In particular, it is shown that a stepsize criterion given by the authors (the so called "restrictive monotonicity test") provides a theoretical justification. Extensions to the related problems of constrained least squares and constrained l_1 parameter estimation problems are suggested. Numerical results for real life applications from aerospace control problems and parameter estimation for chemical processes are given.

2. NEWTON'S METHOD FOR NONLINEAR EQUATIONS

We consider a finite dimensional but possibly large system of highly nonlinear equations

$$F(x) = 0.$$

Starting from an initial guess x^0, Newton's method improves a given estimate x^k iteratively by applying the formula

$$x^{k+1} = x^k + \Delta x^k. \tag{1}$$

The increment Δx^k solves the linear system of equations

$$F(x^k) + J(x^k)\Delta x^k = 0, \tag{2}$$

$$\text{where} \quad J := \left(\frac{\partial F}{\partial x}\right).$$

The local convergence properties of this "full-step" version of Newton's method have been investigated thoroughly and may be formulated as follows.

Theorem 1 (local convergence properties) *Let $F : D \subset R^n \to R^n$ be twice continuously differentiable, $J(x)$ be nonsingular for all $x \in D$, and D be a domain. Assume further that*

$$||J(y)^{-1}(J(x + t\Delta x) - J(x))\Delta x|| \leq \omega t \, ||\Delta x||^2, \qquad \omega \leq \infty, \tag{3}$$

for all $t \in]0,1], x, y = x + \Delta x \in D$ with $\Delta x = -J(x)^{-1}F(x) \neq 0$, i.e. a global bound ω for the "curvature" exists, and that the initial guess x^0 is sufficiently near to a solution:

$$\delta^0 := \frac{\omega}{2}||\Delta x^0|| < 1. \tag{4}$$

Then the following holds:

- *if* $D^0 := B\left(x^0, ||\Delta x^0||/(1-\delta^0)\right) \subset D$, *then the sequence of iterates defined by (1) remains in* D^0,
- *there exists* $x^\star \in D^0$ *with* $F(x^\star) = 0$ *and* $x^k \to x^\star$ $(k \to \infty)$,
- *an a priori error estimate holds*

$$||x^k - x^\star|| \le (\delta^0)^k \frac{||\Delta x^0||}{1-\delta^0},$$

- *and convergence is quadratic with*

$$||\Delta x^{k+1}|| \le \frac{\omega}{2} ||\Delta x^k||^2.$$

The **proof** follows the lines of Banach's Fixed Point Theorem. Contractivity of the sequence of iterates is given by

$$\begin{aligned} ||\Delta x^{k+1}|| &= ||J(x^{k+1})^{-1}\left(F(x^{k+1}) - F(x^k) - J(x^k)(x^{k+1} - x^k)\right)|| \\ &= ||J(x^{k+1})^{-1} \int_0^1 \left(J(x^k + t\Delta x^k) - J(x^k)\right)\Delta x^k dt|| \\ &\le \int_0^1 \omega t \, ||\Delta x^k||^2 \, dt \;=\; \frac{\omega}{2} \, ||\Delta x^k||^2. \end{aligned}$$

Since from here

$$||x^{k+p} - x^k|| \le \sum_{i=0}^{p-1} ||\Delta x^{k+i}|| \le (\delta^0)^k \frac{||\Delta x^0||}{1-\delta^0},$$

we can conclude by induction that the sequence $\{x^k\}$ remains in D^0 and is a Cauchy sequence, so $x^\star$ exists. Finally, $F(x^\star) = 0$ follows from continuity and boundedness of $J(x)^{-1}$ on D^0. □

The local convergence theorem allows some interpretations.

1 The constant ω in (3) is a bound on the nonlinearity of the problem, and its inverse ω^{-1} characterizes the size of the region in which the linearization (2) is an acceptable approximation of F. Hence, condition (4) can also be read as

$$||\Delta x^0|| \le \frac{\eta}{\omega}, \tag{5}$$

for some constant $\eta < 2$, i.e. the increment step should not exceed this region.

2 In the literature, condition (3) is typically replaced by the two conditions $||J(x)^{-1}|| \leq \beta < \infty$, $||J(y) - J(x)|| \leq \gamma ||y - x||$, $\gamma < \infty$. However, $\beta\gamma$ grossly over-estimates the weaker bound ω.

3 In highly nonlinear problems, though, even for the weaker bound ω one cannot expect the initial guess to be close enough to a solution for condition (4) or (5) to hold. One may rather expect

$$||\Delta x^0|| >> \frac{1}{\omega},$$

in which case convergence of the "full-step" Newton method from x^0 cannot be hoped for.

3. GLOBALIZATION BY UNDERRELAXATION

One way to globalize the convergence of Newton's method is by damping or underrelaxation. The iterates are then defined by

$$x^{k+1} = x^k + t^k \Delta x^k, \quad t^k \in]0, 1],$$

where t^k is a relaxation factor, also called the stepsize. The stepsize t^k is chosen such that the next iterate x^{k+1} is "better" than x^k. It is determined by a line search with respect to an appropriate "merit function" or "level function" $T(x)$.

Any piecewise continuously differentiable level function which satisfies the compatibility condition

$$\Delta x^k \neq 0 \;\Rightarrow\; \frac{d}{d\varepsilon} T(x^k + \varepsilon \Delta x^k)\Big|_{\varepsilon=0+} < 0$$

is appropriate to ensure global convergence when the Jacobians are bounded away from singularity. (Note, that at a minimum $x^\star$ of a compatible level function $\Delta x^\star = 0$, hence $F(x^\star) = 0$.) The classical choice of a merit function for the underrelaxed Newton method is

$$T(x) := ||F(x)||_2^2$$

in any suitably scaled Euclidean norm of $F(x)$. For an exact or approximate line search for this level function one easily shows the property:

Lemma 1 (global convergence of damped Newton) *Assume that the level set*

$$N_\alpha := \{x \;\; | \;\; ||F(x)||_2 \leq \alpha\}$$

is compact and is contained in D, that F is twice continuously differentiable and that $J(x)$ is nonsingular on N_α. Then for all $x^0 \in N_\alpha$ there exist a stepsize sequence $\{t^k\}$ and $x^\star \in N_\alpha$ such that $x^k \to x^\star$ $(k \to \infty)$ with $F(x^\star) = 0$.

However, it is well known that already in mildly ill-conditioned problems such a stepsize strategy may be very inefficient since it may produce small stepsizes even in the domain where the full-step Newton's method converges according to Theorem 1. The reason is the following. In ill-conditioned cases, the Newton increment

$$\Delta x^k = -J(x^k)^{-1}F(x^k)$$

may be nearly orthogonal to the steepest descent direction

$$-\nabla T(x^k) = -2J(x^k)^T F(x^k),$$

so that enforcing descent of the level function leads to very small stepsizes. This is due to the fact that with high probability the cosine of the angle between two directions

$$\cos(\Delta x^k, -\nabla T) = \frac{F(x^k)^T F(x^k)}{||J(x^k)^{-1}F(x^k)||\ ||J(x^k)^T F(x^k)||} \geq \frac{1}{\text{cond } J(x^k)}$$

will actually be near its lower bound $(\text{cond } J(x^k))^{-1}$.

Example (Rosenbrock-type)
Let us consider the system of two nonlinear equations

$$F(x) = 0,\ F(x) = \begin{pmatrix} x_1/\sigma_1 \\ \left(x_2 + \frac{1}{200}(x_1 - 50)^2\right)/\sigma_2 \end{pmatrix}, \quad \sigma_1 = 1,\ \sigma_2 = \frac{1}{50},$$

with the initial estimate $x^0 = (50, 1)^T$ and the solution $x^\star = (0, -12.5)^T$. In this example, the condition number of the Jacobian $J(x)$ is near 50 for all $x \in R^2$, which is very moderate compared to the practical nonlinear equations appearing typically in BVP. One can easily check that the conditions of Theorem 1 hold. For $D = R^2$ the curvature is bounded by $\omega \leq 0.01$. Convergence for the full step method is guaranteed in $[-100, 100] \times [-100, 100]$, where $\delta(x) := ||J(x)^{-1}F(x)||\omega/2 < 1$. For the initial point x^0 we have $\Delta x^0 = -(50, 1)^T$, and the estimate $\delta^0 \leq 50.01/200$ holds. The first iteration provides $x^1 = (0, 0)^T$ and the a priori estimate $||x^1 - x^\star|| \leq 17$ holds. The application of damped Newton with $||F(x)||_2^2$ as level function, however, gives the stepsize $t^0 \approx 0.077$ as the optimal relaxation factor. Indeed, the direction of the steepest descent of the level function T at x^0, namely $-\nabla T(x^0)^T = -2F(x^0)^T J(x^0) = -100(1, 50)$, is almost orthogonal to the search direction Δx^0, the cosine of the angle between them being:

$$\frac{-\nabla T(x^0)^T \Delta x^0}{||\nabla T(x^0)|| \; ||\Delta x^0||} \approx 0.040 \text{ (which corresponds to 87.71 degrees).}$$

Thus, although the local contraction conditions are fullfilled quite well, the slight nonlinearity together with the mild ill-conditioning of the problem leads to very small stepsizes.

3.1. NATURAL LEVEL FUNCTIONS

To avoid this effect, two different ways can be followed.

Modification of the search direction. The classical way is the Levenberg-Marquardt or trust region variant. Here, the search direction is replaced by

$$\Delta x^k(\gamma) = -(J(x^k)^T J(x^k) + \gamma I)^{-1} J(x^k)^T F(x^k),$$

i.e. it is turned towards $-\nabla T(x^k)$ for large γ.

Modification of the level function. Recall that any level function of the type

$$T_A(x) = ||AF(x)||_2^2, \; A \text{ nonsingular},$$

is compatible with Newton's method. The special choice $A := J(x^k)^{-1}$ yields a level function, called the "natural level function" by Deuflhard [8], [9], with some distinctive properties.

Lemma 2 (Deuflhard [8], [9])

1 *At the iterate x^k, the Newton direction Δx^k is the steepest descent direction of $T_A(x)$ with the choice $A := J(x^k)^{-1}$.*

2 *The level function is "affine invariant", i.e. invariant with respect to any affine transformation $F(x) \to BF(x)$, where B is a nonsingular matrix.*

Proof. *For $A = J(x^k)^{-1}$ the two vectors*

$$-\nabla T_A = -2J(x^k)^T A^T AF(x^k) \quad \textit{and} \quad \Delta x^k = -J(x^k)^{-1} F(x^k)$$

are obviously collinear, and for all B

$$||J(x^k)^{-1}F(x)||_2^2 = ||(BJ(x^k))^{-1}BF(x)||_2^2. \qquad \square$$

3 If the sequence $\{x^k\}$ converges to a solution $x^\star$, then we have

$$||J(x^k)^{-1}F(x)||_2 = ||x - x^\star||_2 \left(1 + O(||x - x^\star||_2) + O(||x^k - x^\star||_2)\right).$$

From the properties of Lemma 2 one may expect, that the "natural level function" approach should not suffer from the drawbacks demonstrated by the example. It may in fact be viewed as a (local) rescaling of F to AF, such that the condition number of $AJ(x)$ is optimal – namely 1. Figure 1 shows the steepest descent directions and contour lines for both the classical and natural level functions in the case of the Rosenbrock-type example.

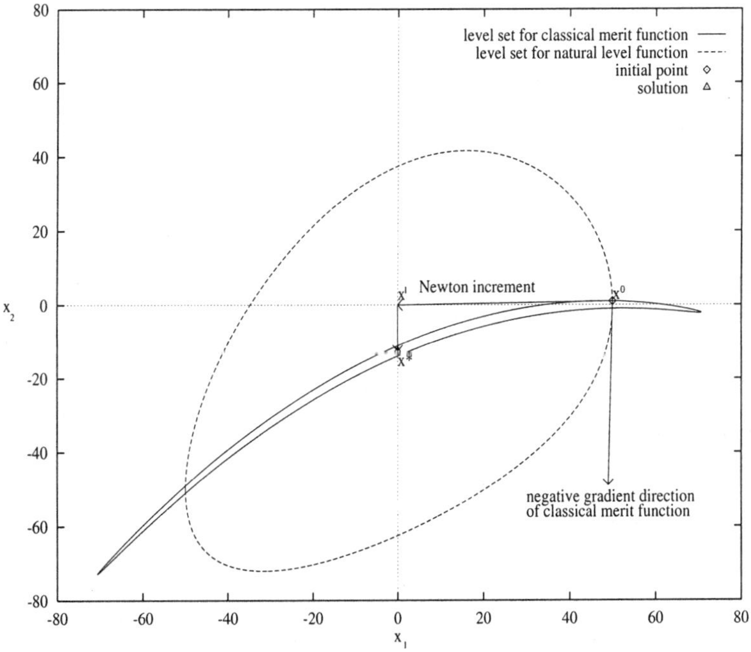

Figure 1 Rosenbrock-type example: natural level function vs classical merit function

Step size procedures based on natural level functions were first introduced by Deuflhard [8], based on a Goldstein type strategy. In Deuflhard [9], refined predictor-corrector strategies for approximate line searches were introduced, and combinations with rank-reduction strategies and quasi-Newton modifications of the Jacobian were studied. The numerical results given by Deuflhard [8], [9], and also by other authors who adopted and modified the natural level function approach, showed very good practical results in difficult applications (e.g., Ascher et al [1], Bock [3], [4], [5], and Nowak and Weimann [10]).

A major deficiency of the approach, however, is the fact that the change of level functions in each step prevents the classical descent arguments of global convergence proofs to hold. Hence no global convergence proof has been given up to now. On the contrary, similar to the Chamberlain [6] result on cycling for SQP methods using the l_1-penalty level function, examples were constructed by Ascher and Osborne [2] and by Plitt [12] that even showed the existence of two-cycles.

4. THE RESTRICTIVE MONOTONICITY TEST (RMT)

In the following, we will derive a more restrictive stepsize strategy than exact or approximate line searches on the natural level function, which is a slight modification of techniques already successfully used in practice [5].

We will first show that these techniques may be interpreted as stepsize strategies, analogous to those used in numerical methods for the discretization of ODE with invariants, thus offering a global convergence argument which is not based on descent properties. In a second step, we will discuss modifications and extensions of damped Newton method that ensure global convergence as well as efficiency.

For the natural level function in step k, we establish the following quadratic bound that will provide a descent property.

Lemma 3 (Quadratic Upper Bound, where Δx^k is the Newton direction)

$$||J(x^k)^{-1}F(x^k+t\Delta x^k)|| \le \left(1-t+\frac{t^2}{2}\omega_1(t)||\Delta x^k||\right)||J(x^k)^{-1}F(x^k)||, \tag{6}$$

where $\omega_1(t) = \sup\limits_{0<s\le t} \dfrac{||J(x^k)^{-1}\left(J(x^k+s\Delta x^k)-J(x^k)\right)||}{s||\Delta x^k||}$.

Proof.

$$\begin{aligned} & ||J(x^k)^{-1}F(x^k+t\Delta x^k)|| - (1-t)||J(x^k)^{-1}F(x^k)|| \\ \le\ & ||J(x^k)^{-1}\Big(F(x^k+t\Delta x^k)-F(x^k)+tF(x^k) \\ & \qquad -tJ(x^k)\Delta x^k + tJ(x^k)\Delta x^k\Big)|| \\ =\ & ||J(x^k)^{-1}\Big(F(x^k+t\Delta x^k)-F(x^k)-tJ(x^k)\Delta x^k\Big)|| \end{aligned}$$

$$
\begin{aligned}
&= \; ||J(x^k)^{-1} \int_0^t \left(J(x^k + s\Delta x^k) - J(x^k) \right) \Delta x^k \, ds|| \\
&\leq \; \omega_1(t) \, ||\Delta x^k||^2 \int_0^t s \, ds = \frac{t^2 \omega_1(t) \, ||\Delta x^k||^2}{2}.
\end{aligned}
$$

From the previous relations it follows that

$$
\begin{aligned}
||J(x^k)^{-1} F(x^k + t\Delta x^k)|| \; &\leq \; (1-t)||\Delta x^k|| + \frac{t^2}{2}\omega_1(t) \, ||\Delta x^k||^2 \\
&= \; \left(1 - t + \frac{t^2}{2}\omega_1(t) \, ||\Delta x^k||\right) ||J(x^k)^{-1} F(x^k)||. \qquad \square
\end{aligned}
$$

Since $\omega_1(t)$, $t \geq 0$, is monotonically nondecreasing, one may choose the damping factor $t^k = t^k(\eta)$ in terms of this quadratic upper bound such that

$$
t = \max! \quad \text{s.t.} \quad t \leq 1, \quad t\omega_1(t)||\Delta x^k|| \leq \eta,
$$

for some prescribed $\eta < 2$. This means that we choose $t^k \leq 1$ maximal such that the "Restricted Monotonicity Test" (RMT)

$$
t^k||\Delta x^k|| \leq \min\left(\frac{\eta}{\omega_1(t^k)}, ||\Delta x^k||\right), \tag{7}
$$

is fulfilled. RMT (7) together with Lemma 3 ensures that the weaker traditional Armijo-type descent condition

$$
||J(x^k)^{-1} F(x^k + t^k(\eta)\Delta x^k)|| \leq \left(1 - t^k(\eta)(1 - \frac{\eta}{2})\right) ||J(x^k)^{-1} F(x^k)|| \tag{8}
$$

holds. Note, that $t^k(\eta)$ for $\eta = 1$ would minimize the right hand side of QUB (6) with respect to t if $\omega_1(t)$ were constant, or replaced by an upper bound.

Remark The RMT ensures that the actual length $t\Delta x$ does not exceed the $1/\omega$ region in which $J(x^k)$ is a valid approximation of $J(x)$ according to the definition of ω!

Similar to Lemma 3, one can show more generally, that

Lemma 4

$$
||AF(x^k + t\Delta x^k)|| \leq \left(1 - t + \frac{t^2}{2}\omega_2^A(t) \, ||\Delta x^k||\right) ||AF(x^k)||,
$$

where $\omega_2^A(t) = \sup_{0<s\le t} \frac{||A\left(J(x^k+s\Delta x^k) - J(x^k)\right)||}{s||AF(x^k)||}$.

Lemma 5 *Assume we choose $\eta < 1$ and t^k such that*

$$t^k\omega_1(t^k)\,||\Delta x^k|| \le \eta < 1. \qquad (9)$$

Then

$$t^k\omega_2^A(t^k)\,||\Delta x^k|| \le \eta\frac{1+\eta}{1-\eta}$$

for all $A \in \{J(x^k+s\Delta x^k)^{-1} \mid s \le t^k\}$.

Proof. Let $J_0 = J(x^k)$ and $A = J(x^k+s\Delta x^k)^{-1}$ for some s, $0 < s \le t^k$. By definition

$$\begin{aligned}
\omega_2^A(t^k) &= \sup_{0<\tau\le t^k}\left(\frac{||A\left(J(x^k+\tau\Delta x^k)-J_0\right)||}{s||AF(x^k)||}\,\frac{||J_0^{-1}F(x^k)||}{||J_0^{-1}F(x^k)||}\right)\\
&\le \sup_{0<\tau\le t^k}\left(\frac{||AJ_0||\;||J_0^{-1}\left(J(x^k+\tau\Delta x^k)-J_0\right)||}{s||AF(x^k)||}\,\frac{||J_0^{-1}A^{-1}||\;||AF(x^k)||}{||J_0^{-1}F(x^k)||}\right)\\
&\le \omega_1(t^k)\,||AJ_0||\;||(AJ_0)^{-1}||.
\end{aligned}$$

Moreover, the choice of A, the definition of $\omega_1(t^k)$ and condition (9) imply

$$||J_0^{-1}(A^{-1}-J_0)|| \le s\omega_1(s)\,||\Delta x^k|| \le \eta < 1,$$

which gives the classical estimates

$$\begin{aligned}
||AJ_0|| &= ||\left(I - J_0^{-1}(J_0 - A^{-1})\right)^{-1}|| &\le \frac{1}{1-\eta},\\
||J_0^{-1}A^{-1}|| &= ||I + J_0^{-1}(A^{-1}-J_0)|| &\le 1+\eta.
\end{aligned}$$

It follows that

$$\omega_2^A(t^k) \le \omega_1(t^k)\frac{1+\eta}{1-\eta},$$

so another application of (9) provides the required result. □

Lemma 4 and Lemma 5 imply that, if η is chosen to satisfy

$$\eta\frac{1+\eta}{1-\eta} < 2, \quad \text{i.e.} \quad \eta < \frac{1}{2}\left(\sqrt{17}-3\right),$$

then all level functions for intermediate choices of Jacobians also descend. In particular, two-cycles are impossible. For example, $\eta = 1/2$ yields the property

Lemma 6 (No Two-Cycles when $\eta = 1/2$) *For all k*

$$||J(x^k)^{-1}F(x^{k+1})|| \;\leq\; (1 - \frac{3t^k}{4})||J(x^k)^{-1}F(x^k)|| \qquad (10)$$

$$||J(x^{k+1})^{-1}F(x^{k+1})|| \;\leq\; (1 - \frac{t^k}{4})||J(x^{k+1})^{-1}F(x^k)|| \qquad (11)$$

hence

$$||J(x^{k+1})^{-1}F(x^{k+2})|| \;\leq\; (1 - \frac{3t^{k+1}}{4})(1 - \frac{t^k}{4})||J(x^{k+1})^{-1}F(x^k)||,$$

so that $x^k \neq x^{k+2}$.

Proof. Due to the choice $\eta = 1/2$, inequality (10) follows immediately from (8), and inequality (11) from Lemmas 4 and 5 in the case $A = J(x^{k+1})^{-1}$. □

Note, however, that this result still does not prove global convergence.

5. PRACTICAL REALIZATION OF THE RMT

The costly evaluation of $\omega_1(t)$ can be avoided. In the quadratic upper bound of Lemma 3, one can replace $\omega_1(t)$ by the weaker estimate for the curvature

$$\begin{aligned}\omega_3(t) \;&:=\; \frac{2||J(x^k)^{-1}\left(F(x^k + t\Delta x^k) - (1-t)F(x^k)\right)||}{t^2||\Delta x^k||^2} \\ &=\; \frac{2||J(x^k)^{-1}\int_0^t \left(J(x^k + s\Delta x^k) - J(x^k)\right)\Delta x^k\, ds||}{t^2||\Delta x^k||^2}.\end{aligned}$$

The estimate $\omega_3(t)$ is easy to evaluate. Indeed, it involves only the calculation of

$$\Delta x^k \;=\; -J(x^k)^{-1}F(x^k),$$

which is necessary anyway, and of

$$\bar{\Delta x}^k \;=\; -J(x^k)^{-1}F(x^k + t\Delta x^k),$$

just as in a line search procedure for the natural level function. However, instead of a one dimensional minimization, and analogously to the more restrictive test (7), we require the conditions

$$\eta_\star \leq t^k\omega_3(t^k)||\Delta x^k|| \leq \eta^\star \qquad (12)$$

with $\eta_\star < \eta < \eta^\star$. In the numerical tests of Section 9, $\eta = 1$, $\eta_\star = 0.8\eta$ and $\eta^\star = 1.2\eta$ were used.

As $\omega_3(t)$ is continuous, a simple rootfinding procedure for $t\omega_3(t)||\Delta x^k|| - \eta = 0$ is applied to satisfy (12). A good starting value for t^k is provided by the curvature estimate $\omega_3(t^{k-1})$ of the previous iteration, namely

$$t^k_{\text{start}} := \min\left(1, \frac{\eta}{\omega_3(t^{k-1})||\Delta x^k||}\right).$$

Thus, according to our experience, at most two F-evaluations per iteration are required.

This restrictive monotonicity test works very well in practical applications. Although the rigorous proof of Lemma 6 does not hold for the weaker curvature measure used here, cycling does not occur for the Ascher–Osborne example, and has not been observed in practical applications. It is hoped that a similar proof of non-cycling, possibly for sharper η, can be found.

Nevertheless, we keep in mind that even Lemma 6 does not provide a global convergence proof for either version (7) or (12) of the RMT based on descent arguments.

6. A DIFFERENT INTERPRETATION OF THE RMT

Let us consider, instead of a single mapping F, a family $H : D\times[0,1] \in R^{n+1} \to R^n$ depending on a parameter λ such that

$$H(x^0, 0) = 0,\ H(x, 1) = F(x),$$

where x^0 is some initial solution ($\lambda = 0$). For example, we can set

$$H(x, \lambda) = F(x) - (1-\lambda)F(x^0). \tag{13}$$

Let us consider the equation

$$H(x(\lambda), \lambda) \equiv 0, \quad \lambda \in [0,1]. \tag{14}$$

Under certain assumptions (see [11]), e.g. those of Lemma 1, (14) defines a unique continuously differentiable function $x(\lambda)$ (homotopy path), and $x(\lambda)$, $\lambda \in [0,1]$, satisfies the implicit, so-called Davidenko differential equations [7]

$$\left[\frac{\partial H(x(\lambda),\lambda)}{\partial x}\right]\dot{x} = -\frac{\partial H(x(\lambda),\lambda)}{\partial \lambda}, \quad \forall\lambda \in [0,1],\ x(0) = x^0.$$

For the choice (13), the last expression takes the form

$$\dot{x} = -J(x)^{-1}F(x^0) = -\frac{1}{1-\lambda}J(x)^{-1}F(x), \quad \forall\lambda \in [0,1], \ x(0) = x^0. \quad (15)$$

Let us introduce the change of variable $\lambda = 1 - e^{-\tau}$. Then τ varies from 0 to $+\infty$ as λ varies from 0 to 1, and (13) has the form

$$H(x,\tau) = F(x) - e^{-\tau}F(x^0), \quad \tau \in [0,+\infty). \quad (16)$$

The differential equation corresponding to (16) is given by

$$\frac{dx}{d\tau} = -J(x)^{-1}F(x), \quad \forall\tau \in [0,+\infty), \ x(0) = x^0. \quad (17)$$

We can consider the problem of constructing the trajectory $x(\lambda)$ of (15) (or $x(\tau)$ of (17)) as numerical integration of the ODE (15) (or (17)). If we integrate (17) by Euler's method with stepsizes t^k we obtain

$$x^{k+1} = x^k - t^k J(x^k)^{-1}F(x^k), \quad k = 0, 1, ...,$$

which is the damped Newton method. Thus, we can view the damped Newton method as an Euler approximation of the continuous Newton equation (17).

Next we will show that the RMT is nothing but a stepsize control for Euler's method applied to the Davidenko differential equations. First we estimate the local integration error. For simplicity we only consider the first iteration step.

Lemma 7 *The local integration error of Euler's method applied to (15) is given by*

$$\varepsilon(t) := -J(x^0)^{-1}\left(F(x^0 + t\Delta x^0) - (1-t)F(x^0)\right) + O(t^3).$$

Proof. The local error is defined as

$$\varepsilon(t) = x(t) - x^0 - t\Delta x^0,$$

where $x(t)$ satisfies the invariant

$$F(x(t)) = (1-t)F(x^0). \quad (18)$$

From a Taylor series expansion of (18) we have

$$\begin{aligned}
\dot{x}(0) &= -J(x^0)^{-1}F(x^0) = \Delta x^0, \\
\ddot{x}(0) &= -J(x^0)^{-1}\left(\frac{\partial}{\partial x}J(x^0)\dot{x}\right)\dot{x} \\
&= -\frac{2}{t^2}J(x^0)^{-1}\left(F(x^0 + t\Delta x^0) - (1-t)F(x^0)\right) + O(t),
\end{aligned}$$

and $x(t) = x^0 + t\Delta x^0 + \frac{t^2}{2}\ddot{x}(0) + O(t^3)$, which imply the required result. □

From the new point of view, the RMT

$$\begin{aligned} ||t\Delta x^0|| \frac{\eta_\star}{2} &\le ||J(x^0)^{-1}\left(F(x^0 + t\Delta x^0) - (1-t)F(x^0)\right)|| \\ &\le \frac{\eta^\star}{2} ||t\Delta x^0|| \end{aligned} \tag{19}$$

is simply a stepsize control for Euler's method. By Lemma 7 the term controlled in formula (19) is an asymptotically correct estimate of the local integration error. It is kept small compared to the increment norm, which is controlled by the choice of η, in order to ensure that the Newton path is followed with a desired accuracy.

We can go one step further, if we take into account that (15) is an implicit ordinary differential equation with known invariant given by equation (18). Similar to techniques used in discretization methods for ODE or DAE with invariants, e.g. [13], we can therefore exploit the invariant for a stabilization step, which is a "back projection" of

$$x^{1,0} := x^0 + t\Delta x^0$$

to the invariant manifold, which is a curve in our case. This step can be performed by adding the correction term already computed for the RMT

$$\begin{aligned} x^{1,1} &:= x^{1,0} + \tilde{\Delta}x^0 = x^0 + t\Delta x^0 + \tilde{\Delta}x^0, \\ \tilde{\Delta}x^0 &:= -J(x^0)^{-1}\left(F(x^0 + t\Delta x^0) - (1-t)F(x^0)\right). \end{aligned}$$

Unlike Euler's method, which is of first order, the combined two step method is of second order. Note, that as soon as $t = 1$ is reached, the additional back projection step extends the quadratically convergent Newton's method to a well-known cubically convergent modification.

In terms of Newton's method, the combined scheme can be interpreted as the first two steps of a simplified full step Newton method to solve $F(x) - (1-t)F(x^0) = 0$, starting from x^0. The RMT then plays the role of a monitoring test to choose t sufficiently small, in order that contractivity (by $\eta^\star/2$) of this scheme is guaranteed. This projection could of course be repeated until the Newton path is met with a desired accuracy. A natural and necessary extension of the RMT is then to check sufficient contractivity of the iterations e.g.

$$\frac{||J(x^0)^{-1}\Big(F(x^{1,i+1}) - (1-t)F(x^0)\Big)||}{||J(x^0)^{-1}\Big(F(x^{1,i}) - (1-t)F(x^0)\Big)||} \le \frac{\eta^{\star\star}}{2}, \quad \eta^\star \le \eta^{\star\star} < 2.$$

If insufficient contractivity occurs, then starting over with reduced η (hence t) seems to be preferable to additional damping during "back projections" or to re-computing the Jacobian.

Remark. It can be shown that the stepsize strategy given here eventually leads to a full step method when the local convergence conditions of Theorem 1 are satisfied.

7. VARIATIONS OF THE DAMPED NEWTON METHOD

The interpretation as an error controlled integration method for an ODE with invariants allows various modifications and variations of the basic method.

Basic strategy

The basic strategy, which was used for the numerical computations presented below, is as follows:

1 compute the Newton direction Δx^k,

2 compute $x^{k+1} = x^k + t^k \Delta x^k$, where t^k satisfies the RMT as error control,

3 restart the homotopy path (equivalently, continue integration) from x^{k+1}.

Basic strategy with back projections

A more expensive strategy which needs one (or more) additional F-evaluation(s) is:

1 compute the Newton direction Δx^k,

2 compute $x^{k+1,0} = x^k + t^k \Delta x^k$, where t^k satisfies the RMT,

3 add one (or more) back projection step(s)

$$\begin{aligned} x^{k+1,i+1} &= x^{k+1,i} + \tilde{\Delta} x^{k,i}, \\ \tilde{\Delta} x^{k,i} &= -J(x^k)^{-1} \left(F(x^{k+1,i}) - (1 - t^k) F(x^k) \right), \end{aligned}$$

4 restart the homotopy path from the last $x^{k+1,i+1}$.

In this variant we first have to ensure local convergence to the Newton path using the convergence behaviour of the back projections for a reduction strategy for η, hence also for the stepsizes t. From this then follows global convergence under certain assumptions like nonsingularity of Jacobians along the Newton path, since along this path all level functions $||AF(x)||_2^2$, with A nonsingular, descend by a factor of $1-t$. A termination criterion could be based on the latter property. In our numerical tests, however, repeated back projections were not found to be superior to the basic strategy. Apparently, the extra effort to iterate back to the continuous Newton path does not necessarily lead to a better iterate x^{k+1}, even though it guarantees global convergence.

Similar to techniques used in discretization methods for ODE with invariants, one can of course consider constructing higher order methods for integration of the Newton path. A few comments from the numerical ODE point of view may be made.

1 Since a highly accurate solution of the Newton path is unlikely to be necessary except maybe in extreme cases of ill-conditioning, low order methods should be most efficient.

2 The Davidenko equation is an implicit ODE and should be treated as such. In order to avoid frequent expensive unnecessary and possibly inaccurate evaluations of $J(x)^{-1}$, the use of higher order Runge-Kutta methods is not recommended.

3 Since back projection to the invariant curve effectively inhibits error propagation, consistency error considerations are sufficient for the construction of integration methods. Suitable candidates may be multistep methods based on solution values and occasional derivative evaluations.

A modification worth investigating may be to vary the steps t^k with the index i, e.g., as in the second order variant

$$\begin{aligned} x^{k+1,0} &= x^k + t^{k,0}\Delta x^k \\ \tilde{\Delta} x^{k+1,0} &= -J(x^k)^{-1}\left(F(x^{k+1,0}) - (1-t^{k+1,0})F(x^k)\right), \\ x^{k+1,1} &= x^k + t^{k,1}\Delta x^k + \frac{(t^{k,1})^2}{(t^{k,0})^2}\tilde{\Delta} x^{k+1,0}. \end{aligned}$$

8. EXTENSIONS TO L_2- AND L_1-PARAMETER ESTIMATION

Parameter estimation problems in dynamic processes can be expressed as nonlinearly constrained optimization problems in the general form

$$\begin{aligned} \min_x \quad & ||F_0(x)||_\nu, \quad \nu \in \{1,2\}, \\ & F_1(x) = 0, \quad F_2(x) \geq 0, \end{aligned} \tag{20}$$

where the cost functional is the l_2- or l_1-norm of the vector function $F_0(x)$. The vector of variables $x = (y, p)$ consists of "state" variables $y \in R^{n_y}$, typically discretization variables for underlying initial or boundary value problems in ODEs or DAEs, and unknown parameters p to be estimated.

Traditionally, the problem (20) is solved by the constrained Gauss–Newton method [3], according to which a new iterate is given by

$$x^{k+1} = x^k + t^k \Delta x^k,$$

where the increment Δx^k solves the linearly constrained problem

$$\begin{aligned} \min_x \quad & \|F_0(x^k) + J_0(x^k)\Delta x^k\|_\nu, \quad \nu \in \{1, 2\}, \qquad (21) \\ & F_1(x^k) + J_1(x^k)\Delta x^k = 0, \\ & F_2(x^k) + J_2(x^k)\Delta x^k \geq 0. \end{aligned}$$

In both cases the solution Δx^k of (21) can be represented in the form $\Delta x^k = -J^+(x^k)F(x^k)$, where $J(x^k)^+$ is a generalized inverse, i.e. it satisfies the condition

$$J^+ J J^+ = J^+, \quad J = \frac{\partial F}{\partial x}, \quad F = \begin{pmatrix} F_0 \\ F_1 \\ F_2 \end{pmatrix}.$$

For example, for the unconstrained l_2-case, $J^+(x^k)$ is the Moore–Penrose inverse of $J(x^k)$. The l_1-solution interpolates some of the measurements, i.e. some components of the linearized function

$$F_0(x) + J_0(x)\Delta x$$

are equal to zero ("active"). Under certain conditions, the generalized inverse of J is then the inverse of a projection of $J(x^k)$, which contains the active measurements, equality constraints and active inequality constraints.

Using the generalized inverse, the Gauss–Newton method becomes

$$x^{k+1} = x^k + t^k \Delta x^k, \quad \Delta x^k = -J^+(x^k)F(x^k). \qquad (22)$$

With these preparations, we can formally extend our quadratic upper bound and restrictive monotonicity test to $\|J^+(x^k)F(x^k + t\Delta x^k)\|$. The Gauss–Newton method (22) can be then interpreted as a stepsize control for the Euler method applied to the continuous Gauss–Newton method. Note, however, that the active inequality constraints incorporated in $J^+(x^k)$ as well as the active measurements are changing along the solution of the continuous problem, so that the stepsize strategy must be combined with an additional monitoring of the changing active sets.

9. NUMERICAL RESULTS

With the new stepsize strategies two challenging test problems were treated. The optimal control problem of the re-entry of an Apollo spacecraft is known for its hard nonlinearities, due to the aerodynamic forces and has a very small region of feasible solutions [14]. In the estimation of the reaction constants in the nonlinear differential equation modelling the denitrogenization of pyridine, ill-conditioning and complicated stability problems for poor initial guesses of the parameters occur. The results show the potential of the new approach.

9.1. RE-ENTRY PROBLEM [14]

In this problem a control has to be chosen to minimize the heating of a space vehicle during the flight through the earth's atmosphere on the way back from the outer space. Numerical difficulties are caused by extreme instability properties due to the aerodynamic forces when entering the atmosphere. Convergence using Newton's method can be expected only if the initial guess is fairly close to the solution.

Applying the maximum principle to this optimal control problem results in a boundary value problem in the states v, γ, ξ, the adjoints λ_v, λ_γ, λ_ξ and the free final time T:

$$
\begin{aligned}
\dot{v} &= \left(-\frac{S\rho v^2}{2m}C_W(u) - \frac{g\sin\gamma}{(1+\xi)^2}\right)T,\\
\dot{\gamma} &= \left(\frac{S\rho v}{2m}C_A(u) + \frac{v\cos\gamma}{R(1+\xi)} - \frac{g\cos\gamma}{v(1+\xi)^2}\right)T,\\
\dot{\xi} &= \frac{v\sin\gamma}{R}T,\\
\dot{\lambda}_v &= \Bigg(30v^2\sqrt{\rho} + \lambda_v\frac{S\rho v}{m}C_W(u)\\
&\qquad -\lambda_\gamma\left\{\frac{S\rho}{2m}C_A(u) + \frac{\cos\gamma}{R(1+\xi)} + \frac{g\cos\gamma}{v^2(1+\xi)^2}\right\} - \lambda_\xi\frac{\sin\gamma}{R}\Bigg)T,\\
\dot{\lambda}_\gamma &= \left(\lambda_v\frac{g\cos\gamma}{(1+\xi)^2} + \lambda_\gamma\left\{\frac{v\sin\gamma}{R(1+\xi)} - \frac{g\sin\gamma}{v(1+\xi)^2}\right\} - \lambda_\xi\frac{v\cos\gamma}{R}\right)T,\\
\dot{\lambda}_\xi &= \Bigg(-5\beta R v^3\sqrt{\rho} - \lambda_v\left\{\frac{\beta R S\rho v^2}{2m}C_W(u) + \frac{2g\sin\gamma}{(1+\xi)^3}\right\}\\
&\qquad +\lambda_\xi\left\{\frac{\beta R S\rho v}{2m}C_A(u) + \frac{v\cos\gamma}{R(1+\xi)^2} - \frac{2g\cos\gamma}{v(1+\xi)^3}\right\}\Bigg)T,
\end{aligned}
$$

with boundary conditions

$$v(0) = 0.36, \quad \gamma(0) = -8.1\pi/180, \quad \xi(0) = 4/R,$$
$$v(T) = 0.27, \quad \gamma(T) = 0, \quad \xi(0) = 2.5/R,$$
$$-10v(T)^3\sqrt{\rho} + \dot{v}(T)\lambda_v(T) + \dot{\gamma}(T)\lambda_\gamma(T) + \dot{\xi}(T)\lambda_\xi(T) = 0,$$

where $\rho_0 = 0.002704$, $R = 209.$, $\beta = 4.26$, $C_W(u) = 1.174 - 0.9\cos u$, $C_A(u) = 0.6\sin u$, $S/2m = 26.600$, $g = 3.2172 \times 10^{-4}$, and the control u is given by

$$\sin u = \frac{0.6\lambda_\gamma}{w}, \quad \cos u = \frac{0.9v\lambda_v}{w}, \quad w = \sqrt{(0.6\lambda_\gamma)^2 + (0.9v\lambda_v)^2}.$$

We parametrized this boundary value problem with a multiple shooting technique, using 6 equidistantly distributed nodes. The initial guesses were generated according to a technique described in [14]. For the solution of the resulting system of nonlinear equations with 37 variables Newton's method using the basic stepsize strategy was used. The solution (relative accuracy 10^{-4}) was achieved after 8 iterations, 4 with damped steps. Figure 2 shows the stepsizes in every iteration.

It is difficult to compare these results with ones documented in the literature [14], [8], [9], because there Broyden approximations and finite difference approximations were used. One may say, however, that our results are very competetive with the fastest results so far published.

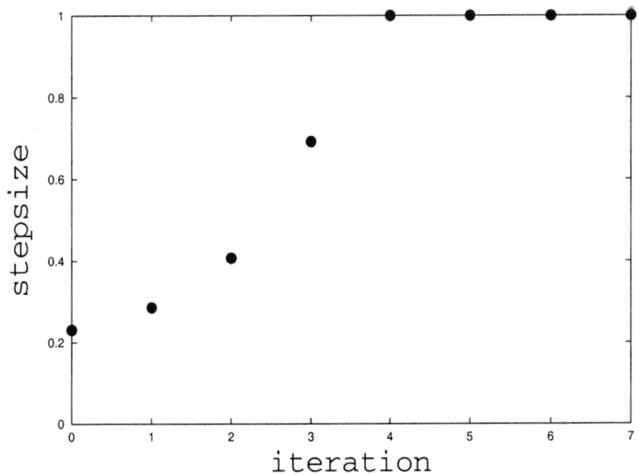

Figure 2 Stepsize history for the re-entry problem

9.2. PARAMETER ESTIMATION IN THE DENITROGENIZATION OF PYRIDINE

This problem (originally due to Zwaga [15]) was investigated in [3]. At first only pyridine is present which initiates a reaction process that can be described by ODEs with 7 state variables, the concentrations of the species:

Pyridine:	$\dot{A} = -p_1 A + p_9 B$
Piperidine:	$\dot{B} = p_1 A - p_2 B - p_3 CB + p_7 D - p_9 B + p_{10} DF$
Pentylamine:	$\dot{C} = p_2 B - p_3 BC - 2p_4 CC - p_6 C + p_8 E$ $+ p_{10} DF + 2p_{11} EF$
N-Pentylpiperidine:	$\dot{D} = p_3 BC - p_5 D - p_7 D - p_{10} DF$
Dipentylamine:	$\dot{E} = p_4 CC + p_5 D - p_8 E - p_{11} EF$
Ammonia:	$\dot{F} = p_3 BC + p_4 CC + p_6 C - p_{10} DF - p_{11} EF$
Pentane:	$\dot{G} = p_6 C + p_7 D + p_8 E.$

The rate constants $p_1, p_2, \dots, p_{11}$ of these ODEs are unknown and have to be estimated from 77 measurements of the states at the times 0.5, 1, ..., 5.5.

We treated this parameter estimation problem with the multiple shooting code PARFIT [3], which has a generalized Gauss–Newton method as a core routine for the solution of the structured constrained least squares problems. For globalization we implemented the basic strategy ($\eta = 1$), replacing the inverse by the generalized inverse, that is the solution operator for the constrained linear least squares problems.

We performed 8 experiments with widely varying initial guesses for the 11 parameters. As initial guesses for the states we chose the measurements. The initial guesses for the parameters $p_i = \alpha, i = 1, ..., 7$, for different α were rather poor guesses, since the true values vary between 0.201 and 29.4. In all cases the algorithm safely converged. The history of stepsizes is shown in Figure 3. The number of iterations, the number of damped steps and the achieved accuracy are given in Table 1.

10. CONCLUSIONS

It is well known that Newton's method for nonlinear equations can be forced to converge globally in a domain where the Jacobian is nonsingular. However, the price one has to pay is unnecessarily small stepsizes in the local convergence domain of the full step method if the problem

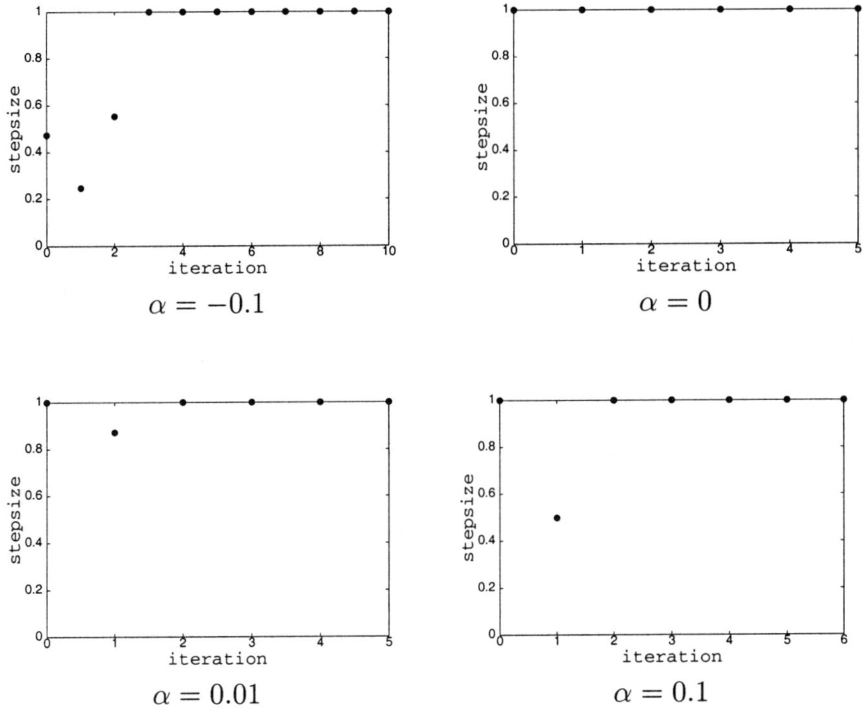

Figure 3 Stepsize histories for the pyridine problem

α	$\|\|\Delta x^0\|\|/\sqrt{n}$	$\|\|\Delta x^\star\|\|/\sqrt{n}$	iterations	damped steps
−0.1	236.04	4.91×10^{-4}	11	3
0.0	73.27	3.86×10^{-5}	6	0
0.01	661.18	1.28×10^{-4}	6	1
0.1	40.09	5.41×10^{-4}	7	1
0.2	16.70	7.82×10^{-5}	8	2
1.0	2.24	3.76×10^{-5}	9	2
5.0	2.65	3.68×10^{-5}	17	11
10.0	66.83	7.64×10^{-5}	27	22

Table 1 Convergence behaviour of the Gauss–Newton method with RMT for the pyridine problem for different initial guesses

is mildly ill-conditioned and nonlinear and one chooses classical merit functions.

The paper presents a new strategy which combines previously defined "natural level functions" with a restrictive monotonicity test. The new

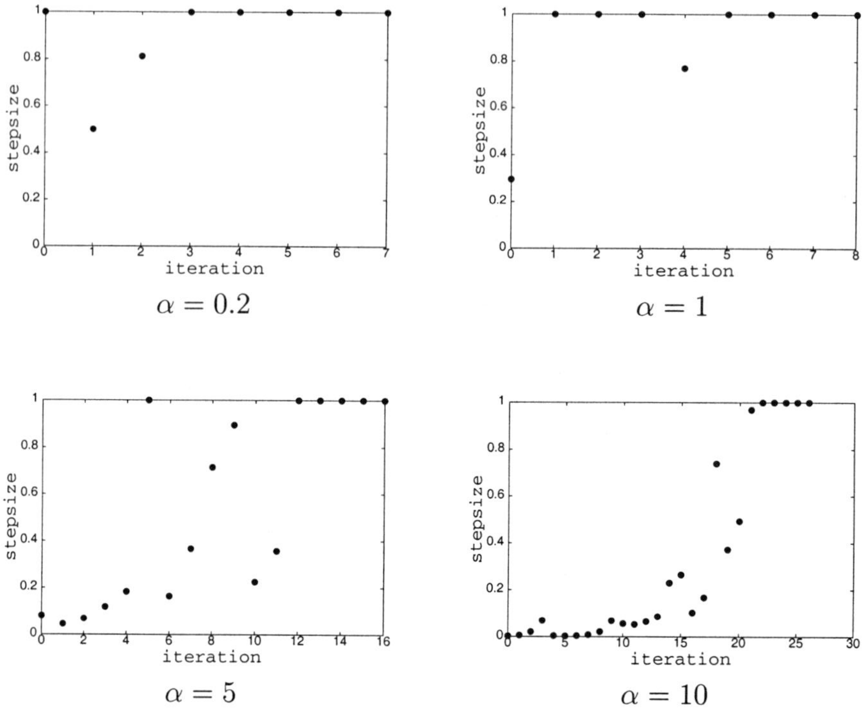

Figure 3 (continued) Stepsize histories for the pyridine problem

global convergence argument is quite different from the classical descent type proof. It is shown that this combination of natural level functions, for which descent proofs do not hold, and the restrictive monotonicity test overcome the problem of choosing too small steps. The new stepsize strategy is viewed as a stepsize control for the continuous Newton method which makes use of an invariant of the Newton path.

Three practical interpretations can be given. The first interpretation is a stepsize control of the continuous Newton method by means of an asymptotically correct estimate of the local error. Secondly, we interprete the damped Newton method as an attempt to solve a relaxed problem with a full step Newton method (with Jacobian kept constant), and the RMT is a test on sufficient contractivity. Thirdly, the stepsize is allowed to go as far as the approximation of the Jacobian is valid.

The second argument suggests generalizations to other stepsize and trust region strategies. Most of them can be interpreted as attempts to solve a relaxed version of the original nonlinear problem. In the spirit of this paper a stepsize (or trust region) strategy can be based on a control whether a sufficient (local) contraction of the method to the solution of the relaxed problem is achieved. It is hoped that this approach proves to be equally effective in these other areas.

Numerical results to two demanding problems from optimal control and parameter estimation in ODE are given, which are notorious for their strong nonlinearities. They show a very nice and reliable convergence behaviour.

Acknowledgments

We would like to thank M.J.D. Powell and an anonymous referee for their valuable comments and suggestions that helped to improve the paper.

Also, we thank the Deutsche Forschungsgemeinschaft (DFG) for financial support through SFB 539.

References

[1] U. Ascher, R. M. M. Mathheij and R. D. Russell (1988), *Numerical Solution of Boundary Value Problems for Ordinary Differential Equations*, Prentice Hall, Englewood Cliffs, New Jersey.

[2] U. Ascher and M. R. Osborne (1987), A note on solving nonlinear equations and the natural criterion function, *J. Optim. Theory Appl.*, 55, no. 1, pp. 147 - 152.

[3] H. G. Bock (1981), Numerical treatment of inverse problems in chemical reaction kinetics, in K. H. Ebert, P. Deuflhard and W. Jäger, eds, *Modelling of Chemical Reaction Systems*, Springer Series in Chemical Physics 18, Heidelberg.

[4] H. G. Bock (1983), Recent advances in parameter identification techniques for O. D. E., in P. Deuflhard and E. Hairer, eds, *Numerical Treatment of Inverse Problems, Progress in Scientific Computing*, 2, Birkhäuser, Boston, pp. 95 - 121.

[5] H. G. Bock (1987), *Randwertproblemmethoden zur Parameteridentifizierung in Systemen nichtlinearer Differentialgleichungen*, Bonner Mathematische Schriften, 187, Bonn.

[6] R. M. Chamberlain (1979), Some examples of cycling in variable metric methods for constrained minimization, *Math. Programming*, 16, pp. 378 - 383.

[7] D. Davidenko (1953), On the approximate solution of a system of nonlinear equations (Russian), *Ukrain. Mat. Zh.*, 5, pp. 196 - 206.

[8] P. Deuflhard (1974), A modified Newton method for the solution of ill-conditioned systems of nonlinear equations with applications to multiple shooting, *Num. Math.*, 22, pp. 289 - 315.

[9] P. Deuflhard (1975), A relaxation strategy for the modified Newton method, in R. Bulirsch, W. Oettli and J. Stoer, eds, *Lecture Notes in Math.*, 447, Springer Verlag, pp. 59 – 73.

[10] U. Nowak and L. Weimann (1991), A family of Newton codes for systems of highly nonlinear equations, Technical Report TR-91-10, Konrad-Zuse-Zentrum für Informationstechnik Berlin.

[11] J. M. Ortega and W. C. Rheinboldt (1970), *Iterative Solution of Nonlinear Equations in Several Variables*, Academic Press, New York.

[12] K. J. Plitt (1983), Private Communication.

[13] V. H. Schulz, H. G. Bock and M. C. Steinbach (1998), Exploiting invariants in the numerical solution of multipoint boundary value problems for DAE, *SIAM J. Sci. Comp.*, 19, no. 2, pp. 440 – 467.

[14] J. Stoer and R. Bulirsch (1973), *Einführung in die Numerische Mathematik II*, Heidelberger Taschenbuch 114, Springer, Berlin – Heidelberg – New York.

[15] P. Zwaga (1977), Private Communication.

COMPUTATIONAL EXPERIENCE WITH A PARALLEL IMPLEMENTATION OF AN INTERIOR-POINT ALGORITHM FOR MULTICOMMODITY NETWORK FLOWS

Jordi Castro *

Statistics and Operations Research Dept.

Universitat Politècnica de Catalunya,

Pau Gargallo 5, 08028 Barcelona

jcastro@eio.upc.es

Abstract A parallel implementation of the specialized interior-point algorithm for multicommodity network flows introduced in [6] is presented. In this algorithm, the positive definite systems of each iteration are solved through a scheme that combines direct factorizations and a preconditioned conjugate gradient (PCG) method. Although this numerical procedure works well in practice, it requires the solution of at least k systems of equations at each iteration of the PCG, k being the number of commodities to be routed through the network.

In order to reduce the time spent by the PCG method, we propose the application of coarse-grained parallel strategies for computing the k linear systems of equations at each PCG iteration. Since the number of arithmetic operations to be performed for each commodity is the same, the load balancing between processors is guaranteed, which avoids unnecessary delays. An extensive set of computational results on a shared memory machine are presented, using problems of up to 2.5 million variables and 260,000 constraints. For the largest PDS (Patient Distribution System) problems, the efficiency of the parallel implementation developed is about 80%, which confirms that it can be a promising tool for very large and difficult multicommodity instances.

Keywords: interior-point methods, linear programming, multicommodity network flows, parallel computing.

*This work has been supported by Spanish CICYT Project TAP96-1044 and by the European Center for Parallelism of Barcelona (CEPBA).

M.J.D. Powell and S. Scholtes (Eds.), *System Modelling and Optimization: Methods, Theory and Applications.*

1. INTRODUCTION

Multicommodity flows are one of the most challenging problems for linear programming solvers. This is partly due to the large size of these models in real world applications (e.g., routing in telecommunications networks). The need to solve very large multicommodity instances has led to the development of both specialized algorithms and parallel implementations. In this work we introduce a parallel implementation of a specialized multicommodity interior-point algorithm. The implementation has two main features. From the multicommodity point of view, it is not based on a decomposition approach, and thus it does not follow the master-slaves (or coordinator-subtasks) parallel scheme. From the interior-point point of view, unlike other parallel interior-point codes [4, 8, 15], the parallelization is not focused on the Cholesky factorization to be performed at each iteration —though it could be included— but on the parallel solution of smaller subsystems related to the various commodities of the problem.

The block angular structure of the multicommodity problem constraints matrix has led to a number of specialized methods. Among the earlier approaches we could mention primal partitioning, and price and resource directive decomposition (see [2, 14] for a general description). Recent variants of price directive decomposition have successfully applied bundle methods [10] and analytic centers [11]. Multicommodity problems, such as the PDS ones, were also used to test the efficiency of the early general interior-point solvers for linear programming (e.g., [1]). Attempts to develop specialized interior-point algorithms for multicommodity flows were presented in [13, 20, 6], the latter being the most successful. This is the algorithm that will be parallelized in this work.

Parallel approaches for multicommodity problems have also been widely studied in the past. As in the sequential case, the parallel implementations make use of several decomposition strategies, such as bundle methods [7, 17], linear-quadratic penalty terms [19], and, more recently, analytic centers [12]. A discussion of these and other parallel decomposition approaches is presented in [7]. A general description of the parallelization of mathematical programming algorithms can be found in [5] and [21].

The paper is organized as follows. Section 2 presents the formulation of the problem to be solved. Section 3 outlines the specialized interior-point algorithm for multicommodity flows, including a brief description of the general path-following method. Section 4 deals with the parallelization issues of the specialized multicommodity algorithm. Finally,

Section 5 gives the computational results obtained with the parallel implementation developed.

2. PROBLEM FORMULATION

In the most general case, the multicommodity network flow problem can be stated as how to obtain the best routing (that which involves the minimum cost) of a set of k commodities through a network of m nodes and n arcs, where the arcs have an individual capacity for each commodity, and a mutual capacity for all the commodities. The resulting problem can be written as

$$\min_{x^{(1)},\ldots,x^{(k)}} \quad \sum_{i=1}^{k} {c^{(i)}}^T x^{(i)} \tag{1}$$

subject to

$$\begin{bmatrix} A_N & \mathbf{0} & \ldots & \mathbf{0} & \mathbf{0} \\ \mathbf{0} & A_N & \ldots & \mathbf{0} & \mathbf{0} \\ \vdots & \vdots & \ddots & \vdots & \vdots \\ \mathbf{0} & \mathbf{0} & \ldots & A_N & \mathbf{0} \\ \mathbb{1}_n & \mathbb{1}_n & \ldots & \mathbb{1}_n & \mathbb{1}_n \end{bmatrix} \begin{bmatrix} x^{(1)} \\ x^{(2)} \\ \vdots \\ x^{(k)} \\ s_{mc} \end{bmatrix} = \begin{bmatrix} b^{(1)} \\ b^{(2)} \\ \vdots \\ b^{(k)} \\ b_{mc} \end{bmatrix} \tag{2}$$

$$0 \le x^{(i)} \le \overline{x}^{(i)}, \quad i = 1, \ldots, k \tag{3}$$

$$0 \le s_{mc} \le b_{mc}. \tag{4}$$

Vectors $x^{(i)} \in \mathbb{R}^n$ and $c^{(i)} \in \mathbb{R}^n$ are the flow and cost arrays for each commodity i, $i = 1, \ldots, k$. $s_{mc} \in \mathbb{R}^n$ denote the slacks of the mutual capacity constraints. $A_N \in \mathbb{R}^{m \times n}$ is the node-arc incidence matrix. We shall assume that A_N is a full row-rank matrix. This can always be guaranteed by removing any of the (redundant) node balance constraints. $b^{(i)} \in \mathbb{R}^m$ is the vector of supplies/demands for commodity i at the nodes of the network. Constraints (3) are simple bounds on the flows, $\overline{x}^{(i)} \in \mathbb{R}^n, i = 1, \ldots, k$, being the upper bounds. $b_{mc} \in \mathbb{R}^n$ are the mutual capacities of the arcs for all the commodities. $\mathbb{1}_n$ denotes the $n \times n$ identity matrix.

Note that the multicommodity flow problem can be formulated as a linear programming one with $\tilde{m} = km + n$ constraints and $\tilde{n} = (k+1)n$ variables.

3. OUTLINE OF THE SPECIALIZED INTERIOR-POINT ALGORITHM FOR MULTICOMMODITY FLOWS

The interior-point algorithm for multicommodity flows introduced in [6] is a specialization of the path-following algorithm for linear programming (see [23] for a thorough description). Let us consider the following linear programming problem in primal form

$$\begin{array}{rl} \min & c^T x \\ \text{subject to} & Ax = b, \\ & x + f = \overline{x} \\ & x, f \geq 0, \end{array} \tag{5}$$

where $x \in \mathbb{R}^{\tilde{n}}$ and $f \in \mathbb{R}^{\tilde{n}}$ are the primal variables, $\overline{x} \in \mathbb{R}^{\tilde{n}}$ are the upper bounds, $c \in \mathbb{R}^{\tilde{n}}$, $b \in \mathbb{R}^{\tilde{m}}$, and $A \in \mathbb{R}^{\tilde{m} \times \tilde{n}}$ is a full row-rank matrix. The dual of (5) is

$$\begin{array}{rl} \max & b^T y - \overline{x}^T w \\ \text{subject to} & A^T y + z - w = c \\ & z, w \geq 0, \end{array} \tag{6}$$

where $y \in \mathbb{R}^{\tilde{m}}$, $z \in \mathbb{R}^{\tilde{n}}$ and $w \in \mathbb{R}^{\tilde{n}}$ are the dual variables.

Replacing the inequalities in (5) by a logarithmic barrier in the objective function, with parameter μ, it can be seen that the KKT first order optimality conditions of this barrier problem are equivalent to the following system of nonlinear equations:

$$\begin{array}{lcrcl} r_{xz} & \equiv & \mu e_{\tilde{n}} - XZe_{\tilde{n}} & = & 0 \\ r_{fw} & \equiv & \mu e_{\tilde{n}} - FWe_{\tilde{n}} & = & 0 \\ r_b & \equiv & b - Ax & = & 0 \\ r_c & \equiv & c - (A^T y + z - w) & = & 0 \\ & & (x, z, w) & \geq & 0 \\ & & \overline{x} & \geq & x, \end{array} \tag{7}$$

where $e_{\tilde{n}}$ is the $\tilde{n}$-dimensional vector of 1's, X, Z, F, and W are diagonal matrices defined as $M \in \mathbb{R}^{\tilde{n} \times \tilde{n}} = \operatorname{diag}(m_1, \ldots, m_{\tilde{n}})$, and the vectors r_* define the left-hand side terms of (7). Note that we did not include the slacks equation $x + f = \overline{x}$ in (7). Instead we replaced the slacks f by $\overline{x} - x$ (thus, $F = \overline{X} - X$ in (7)), reducing by $\tilde{n}$ the number of equations and variables. The solutions of system (7) – considering inequalities as strict inequalities – for different μ values give rise to an arc of strictly feasible points known as the central path. As μ tends to 0, the solutions of (7) converge to that of the original primal and dual problems. A

Algorithm path-following$(A, b, c, \overline{x}, \xi)$

1 Initialize ξ, where $\xi = (x^T, f^T, y^T, z^T, w^T)^T$

2 <u>while</u> ξ is not optimal <u>do</u>

3 $\quad \Theta = (X^{-1}Z + F^{-1}W)^{-1}$

4 $\quad r = F^{-1}r_{fw} + r_c - X^{-1}r_{xz}$

$\quad$ Compute direction:

5 $\quad\quad (A\Theta A^T)dy = r_b + A\Theta r$

6 $\quad\quad dx = \Theta(A^T dy - r)$

7 $\quad\quad dw = F^{-1}(r_{fw} + W dx)$

8 $\quad\quad dz = r_c + dw - A^T dy$

9 $\quad$ Update μ

10 $\quad$ Compute α

11 $\quad \xi \leftarrow \xi + \alpha\, d\xi$

12 <u>end_while</u>

Figure 1 Path-following algorithm.

path-following algorithm attempts to follow the central path. Figure 1 shows a damped version of Newton's iteration applied to the nonlinear system (7). We use it for the multicommodity specialization. Note that the matrix Θ computed at step 3 is a positive definite diagonal matrix, because of the way it is formed from positive definite diagonal matrices. A more comprehensive description of the algorithm can be found in [23].

The main computational burden of the algorithm is the solution of the positive definite system

$$(A\Theta A^T)dy = \bar{b} \tag{8}$$

at step 5 of Figure 1 ($\bar{b}$ in (8) denotes the right-hand side $r_b + A\Theta r$ of the system). General interior-point codes attempt to solve (8) through a Cholesky factorization $LL^T = P(A\Theta A^T)P^T$, where P denotes a permutation matrix obtained by some heuristic. However, even for such good permutation matrices as those obtained by the minimum degree ordering or minimum local fill-in heuristics, when A is the multicommodity constraints matrix defined in (2), the Cholesky factorization LL^T turns out to be fairly dense, making this procedure computationally expensive. This is shown in Figure 2, in which the sparsity patterns of both A and $L + L^T$ are depicted for a multicommodity problem with 64 nodes, 524 arcs and 4 commodities, using the state-of-the-art interior-point code BPMPD [16].

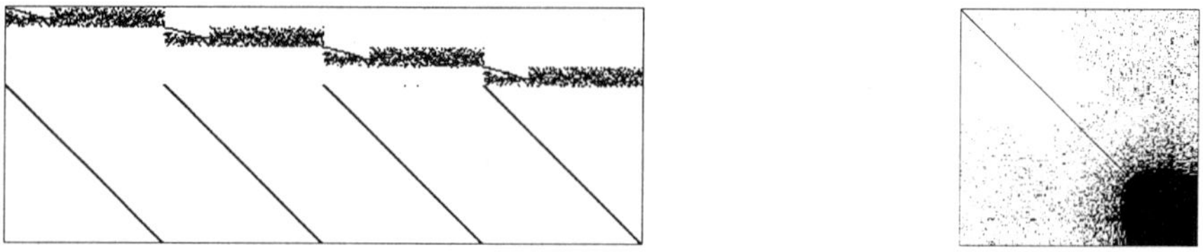

(a) (b)

Figure 2 (a) Sparsity pattern of a multicommodity constraint matrix.
(b) Sparsity pattern of the factorization of $P(A\Theta A^T)P^T$.

The specialized interior-point method suggested in [6] considers the structure of A presented in (2), and the following partitioning for the diagonal matrix Θ

$$\Theta = \begin{bmatrix} \Theta^{(1)} & & & \\ & \ddots & & \\ & & \Theta^{(k)} & \\ & & & \Theta_{mc} \end{bmatrix}, \tag{9}$$

where $\Theta^{(i)} \in \mathbb{R}^{n\times n}$ and $\Theta_{mc} \in \mathbb{R}^{n\times n}$ are related to the flows $x^{(i)}$ of commodity i and the slacks s_{mc} respectively. It is straightforward to see that the structure of $A\Theta A^T$ is

$$\left[\begin{array}{c|c|c|c} A_N\Theta^{(1)}A_N^T & \ldots & \mathbf{0} & A_N\Theta^{(1)} \\ \hline \vdots & \ddots & \vdots & \vdots \\ \hline \mathbf{0} & \ldots & A_N\Theta^{(k)}A_N^T & A_N\Theta^{(k)} \\ \hline \Theta^{(1)}A_N^T & \ldots & \Theta^{(k)}A_N^T & \Theta_{mc} + \sum_{i=1}^{k}\Theta^{(i)} \end{array}\right] = \left[\begin{array}{c|c} B & C \\ \hline C^T & D \end{array}\right], \tag{10}$$

where $B \in \mathbb{R}^{km\times km}$ is the block diagonal matrix

$$B = \mathrm{diag}(A_N\Theta^{(i)}A_N^T,\ i = 1,\ldots,k), \tag{11}$$

each block being a square matrix of dimension m, where $C \in \mathbb{R}^{km\times n}$ is defined as

$$C = \begin{bmatrix}\Theta^{(1)}A_N^T & \ldots & \Theta^{(k)}A_N^T\end{bmatrix}^T, \tag{12}$$

and where $D \in \mathbb{R}^{n\times n}$ corresponds to the lower diagonal submatrix of $A\Theta A^T$:

$$D = \Theta_{mc} + \sum_{i=1}^{k}\Theta^{(i)}. \tag{13}$$

Since Θ is diagonal and positive definite, it follows that D is also a positive definite diagonal matrix.

Using the above structure of $A\Theta A^T$, and partitioning vectors dy and $\bar{b}$ accordingly, the solution of (8) is reduced to

$$\begin{aligned}(D - C^T B^{-1} C) dy_2 &= (\bar{b}_2 - C^T B^{-1} \bar{b}_1) \equiv \beta_2 &\quad (14)\\ B dy_1 &= (\bar{b}_1 - C dy_2) \equiv \beta_1, &\quad (15)\end{aligned}$$

where β_2 and β_1 denote the right-hand sides of (14) and (15) respectively. The matrix

$$S = D - C^T B^{-1} C \quad (16)$$

is known as the Schur complement. To solve (14) and (15) efficiently, we only need to deal with systems involving B and S. Systems with the matrix B can be decomposed into k smaller ones of dimension m with matrices $A_N \Theta^{(i)} A_N^T$, $i = 1, \ldots, k$, according to (11).

The system (14) cannot be solved using a direct method (e.g., factorization of the Schur complement), since this would mean forming the matrix S, which is computationally prohibitive. Instead, we suggest using a conjugate gradient method, in virtue of the following result (see [6] for a proof).

Proposition 1 *The Schur complement matrix* $S = D - C^T B^{-1} C$ *defined in* (16) *is symmetric and positive definite at each iteration of the path-following algorithm.*

The main drawback of the conjugate gradient method is its slow convergence, especially when (5) and (6) are close to their solution point (the Schur complement becomes more ill-conditioned). It seems more reliable to use a preconditioned conjugate gradient (PCG) algorithm. The preconditioner that will be used consists of an approximation of the inverse of S, and it is based on Proposition 2. A proof of this result can be found in [6].

Proposition 2 *The inverse of* $S = D - C^T B^{-1} C$ *can be computed as*

$$S^{-1} = \left(\sum_{i=0}^{\infty} (D^{-1} Q)^i\right) D^{-1}, \quad (17)$$

where

$$Q = C^T B^{-1} C. \quad (18)$$

The preconditioner is then obtained by truncating the power series (17) at the term with index $i = \phi$, say. Clearly, the higher ϕ the better the preconditioning, and the fewer iterations of the PCG will be required. However, each new term in the preconditioner, after the first one, means solving one additional system with matrix B, which increases the cost of each PCG iteration. Therefore, we must balance two objectives: reducing the number of PCG iterations and the number of systems to be solved. Several numerical experiments have shown that the best results are obtained for $\phi = 0$ (in this case the preconditioner is D^{-1}, thus being diagonal) and, in some problems, for $\phi = 1$. The algorithm uses $\phi = 0$ as the default value. The extensive computational experience reported in [6] proved the efficiency of this specialized interior-point algorithm.

4. PARALLELIZATION OF THE ALGORITHM

Computing the direction of the dual variables dy at step 5 of the path-following method in Figure 1 is by far the most costly procedure to be performed by the specialized multicommodity algorithm. Figure 3 summarizes the steps required to compute dy, according to (14) and (15). Looking at Figure 3 we see that all of the steps require either a factorization of B, or a backward and forward substitution with this factorization, or products of vectors with matrices C or C^T. In fact, these will be the only four procedures to be run in parallel, following a coarse-grained scheme. Considering the partitioning of B and C defined in (11) and (12), these four procedures are implemented as follows:

1 Factorization of B. Perform in parallel the k factorizations of $A_N \Theta^{(i)} A_N^T$. Note that the current implementation uses sequential Cholesky solvers. Using parallel implementations of Cholesky decompositions, such as those described in [4, 8, 15], each of the k factorizations could itself be performed in parallel, improving the efficiency of the code.

2 Solution of system $Br = s$, for any $s \in \mathbb{R}^{km}$. Solve in parallel for each diagonal block $A_N \Theta^{(i)} A_N^T$ of B.

3 Computation of $w = Cv$, for any $v \in \mathbb{R}^n$. Using (12), we compute in parallel $w^{(i)} = A_N \Theta^{(i)} v, i = 1, \ldots, k$, so $w^{(i)}$ has the components of w related to commodity i.

4 Computation of $v = C^T w$, for any $w \in \mathbb{R}^{km}$. Using (12), we compute in parallel the temporary vectors $v^{(i)} = \Theta^{(i)} A_N^T w^{(i)}, i = 1, \ldots, k$. We then add the temporary vectors sequentially ($v = \sum_{i=1}^{k} v^{(i)}$), obtaining v. Due to its low computational cost, the addition of the k $v^{(i)}$ vectors has not been parallelized.

Procedure $A\Theta A^T dy = \bar{b}(A, \Theta, \bar{b}, dy)$

1 Factorize the k blocks of B

2 Compute $\beta_2 = \bar{b}_2 - C^T B^{-1} \bar{b}_1$

3 PCG: Solve $(D - C^T B^{-1} C) dy_2 = \beta_2$

$3._1$ <u>while</u> dy_2 is not optimal <u>do</u>

$\vdots$ $\vdots$

$3._i$ Compute $w = (D - C^T B^{-1} C)v$

$\vdots$ $\vdots$

$3._n$ <u>end while</u>

4 Compute $\beta_1 = \bar{b}_1 - C dy_2$

5 Solve $B dy_1 = \beta_1$

6 Return: $dy = (dy_1^T \;\; dy_2^T)^T$

Figure 3 Procedure for computing systems (14) and (15).

From our computational experience, it can be stated that for large problems the above four procedures represent more than 97% of the execution time (see Figure 6 in Section 5). This guarantees that the fraction of the sequential region will be small enough for it not to be a major bottleneck. It should also be noted that, in each of the four parallelized procedures, the number of floating point operations for each commodity (and thus for each processor) will be the same, which guarantees the load balancing between processors and avoids unnecessary delays.

4.1. PARALLEL PROGRAMMING ENVIRONMENT

The parallel implementation of the multicommodity interior-point algorithm was developed on a Silicon Graphics Origin2000 (SGI O2000) server. The SGI O2000 is a shared memory machine, main memory being physically distributed across several processors. In addition, each processor has a first level cache memory of 64Kb (32Kb for instructions, 32Kb for data), and a secondary data cache memory of 4Mb (for both instructions and data).

The main advantage of using a shared memory machine such as the SGI O2000 is the ease with which an existing sequential code can be ported, in comparison with distributed parallel environments. The latter require the use of one of the message passing communication standards (e.g., MPI or PVM), whereas the SGI O2000 provides a more user-friendly system based on including special directives in sequential C or Fortran codes [22]. These directives, usually located at the beginning of loops, create different threads of execution that will run in parallel different sections of the iterative region. Moreover, unlike distributed systems, which force the programmer to allocate data structures between processors and to keep communication low, the parallel environment of the SGI O2000 automatically attempts to perform these tasks. The default data distribution provided, however, can result in an excessive number of cache misses and page faults from the local memory of each processor, the performance of the parallel executions thus being severely impaired. Although advanced directives enable this feature to be controlled, the computational results presented in Section 5 were obtained with the default data distribution across processors provided by the system. This default distribution was also used in [4]. Further details about the use of the parallel directives of the SGI O2000 can be found in [22].

4.2. PERFORMANCE MEASURES

The performance measures presented below will be considered in Section 5 when reporting the computational results obtained. All of these performance measures are widely used in the field of parallel computing [5]. Considering a particular parallel implementation of an algorithm, we will denote the execution time obtained with p processors by T_p. The speedup S_p obtained with p processors can thus be defined as

$$S_p = \frac{T_1}{T_p}.$$

The fraction of the total execution time consumed in the sequential version by the parallel region will be denoted by f. Values of f close to 1 guarantee good theoretical speedups, whereas the bottleneck represented by the sequential region increases with $1 - f$. This is summarized by Amdahl's law, which provides a theoretical upper bound $\overline{S_p}$ for the best possible speedup

$$\overline{S_p} = \frac{1}{f/p + (1-f)} \leq \frac{1}{(1-f)}.$$

Finally, we can define the efficiency with p processors as

$$E_p = \frac{S_p}{p} \leq \overline{E_p} = \frac{\overline{S_p}}{p}.$$

The efficiency represents the fraction that a particular processor (of the p available) is usefully employed during the execution of the algorithm. Note than when $f = 1$, we have $\overline{S_p} = p$ and $\overline{E_p} = 1$.

5. COMPUTATIONAL RESULTS

The sequential code of the algorithm outlined in Section 3 was implemented and named IPM in [6]. The parallel version developed in this work will be denoted as pIPM. It is written mainly in C, with only the Cholesky factorization routines (devised by E. Ng and B. Peyton [18]) coded in Fortran. Both the sequential and parallel versions can be freely obtained for academic purposes from `http://www-eio.upc.es/castro/software.html`. All the runs were carried out on the SGI Origin2000 server located at the European Center for Parallelism of Barcelona (CEPBA), running an IRIX64 6.5 Unix operating system. The main characteristics of the server are shown in Figure 4, as reported by the `hinv` (hardware inventory) command. This computer appears at position 275 of the TOP500 supercomputer sites list [9].

64 250 MHZ IP27 Processors
CPU: MIPS R10000 Processor Chip Revision: 3.4
FPU: MIPS R10010 Floating Point Chip Revision: 0.0
Main memory size: 8192 Mbytes
Instruction cache size: 32 Kbytes
Data cache size: 32 Kbytes
Secondary unified instruction/data cache size: 4 Mbytes

Figure 4 Characteristics of the SGI Origin2000 server used for the executions.

Two sets of multicommodity instances were used for the computational experiments. The first is made up of 18 problems obtained with Ali and Kennington's Mnetgen generator [3]. Table 4 shows the dimensions and optimal solutions of the Mnetgen problems. The parameters used to generate the instances can be found in [10], and can be retrieved from `http://www.di.unipi.it/di/groups/optimize/Data/MMCF.html#MNetGen`. Columns "m", "n", and "k" show the number of nodes, arcs, and commodities. Columns "$\tilde{n}$" and "$\tilde{m}$" give the number of variables and constraints of the linear problem (where $\tilde{n} = (k+1)n$ and $\tilde{m} = km + n$). Finally, column "$c^T x^*$" gives the exact optimal objective function value. For the last two problems no exact objective value has been computed (an approximate solution obtained with IPM is reported in Table 3).

Problem	m	n	k	$\tilde{n}$	$\tilde{m}$	$c^T x^*$
M_{128-8}	128	1089	8	9801	2113	1924133.9
M_{128-16}	128	1114	16	18938	3162	4145079.4
M_{128-32}	128	1141	32	37653	5237	9785961.1
M_{128-64}	128	1171	64	76115	9363	19269824.2
$M_{128-128}$	128	1204	128	155316	17588	40143200.8
M_{256-8}	256	2165	8	19485	4213	9919483.2
M_{256-16}	256	2308	16	39236	6404	20692883.7
M_{256-32}	256	2314	32	76362	10506	45671076.1
M_{256-64}	256	2320	64	150800	18704	92249381.1
$M_{256-128}$	256	2358	128	304182	35126	190137259.9
$M_{256-256}$	256	2204	256	566428	67740	397882591.3
M_{512-8}	512	4373	8	39357	8469	46339269.9
M_{512-16}	512	4620	16	78540	12812	96992237.2
M_{512-32}	512	4646	32	153318	21030	192941834.8
M_{512-64}	512	4768	64	309920	37536	412943158.7
$M_{512-128}$	512	4786	128	617394	70322	828013599.8
$M_{512-256}$	512	4810	256	1236170	135882	—
$M_{512-512}$	512	4786	512	2455218	266930	—

Table 1 Dimensions and optimal solutions of the Mnetgen problems.

The second set consists of ten of the PDS (Patient Distribution System) problems. These problems arise from a logistic model for evacuating patients from a place of military conflict. They can be retrieved from `http://www.di.unipi.it/di/groups/optimize/Data/MMCF.html#Pds`. Their dimensions and optimal objective functions can be found in Table 1. The meaning of the columns is the same as in Table 4.

Before performing all the executions, we studied in detail the performance of pIPM in two particular instances, $M_{128-128}$ for the Mnetgen and PDS30 for the PDS problems. The behavior of the code with these two instances turned out to be fairly representative of the general behavior for each data set. Figure 5 shows the results obtained. Each plot gives the execution time T_p (left-hand vertical scale) and the theoretical best speedup $\overline{S_p}$, observed speedup S_p, and observed efficiency E_p (right-hand vertical scale) for different numbers of processors (horizontal axis). Although both problems have a similar f value (0.88 for $M_{128-128}$, 0.92 for PDS30), pIPM behaved very differently in each case. For $M_{128-128}$, the gap between S_p and $\overline{S_p}$ increases with the number of processors, whereas for PDS30 the best theoretical speedup is almost always achieved. This

Problem	m	n	k	$\tilde{n}$	$\tilde{m}$	$c^T x^*$
PDS1	126	372	11	4464	1758	29083930523.0
PDS10	1399	4792	11	57504	20181	26727094976.0
PDS20	2857	10858	11	130296	42285	23821658640.0
PDS30	4223	16148	11	193776	62601	21385445736.0
PDS40	5652	22059	11	264708	84231	18855198824.0
PDS50	7031	27668	11	332016	105009	16603525724.0
PDS60	8423	33388	11	400656	126041	14265904407.0
PDS70	9750	38396	11	460752	145646	12241162812.0
PDS80	10989	42472	11	509664	163351	11469077462.0
PDS90	12186	46161	11	553932	180207	11087561635.0

Table 2 Dimensions and optimal solutions of the PDS problems.

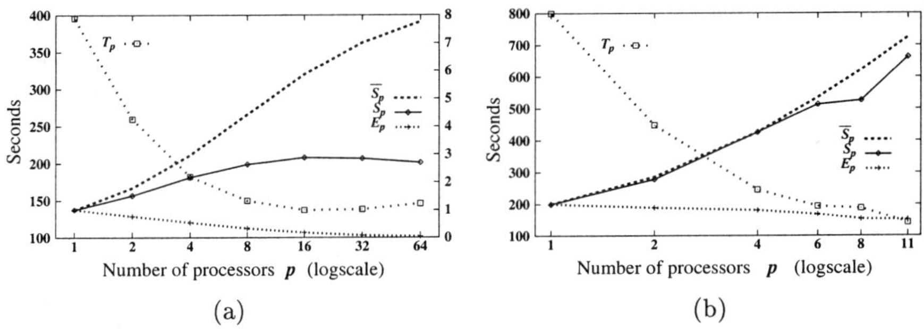

Figure 5 Behavior of the algorithm with the (a) $M_{128-128}$ problem.
(b) PDS30 problem.

fact, together with the different maximum number of processors used in the two problems (64 vs. 11), gives rise to efficiencies of $E_{64} = 0.04$ for $M_{128-128}$ (the best possible value was $\overline{E_{64}} = 0.12$) and $E_{11} = 0.51$ for PDS30 ($\overline{E_{11}} = 0.56$). It can also be observed that the execution time T_p decreases for PDS30 with p, whereas for $M_{128-128}$ it remains almost the same for 8, 16, and 32, and slightly increases for 64 processors. This lack of scalability of a shared memory machine when using a large number of processors was also stated in [4]. However, we believe that these results can be improved by exploiting the data distribution between processors, as suggested in Subsection 4.1. This additional work remains to be done.

On the basis of the results in Figure 5 we decided to execute the Mnetgen problems with 8, 16, and 32 processors (always guaranteeing

$k \geq p$), whereas 6 and 11 where used for the PDS ones. Tables 3 and 4 show the results obtained for the two sets of problems. Column $c^T x^*_{\mathrm{pIPM}}$ gives the optimal solution computed by pIPM. The relative error with respect to the exact optimal solutions of Tables 4 and 1 ranges between 10^{-5} and 10^{-8} for all the cases. Column f gives the f value (fraction represented by the parallel region in the sequential version). Column p is the number of processors used in the execution. T_p denotes the execution time. For $p = 1$, this time means CPU time, as reported by the `times` Unix command, and was obtained by executing the instances on a single processor to reduce context switches. For $p > 1$, T_p denotes wall-clock time, and was obtained by executing pIPM alone on the server to improve the accuracy of the time measures. Columns S_p and E_p give the observed speedups and efficiencies, and, enclosed in parentheses, their best theoretical values $\overline{S_p}$ and $\overline{E_p}$ respectively. Note that, for some of the PDS problems, the observed speedups and efficiencies are greater than their theoretical upper bounds. These superlinear speedups can be explained by: firstly, a lack of accuracy in the measures of both T_1 and T_p, $p > 1$; and secondly, as suggested in [7], a reduction of the number of cache misses in the parallel execution with respect to the sequential one due to the data distribution between processors.

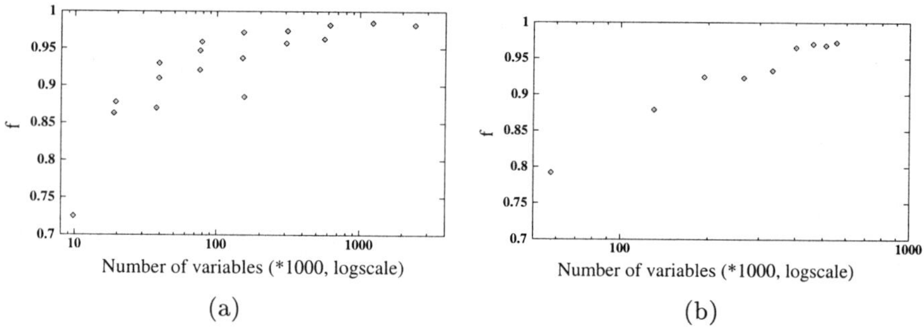

Figure 6 Fraction f represented by the parallel section in
(a) the Mnetgen problems.
(b) the PDS problems.

Some of the information in Tables 3 and 4 is summarized in Figures 6, 7, and 8. Figure 6 shows the evolution of the f value with the number of variables of the problem, for both the Mnetgen and PDS instances. Clearly, this value increases with the size of the problem, and for the largest ones it is greater than 0.97, as stated in Section 4 above. Accordingly, the bottleneck associated with the sequential version is consistently reduced for larger and larger instances, which results in a (theoretical) good behavior of the parallel implementation.

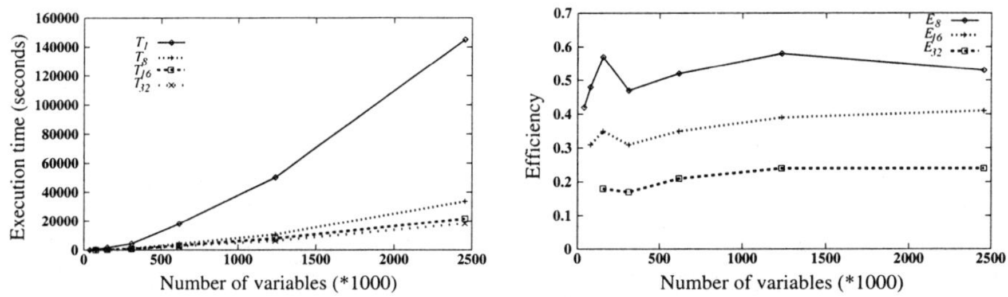

(a) (b)

Figure 7 (a) Execution times for the M_{512-*} problems.
(b) Efficiencies for the M_{512-*} problems.

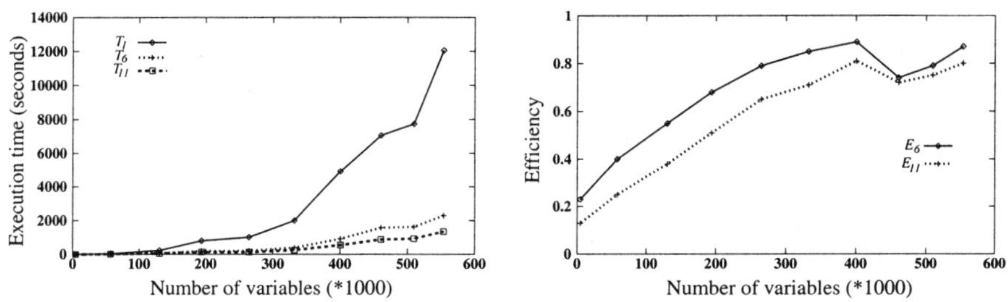

(a) (b)

Figure 8 (a) Execution times for the PDS problems.
(b) Efficiencies for the PDS problems.

Figures 7 and 8 show the execution times and efficiencies for the M_{512-*} and PDS problems, for different number of processors. For the M_{512-*} problems the best improvements are clearly obtained when moving from 1 to 8 processors. It is also clear that efficiencies tend to decrease with the number of processors, and that, for any p, they remain stable with the size of the problem. These results for the Mnetgen problems are not so good as those observed by parallel implementations of decomposition approaches for multicommodity flows (e.g., [7]). pIPM has a better behavior for the PDS problems. For instance, speedups of about 5 are obtained for the largest problems with $p = 6$, and execution times are almost reduced to half when moving from 6 to 11 processors. Figure 8(b) shows that, unlike in the M_{512-*} problems, efficiencies for 6 and 11 processors are almost the same, and that they become better with the dimension of the problem. The scalability of the code with the PDS problems is not observed, in general, with other parallel implementations of interior-point algorithms using a similar number of processors (e.g., [4, 8, 12]).

6. CONCLUSIONS AND FUTURE RESEARCH

The parallel code pIPM introduced in this work can be an efficient and promising tool for the solution of certain types of large and difficult multicommodity problems. We have found that it is especially appropriate for those instances with large networks and few commodities, where a small number of processors is required.

However, it can be improved with many additional refinements, that form part of the further work to be done. Among these we would mention:

- The fraction f of the parallel region should be augmented to guarantee better theoretical speedups. This would mean parallelizing additional routines while keeping overhead costs low.

- It would be worth attempting to use higher order preconditioners (e.g., $\phi > 0$) for the solution of the system with the Schur complement. Although for sequential executions this would reduce the performance of the algorithm, it could augment the fraction f of the parallel region, providing better parallel executions.

- The gap between the observed and theoretical efficiencies for problems with many commodities, such as the Mnetgen ones, should be reduced. This could be attempted by considering and exploiting the data distribution across the various processors. The reduction of the number of cache misses and remote memory access could mean improvements by a factor of two.

Problem	$c^T x^*_{\text{pIPM}}$	f	p	T_p	S_p $(\overline{S_p})$	E_p $(\overline{E_p})$
M_{128-8}	1924113.4	0.72	1	3.1	1.0 (1.0)	1.00 (1.00)
			8	1.7	1.8 (2.7)	0.22 (0.34)
M_{128-16}	4145089.5	0.86	1	16.2	1.0 (1.0)	1.00 (1.00)
			8	7.3	2.2 (4.1)	0.27 (0.51)
			16	7.7	2.1 (5.2)	0.13 (0.32)
M_{128-32}	9785902.8	0.87	1	36.4	1.0 (1.0)	1.00 (1.00)
			8	15.8	2.3 (4.2)	0.28 (0.52)
			16	14.8	2.5 (5.4)	0.15 (0.33)
			32	18.8	1.9 (6.4)	0.06 (0.19)
M_{128-64}	19269830.9	0.92	1	195.4	1.0 (1.0)	1.00 (1.00)
			8	79.9	2.5 (5.2)	0.30 (0.64)
			16	73.9	2.6 (7.3)	0.16 (0.45)
			32	71.9	2.7 (9.3)	0.08 (0.29)
$M_{128-128}$	40143266.0	0.89	1	395.9	1.0 (1.0)	1.00 (1.00)
			8	150.1	2.6 (4.4)	0.33 (0.55)
			16	137.3	2.9 (5.9)	0.18 (0.36)
			32	138.6	2.9 (7.0)	0.08 (0.21)
M_{256-8}	9919478.8	0.88	1	19.4	1.0 (1.0)	1.00 (1.00)
			8	7.7	2.5 (4.3)	0.31 (0.54)
M_{256-16}	20692714.5	0.91	1	69.0	1.0 (1.0)	1.00 (1.00)
			8	25.4	2.7 (4.9)	0.33 (0.61)
			16	23.5	2.9 (6.8)	0.18 (0.42)
M_{256-32}	45671345.2	0.95	1	342.0	1.0 (1.0)	1.00 (1.00)
			8	123.5	2.8 (5.8)	0.34 (0.73)
			16	91.1	3.8 (8.9)	0.23 (0.55)
			32	98.0	3.5 (12.1)	0.10 (0.37)
M_{256-64}	92249411.9	0.94	1	586.2	1.0 (1.0)	1.00 (1.00)
			8	199.0	3.0 (5.5)	0.36 (0.69)
			16	146.5	4.0 (8.2)	0.25 (0.51)
			32	143.8	4.1 (10.8)	0.12 (0.33)
$M_{256-128}$	190138392.4	0.96	1	3352.8	1.0 (1.0)	1.00 (1.00)
			8	627.6	5.3 (6.2)	0.66 (0.76)
			16	558.9	6.0 (9.7)	0.37 (0.60)
			32	511.9	6.5 (13.7)	0.20 (0.42)

Table 3 Results obtained for the Mnetgen problems.

Problem	$c^T x^*_{\text{pIPM}}$	f	p	T_p	S_p $(\overline{S_p})$	E_p $(\overline{E_p})$
$M_{256-256}$	397883691.5	0.96	1	7486.0	1.0 (1.0)	1.00 (1.00)
			8	1597.0	4.7 (6.3)	0.58 (0.79)
			16	1494.0	5.0 (10.3)	0.31 (0.64)
			32	1274.5	5.9 (14.9)	0.18 (0.46)
M_{512-8}	46338411.5	0.93	1	109.5	1.0 (1.0)	1.00 (1.00)
			8	31.9	3.4 (5.4)	0.42 (0.67)
M_{512-16}	96992142.3	0.96	1	550.9	1.0 (1.0)	1.00 (1.00)
			8	140.9	3.9 (6.2)	0.48 (0.77)
			16	111.0	5.0 (9.9)	0.31 (0.61)
M_{512-32}	192941650.0	0.97	1	1820.3	1.0 (1.0)	1.00 (1.00)
			8	395.1	4.6 (6.7)	0.57 (0.83)
			16	318.4	5.7 (11.3)	0.35 (0.70)
			32	313.7	5.8 (17.1)	0.18 (0.53)
M_{512-64}	412943655.4	0.97	1	4473.1	1.0 (1.0)	1.00 (1.00)
			8	1190.8	3.8 (6.8)	0.47 (0.84)
			16	896.2	5.0 (11.5)	0.31 (0.71)
			32	783.5	5.7 (17.7)	0.17 (0.55)
$M_{512-128}$	828014985.2	0.98	1	18156.2	1.0 (1.0)	1.00 (1.00)
			8	4302.4	4.2 (7.1)	0.52 (0.88)
			16	3168.2	5.7 (12.6)	0.35 (0.78)
			32	2667.0	6.8 (20.5)	0.21 (0.64)
$M_{512-256}$	1649358223.7	0.98	1	50178.7	1.0 (1.0)	1.00 (1.00)
			8	10745.5	4.7 (7.2)	0.58 (0.90)
			16	7982.6	6.3 (13.1)	0.39 (0.81)
			32	6346.4	7.9 (21.8)	0.24 (0.68)
$M_{512-512}$	3487594874.0	0.98	1	145018.9	1.0 (1.0)	1.00 (1.00)
			8	33655.5	4.3 (7.1)	0.53 (0.88)
			16	21630.5	6.7 (12.6)	0.41 (0.78)
			32	18312.8	7.9 (20.5)	0.24 (0.64)

Table 3 (continued) Results obtained for the Mnetgen problems.

Problem	$c^T x^*_{\text{pIPM}}$	f	p	T_p	S_p $(\overline{S_p})$	E_p $(\overline{E_p})$
PDS1	29083850483.5	0.57	1	0.7	1.0 (1.0)	1.00 (1.00)
			6	0.5	1.4 (1.9)	0.23 (0.31)
			11	0.5	1.5 (2.1)	0.13 (0.18)
PDS10	26726869329.4	0.79	1	46.2	1.0 (1.0)	1.00 (1.00)
			6	19.3	2.4 (2.9)	0.40 (0.49)
			11	16.6	2.8 (3.6)	0.25 (0.32)
PDS20	23820311896.6	0.88	1	234.4	1.0 (1.0)	1.00 (1.00)
			6	70.1	3.3 (3.7)	0.55 (0.62)
			11	55.4	4.2 (5.0)	0.38 (0.45)
PDS30	21385482088.8	0.92	1	799.1	1.0 (1.0)	1.00 (1.00)
			6	193.7	4.1 (4.3)	0.68 (0.72)
			11	141.9	5.6 (6.2)	0.51 (0.56)
PDS40	18852465159.0	0.92	1	1017.7	1.0 (1.0)	1.00 (1.00)
			6	213.4	4.8 (4.3)	0.79 (0.72)
			11	141.3	7.2 (6.2)	0.65 (0.56)
PDS50	16601676244.4	0.93	1	2003.1	1.0 (1.0)	1.00 (1.00)
			6	390.4	5.1 (4.5)	0.85 (0.74)
			11	254.1	7.9 (6.6)	0.71 (0.59)
PDS60	14265869776.2	0.96	1	4924.1	1.0 (1.0)	1.00 (1.00)
			6	917.5	5.4 (5.1)	0.89 (0.85)
			11	551.5	8.9 (8.2)	0.81 (0.74)
PDS70	12240890481.6	0.97	1	7055.5	1.0 (1.0)	1.00 (1.00)
			6	1574.4	4.5 (5.2)	0.74 (0.87)
			11	881.5	8.0 (8.5)	0.72 (0.76)
PDS80	11468486724.2	0.97	1	7737.3	1.0 (1.0)	1.00 (1.00)
			6	1628.6	4.8 (5.2)	0.79 (0.86)
			11	928.4	8.3 (8.3)	0.75 (0.75)
PDS90	11087270971.3	0.97	1	12059.8	1.0 (1.0)	1.00 (1.00)
			6	2293.7	5.3 (5.3)	0.87 (0.87)
			11	1355.9	8.9 (8.6)	0.80 (0.78)

Table 4 Results obtained for the PDS problems.

References

[1] I. Adler, M.G.C. Resende and G. Veiga (1989), An implementation of Karmarkar's algorithm for linear programming, *Mathematical Programming*, 44, pp. 297–335.

[2] R.K. Ahuja, T.L. Magnanti and J.B. Orlin (1993), *Network Flows*, Prentice Hall, Englewood Cliffs, NJ.

[3] A. Ali and J.L. Kennington (1977), Mnetgen program documentation, Technical Report 77003, Dept. of Ind. Eng. and Operations Research, Southern Methodist University, Dallas.

[4] E.D. Andersen and K.D. Andersen (1998), A parallel interior-point algorithm for linear programming on a shared memory machine, CORE Discussion Paper 9808, CORE, Louvain-La-Neuve, Belgium.

[5] D.P. Bertsekas and J.N. Tsitsiklis (1995), *Parallel and Distributed Computation*, Prentice-Hall, Englewood Cliffs.

[6] J. Castro, A specialized interior-point algorithm for multicommodity network flows, *SIAM J. Optim.* (to appear). Available from `http://www-eio.upc.es/jcastro`.

[7] P. Cappanera and A. Frangioni (1996), Symmetric and asymmetric parallelization of a cost-decomposition algorithm for multicommodity flow problems, Technical Report TR-96-36, Dip. di Informatica, Università di Pisa, Italy.

[8] T.F. Coleman, J. Czyzyk, C. Sun, M. Wagner and S.J. Wright (1997), pPCx: parallel software for linear programming, in *Proceedings of the Eight SIAM Conference on Parallel Processing in Scientific Computing*, SIAM.

[9] J.J. Dongarra, H.W. Meuer and E. Strohmaier (1998), TOP500 supercomputer sites, Technical Report UT-CS-98-404, Computer Science Dept., University of Tennessee.

[10] A. Frangioni and G. Gallo (1999), A bundle type dual-ascent approach to linear multicommodity min cost flow problems, to appear in *INFORMS Journal on Computing*.

[11] J. Gondzio, R. Sarkissian and J.-P. Vial (1998), Parallel implementation of a central decomposition method for solving large scale planning problems, HEC Technical Report 98.1.

[12] J.-L. Goffin, J. Gondzio, R. Sarkissian and J.-P. Vial (1996), Solving nonlinear multicommodity flow problems by the analytic center cutting plane method, *Mathematical Programming*, 76, pp. 131–154.

[13] A.P. Kamath, N.K. Karmarkar and K.G. Ramakrishnan (1993), Computational and complexity results for an interior point algorithm on multicommodity flow problems, Technical Report TR-21/93, Dip. di Informatica, Università di Pisa, Italy, pp. 116-122. Extended abstracts of Netflow'93.

[14] J.L. Kennington and R.V. Helgason (1980), *Algorithms for Network Programming*, Wiley, New York.

[15] I.J. Lustig and E. Rothberg (1996), Gigaflops in linear programming, *Operations Research Letter*, 18(4), pp. 157–165.

[16] Cs. Mészáros (1996), *The Efficient Implementation of Interior Point Methods for Linear Programming and their Applications*, Ph.D. Thesis, Eötvös Loránd University of Sciences.

[17] D. Medhi (1990), Parallel bundle-based decomposition for large-scale structured mathematical programming problems, *Annals of Operations Research*, 22, pp. 101–127.

[18] E. Ng and B.W. Peyton (1993), Block sparse Cholesky algorithms on advanced uniprocessor computers, *SIAM J. Sci. Comput.*, 14, pp. 1034–1056.

[19] M.C. Pinar and S.A. Zenios (1992), Parallel decomposition of multicommodity network flows using a linear-quadratic penalty algorithm, *ORSA Journal on Computing*, 4, pp. 235–249.

[20] L. Portugal, M.G.C. Resende, G. Veiga, G. and J. Júdice (1997), A truncated interior-point method for the solution of minimum cost flow problems on an undirected multicommodity flow network (in Portuguese), in *Proceedings of First Portuguese National Telecommunications Conference*, pp. 381–384.

[21] J.B. Rosen, editor, (1990), Supercomputers and large-scale optimization: algorithms, software, applications, *Annals of Operations Research*, 22.

[22] Silicon Graphics Inc. (1998), *C Language Reference Manual.*

[23] S.J. Wright (1997), *Primal-Dual Interior-Point Methods*, Society for Industrial and Applied Mathematics, Philadelphia, PA.

MODELLING THE BEHAVIOR OF SYSTEMS: BASIC CONCEPTS AND ALGORITHMS

Tommaso Cotroneo
Mathematics Institute,
University of Groningen,
P.O. Box 800, 9700 AV Groningen, The Netherlands.
T.Cotroneo@math.rug.nl

Jan C. Willems
As above
J.C.Willems@math.rug.nl

Abstract In this paper we introduce the behavioral approach as a mathematical language for describing dynamical systems, in particular systems modeled by high order constant coefficient linear differential equations. We investigate what data have to be added in order to express the influence of the environment and the initial conditions on the system. We give an algorithm to check whether these additional constraints are satisfied by a (unique) trajectory. We define the concepts of observability and controllability, and present algorithms which provide a constructive verification of such properties.

1. INTRODUCTION

The purpose of this paper is two-fold. Firstly, we introduce some of the main ideas of behavioral systems theory as an abstract framework for modeling and analysis of dynamical systems. Secondly, we show how algorithms can be developed which allow the analysis of system properties, even at the high level of generality at which we shall be working.

The starting point of our discussion is the definition of a system as the set of feasible trajectories of the variables whose dynamics we are modeling. Such a definition has two crucial aspects: on the one hand

M.J.D. Powell and S. Scholtes (Eds.), *System Modelling and Optimization: Methods, Theory and Applications.*

it abandons the idea of a system as an input/output map, i.e. as a signal processor, and simply makes it a relation between variables which evolve in time, but for which no hierarchical or cause/effect structure is given a priori. On the other hand, it distinguishes clearly between the system (i.e. the feasible trajectories) and its representations (e.g. equations, graphs, grammar rules, etc.).

Crucial definitions such as controllability and observability are also given in a representation-free fashion.

At the level of representations, we concentrate on linear differential systems, i.e. systems whose trajectories can be described as solutions to a set of linear differential equations. For such systems we describe algorithms which allow one to check controllability and observability, and which allow one to verify whether trajectories satisfying given constraints can be simulated.

One crucial issue we shall overlook for lack of space is that of describing systems as interconnections of smaller subsystems. Such a concept, very typical in engineering thinking, finds in the behavioral framework a nice formal description. It also motivates the introduction of the concept of latent variables, namely variables which we are not interested in modeling, but which we have to take into account, in order to describe the subsystems and the interconnections that provide the final model.

The main references to the ideas we will discuss are [7, 9]; in [10], control issues are also addressed from this point of view; finally, [8, 6] use this same framework to study properties of systems described by partial difference and differential equations.

2. MODELING A DYNAMICAL SYSTEM

If one had to define what the purpose of modeling a dynamical system is, one could reasonably say it is describing how a set of variables of interest, call them w, evolve as a function of time. If we indicate by $\mathbb{T}$ the time axis of interest (typically $\mathbb{T} = \mathbb{R}$ for continuous time models and $\mathbb{T} = \mathbb{Z}$ for discrete time ones), and by $\mathbb{W}$ the space in which the variables of interest take on their values (e.g. $\mathbb{R}^q$ if there are q real valued variables), then the w's are elements of $\mathbb{W}^{\mathbb{T}}$, with $\mathbb{W}^{\mathbb{T}}$ denoting the set of maps from $\mathbb{T}$ to $\mathbb{W}$. The model of the system tells us that only a subset of such trajectories can actually happen, namely the subset that complies with the laws of the system. We will indicate this set of admissible trajectories as $\mathfrak{B}$ and refer to it as the *behavior* of the dynamical system.

Formalizing the above discussion, we define a *dynamical system* as a triple $\S = (\mathbb{W}, \mathbb{T}, \mathfrak{B})$, with $\mathbb{W}$ the *signal space*, $\mathbb{T}$ the *time axis* and $\mathfrak{B} \subseteq \mathbb{W}^{\mathbb{T}}$ the *behavior* of the system.

The above definition is the cornerstone of behavioral systems theory, and in essence it defines a model as an exclusion law, a rule that allows us to pick a subset of feasible trajectories out of a set of possible ones. Given its crucial importance, we illustrate it with two examples.

1. *Newton's second law* imposes a restriction that relates the position $\vec{q}$ of a point mass to the force $\vec{F}$ acting on it. This relation is $\vec{F} = m\frac{d^2}{dt^2}\vec{q}$, with m the mass. This is a dynamical system with $\mathbb{T} = \mathbb{R}$, $\mathbb{W} = \mathbb{R}^3 \times \mathbb{R}^3$, and behavior $\mathfrak{B}$ consisting of all maps $t \in \mathbb{R} \mapsto (\vec{q}, \vec{F})(t) \in \mathbb{R}^3 \times \mathbb{R}^3$ that satisfy $\vec{F} = m\frac{d^2}{dt^2}\vec{q}$.

2. *Kepler's laws* describe the possible motions of the planets in the solar system. They define a dynamical system with $\mathbb{T} = \mathbb{R}$, $\mathbb{W} = \mathbb{R}^3$, and $\mathfrak{B}$ the set of maps $w : \mathbb{R} \to \mathbb{R}^3$ that satisfy the following laws. The paths w must be ellipses in $\mathbb{R}^3$ with the sun (assumed in fixed position) at one of the foci, the radius vector from the sun to the planet must sweep out equal areas in equal time, and the ratio of the period of revolution around the ellipse to the major axis must be the same for all w's in $\mathfrak{B}$.

Classical notions such as linearity and time-invariance are also introduced very naturally, starting from the above formal definition of a system. In particular, we talk about a *linear* system if $\mathbb{W}$ is a vector space and $\mathfrak{B}$ a linear subspace of $\mathbb{W}^{\mathbb{T}}$, and about a *time-invariant* one (assuming $\mathbb{T} = \mathbb{R}$ or $\mathbb{Z}$) if $\sigma^t\mathfrak{B} = \mathfrak{B}$ for all $t \in \mathbb{T}$, where σ^t denotes the t-shift defined by $(\sigma^t f)(t') := f(t' + t)$, $t' \in \mathbb{R}$.

3. DIFFERENTIAL SYSTEMS

As discussed in the above section, when it comes to modeling a dynamical system, what we are really after is the behavior $\mathfrak{B}$, the set of admissible trajectories. Of course such a set can be described in many possible ways, for example through differential equations as in Newton's second law, or through formal descriptions, such as in Kepler's laws. It is therefore conceptually misleading to identify the idea of a system with that of a set of differential equations, because, as we have pointed out, equations are just one of many possible instruments that can be used to specify behaviors. As further examples, think of finite state automata whose behavior is typically described graphically, or non-linear electronic components that are often described by graphs in the I-V plane.

Although identifying systems with equations is not appropriate, the class of systems whose behavior is specified by differential equations deserves special attention, because it plays such a prominent role in physical and engineering applications. We define a *differential system* as a system with $\mathbb{T} = \mathbb{R}$, whose behavior $\mathfrak{B}$ consists of all solutions of a set of differential equations of the form

$$f\left(t, w, \frac{d}{dt}w, \ldots, \frac{d^L}{dt^L}w\right) = 0.$$

Of even greater interest to us is the subclass of differential systems for which $\mathbb{W}$ is a finite dimensional vector space, and the defining equations are not only linear but also the variable t does not appear explicitly in the above equation. In this case we talk about a *linear time-invariant differential system.* If $\mathbb{W}$ is q-dimensional, say, the behavior is then specified as all solutions to

$$R_0 w + R_1 \frac{d}{dt} w + \cdots + R_L \frac{d^L}{dt^L} w = 0,$$

with $R_i \in \mathbb{R}^{p \times q}$, $i = 0, 1, \ldots, L$, where p denotes the number of rows in the above system. Notice how algebraic constraints (i.e. differential equations of order 0) are automatically included in this class. To avoid technical issues, in the following we regard $\mathfrak{B}$ as the set of $\mathfrak{C}^\infty$ solutions of the above set of equations, in other words we take $\mathfrak{B} \subseteq \mathfrak{C}^\infty(\mathbb{R}, \mathbb{R}^q)$.

To the above system of differential equations, we can associate in a natural way the polynomial matrix $R(\xi) = R_0 + R_1\xi + \cdots + R_L\xi^L \in \mathbb{R}[\xi]^{p \times q}$, the elements of the matrix being polynomials in $\xi \in \mathbb{R}$. Given this association, we often write the set of equations as

$$R\left(\frac{d}{dt}\right) w = 0. \tag{1}$$

For obvious reasons, we may refer to the above as a *kernel representation* of the behavior of our linear time-invariant differential system, and write $\mathfrak{B} = \ker(R\left(\frac{d}{dt}\right))$.

If we are given a linear subspace $V \subseteq \mathbb{R}^q$, we know from basic linear algebra that we can always find a matrix R such that $V = \ker(R)$; we also know that such a matrix is not uniquely defined. Something very similar happens when looking at kernel representations of linear differential behaviors, which are subspaces of $\mathfrak{C}^\infty(\mathbb{R}, \mathbb{R}^q)$; in order to investigate this aspect we first recall the definition of a module over the polynomial ring $\mathbb{R}[\xi]$.

The *module* spanned by a set $v_1, \ldots, v_p \in \mathbb{R}^n[\xi]$ of polynomial vectors, denoted by $\langle v_1, \ldots, v_p \rangle$, is defined as the set of all linear combinations with polynomial coefficients of the given vectors. Thus we have

$$\langle v_1, \ldots, v_p \rangle = \left\{ \sum_{i=1}^{p} h_i v_i : h_i \in \mathbb{R}[\xi] \right\} \subseteq \mathbb{R}^n[\xi].$$

Further, $\mathbb{R}^n[\xi]$ itself is a module, trivially obtained as $\langle e_1, \ldots, e_n \rangle$ with e_i the i-th unit vector. Modules of the form $\langle v_1, \ldots, v_p \rangle \subseteq \mathbb{R}^n[\xi]$ are *submodules* of $\mathbb{R}^n[\xi]$.

The set of generators of such a submodule is not unique. In other words there can exist elements $u_1, \ldots, u_r \in \mathbb{R}^n[\xi]$ such that $\langle v_1, \ldots, v_p \rangle = \langle u_1, \ldots, u_r \rangle$. Because in general $p \neq r$, the cardinality of the generating set of a given module is also not uniquely determined. The minimal cardinality is unique, however; in other words, for any submodule $\mathfrak{M} \subseteq \mathbb{R}^n[\xi]$, there is a greatest integer c such that any generating set of the given submodule must contain at least c elements. Generating sets with exactly c elements are called *minimal generating sets* for $\mathfrak{M}$.

Given a polynomial matrix $R \in \mathbb{R}^{p \times q}[\xi]$, we will now indicate by $\langle R \rangle$ the submodule spanned by its rows. It turns out that, if R' is also a polynomial matrix with q columns, then

$$\mathfrak{B} = \ker\left(R\left(\frac{d}{dt}\right)\right) = \ker\left(R'\left(\frac{d}{dt}\right)\right) \quad \Leftrightarrow \quad \langle R \rangle = \langle R' \rangle.$$

In other words, any behavior will admit many different kernel representations, but is associated with one and only one submodule of $\mathbb{R}^q[\xi]$, namely the module generated by the rows of one, and therefore all, of its possible kernel representations.

This non-uniqueness in the representation of a behavior is a consequence of our definition of a system as a set of trajectories, rather than as a set of equations. It also has practical relevance, because it enables us to use in each situation the representation which we find most appropriate for the purpose at hand.

The non-uniqueness in the cardinality of generating sets for modules implies not only that several matrices R satisfy $\mathfrak{B} = \ker(R\left(\frac{d}{dt}\right))$, but also that the row dimension of R is not uniquely defined (i.e. the number of equations we need to specify a behavior is not unique). The above discussion, however, provides a minimal number of rows, say c, that such a matrix must contain. Any $R \in \mathbb{R}^{c \times q}[\xi]$ such that $\mathfrak{B} = \ker(R\left(\frac{d}{dt}\right))$ will be called a *minimal representation* of $\mathfrak{B}$. The minimal representations correspond to polynomial matrices R which are of full row rank over the ring $\mathbb{R}[\xi]$ (that is, $R \in \mathbb{R}^{c \times q}[\xi]$ has a nonsingular $c \times c$ submatrix).

4. MODELING THE INFLUENCE OF THE ENVIRONMENT

When modeling systems, we will usually be dealing with "open" systems. This means that the systems interact with the environment around them, so the given model includes some freedom in the w's for the influence of the environment. From the mathematical point of view, this will show up in the fact that, if $\mathfrak{B} = \ker(R\left(\frac{d}{dt}\right))$, and if $R \in \mathbb{R}^{p\times q}[\xi]$ is a full row rank (equivalently, a minimal) representation of $\mathfrak{B}$, then $p < q$. In other words the system is underdetermined, having more variables than equations.

A special case of the situation described above is obtained by looking at systems in the traditional input-output form, which are described by the equations

$$P\left(\frac{d}{dt}\right)y = Q\left(\frac{d}{dt}\right)u, \tag{2}$$

with P square, $\det(P) \neq 0$, and $P^{-1}Q$ a matrix of proper rational functions. Such systems correspond in the notation (1) to $R = [P \quad -Q]$ and $w = \begin{pmatrix} y \\ u \end{pmatrix}$. It can be shown that the u's are free in (2), meaning that, for any given u, there exists a y that satisfies the equations (2). In addition, the y's are bounded, meaning that they are uniquely specified by u and by the initial values $y(0)$, $\frac{d}{dt}y(0)$, ...

5. SIMULATING TRAJECTORIES

As discussed above we are dealing mainly with underdetermined systems of equations. Suppose now that we are interested in simulating a possible system trajectory, in other words in reproducing one solution of the system of equations (1). It is necessary to deal with the underdetermination of the original system. A very general way of doing so is by specifying an additional set of equations of the form

$$U\left(\frac{d}{dt}\right)w = V\left(\frac{d}{dt}\right)f, \tag{3}$$

with $f \in \mathfrak{C}^\infty(\mathbb{R}, \mathbb{R}^f)$. Typically, the choice of U, V and f corresponds to fixing the external influences that act on the system. As a special case of additional equations (3), we may choose $U = [0 \ I]$ and $V = I$. It follows from $w = \begin{pmatrix} y \\ u \end{pmatrix}$ that (3) takes up the freedom in u by the classical assignment $u = f$.

Another possibility is that one might want or have to impose a set of conditions that the system variables and their derivatives should satisfy at a given instant in time, say $t = 0$. The conditions may take the form

$$S\left(\frac{d}{dt}\right) w(0) = a, \tag{4}$$

where a is a given real vector of suitable dimension.

Thus the problem becomes one of first investigating existence and uniqueness of a solution w, which satisfies both the system equations (1) and the extra constraints (3) and (4), and then providing an algorithm to compute a (unique) solution. In the next pages we shall address the first issue in detail and overlook the second one.

Notice that, by setting $K = \begin{pmatrix} R \\ U \end{pmatrix}$ and $J = \begin{pmatrix} 0 \\ V \end{pmatrix}$, equations (1) and (3) can be written together as $K\left(\frac{d}{dt}\right) w = J\left(\frac{d}{dt}\right) f$. Slightly generalizing the situation described above, we will therefore look at the following problem. We are given polynomial matrices $K \in \mathbb{R}[\xi]^{p\times q}$, $S \in \mathbb{R}[\xi]^{s\times q}$, $J \in \mathbb{R}[\xi]^{p\times m}$, a real vector $a \in \mathbb{R}^{s\times 1}$ and a function vector $f \in \mathfrak{C}^\infty(\mathbb{R}, \mathbb{R}^m)$. These matrices and vectors provide the following system of differential equations with initial conditions:

$$\left.\begin{aligned} K\left(\frac{d}{dt}\right) w = J\left(\frac{d}{dt}\right) f \\ S\left(\frac{d}{dt}\right) w(0) = a \end{aligned}\right\}. \tag{5}$$

The question is to determine conditions under which there exists a solution $w \subset \mathfrak{C}^\infty(\mathbb{R}, \mathbb{R}^q)$ of these equations. The question is answered by the following theorem, already presented in [2].

Theorem 1 : *Let $K \in \mathbb{R}[\xi]^{p\times q}$, $S \in \mathbb{R}[\xi]^{s\times q}$, $J \in \mathbb{R}[\xi]^{p\times m}$, $a \in \mathbb{R}^{s\times 1}$ and $f \in \mathfrak{C}^\infty(\mathbb{R}, \mathbb{R}^m)$ be given. The system of differential equations with initial conditions (5) has a solution $w \in \mathfrak{C}^\infty(\mathbb{R}, \mathbb{R}^q)$ if and only if the data have the two properties*

1 : $n \in \mathbb{R}^{1\times p}[\xi]$ *and* $nK = 0 \;\Rightarrow\; n\left(\frac{d}{dt}\right) J\left(\frac{d}{dt}\right) f = 0,$

2 : $\ell \in \mathbb{R}^{1\times s}$, $b \in \mathbb{R}^{1\times p}[\xi]$ *and* $\ell S = bK$

$$\Rightarrow\; \ell a = \left(b\left(\frac{d}{dt}\right) J\left(\frac{d}{dt}\right) f\right)(0).$$

The first of these conditions states that, in order to have a solution w to $K\left(\frac{d}{dt}\right) w = J\left(\frac{d}{dt}\right) f$ for a given f, any differential relationships which hold for the rows of K must also hold for the corresponding components of the vector $J\left(\frac{d}{dt}\right) f$. Moreover, the other condition states that, if a solution satisfies $S\left(\frac{d}{dt}\right) w(0) = a$, then, whenever a linear combination of the left hand side of the initial conditions can be written as a consequence of the left hand side of the equations, then the right hand side of the initial conditions must be in the same way a consequence of the right hand side of the equations. The first condition, therefore, expresses consistency of the set of equations $K\left(\frac{d}{dt}\right) w = J\left(\frac{d}{dt}\right) f$, while the second one expresses consistency of the initial conditions with respect to the given equations. Both conditions can be seen as generalizing the well known rank condition for solvability of systems of algebraic equations $Ax = y$, say.

A MATLAB pseudocode will be given at the end of this section that sketches an algorithm that checks these conditions. It requires the following useful concepts.

- For any $K \in \mathbb{R}[\xi]^{p\times q}$, the *annihilators* for the rows of K are defined to be the elements of the set

$$\mathfrak{H}_K = \left\{ n \in \mathbb{R}^{1\times p}[\xi] \ : \ nK = 0 \right\}.$$

  Such a set is a submodule of $\mathbb{R}^p[\xi]$; in algebraic literature it is called the *syzygy module* of the rows of K (see [4, 1]). One can always find a polynomial matrix N such that its rows generate $\mathfrak{H}_K$, in other words such that $\mathfrak{H}_K = \langle N \rangle$. In [1] an algorithm is presented which constructs a suitable N from K; such an algorithm is part of most computer algebra packages; in our pseudocode we assume that a procedure **SYZYGY** is available which performs this computation. Condition 1 of the above theorem can then be checked by verifying $N(\frac{d}{dt}) J\left(\frac{d}{dt}\right) f = 0$.

- Given $K \in \mathbb{R}[\xi]^{p\times q}$, we define its highest row coefficient matrix K_{hc} to be the real matrix whose i-th row contains the coefficients of the highest power of ξ that occurs in the i-th row of K. For example,

$$K = \begin{bmatrix} 3\xi^2 + \xi + 1 & 2 \\ 2\xi & \xi + 1 \end{bmatrix} \quad \Rightarrow \quad K_{hc} = \begin{bmatrix} 3 & 0 \\ 2 & 1 \end{bmatrix}.$$

  The matrix K is defined to be *row proper* if the rows of K_{hc} are linearly independent. If K is not row proper, its *row proper form* is defined to be any matrix K' such that the modules $\langle K \rangle$ and $\langle K' \rangle$ are the same, and such that K' is row proper. Of course K' is not uniquely defined, but any row proper form can be obtained from another by taking linear combinations of the rows. For example, if $K = \begin{bmatrix} \xi^2+\xi+1 & \xi^2 \\ \xi & \xi+1 \end{bmatrix}$, then

$$K_1' = \begin{bmatrix} \xi+1 & -\xi \\ \xi & \xi+1 \end{bmatrix} \quad \text{and} \quad K_2' = \begin{bmatrix} 2\xi+1 & 1 \\ \xi & \xi+1 \end{bmatrix}$$

are both row proper forms of K. Because $\langle K \rangle = \langle K' \rangle$, it follows that there exists a polynomial matrix U such that $K' = UK$. Our pseudocode uses a procedure $[K', U] =$**ROWPROP**(K) that returns a row proper form of K and the transformation matrix U. Classical algorithms for doing so exist in the literature (e.g. in [5]).

- The leading monomial matrix K_{lm} of $K \in \mathbb{R}[\xi]^{p\times q}$ is obtained by taking for each row the leftmost occurence of the highest power of ξ appearing in the given row. For example,

$$K = \begin{bmatrix} \xi^2+\xi+1 & 2 & 3\xi^2 \\ 4 & 2\xi+1 & \xi \end{bmatrix} \quad \Rightarrow \quad K_{lm} = \begin{bmatrix} \xi^2 & 0 & 0 \\ 0 & 2\xi & 0 \end{bmatrix}.$$

A matrix P with q columns is said to be *reduced* with respect to K if there exists no polynomial matrix V such that $P_{lm} = VK_{lm}$. For example, the vector $P = [\xi\ 0\ 0]$ is reduced with respect to the above K while $P = [0\ \xi^2\ 0]$ is not.

Given any two polynomial matrices K and S with the same number of columns, a *division algorithm* can be designed that generates a quotient matrix B and a remainder matrix P such that $S = P + BK$, where P is *reduced* with respect to K. Our pseudocode uses a procedure $[B, P] =$**DIVIDE**(S, K), which returns the quotient and remainder matrix of the division of S by K. An algorithm to perform this kind of division is described in [1], using generic orderings of vector polynomials that are more generic than our concept of leading monomials. Actually, all we are doing in this section could be recast in the more abstract language of canonical forms for polynomial matrices as in [2], but we prefer not to do so.

- Let K' be a row proper form of K and let S be as in the previous paragraph. Using the division algorithm we can find a quotient matrix B' and a remainder matrix P such that $S = P + B'K'$. Because $K' = UK$ for some matrix U, these remarks imply $S = P + BK$ with $B = B'U$. We note, however, that the P and B obtained in this way are not necessarily the same as those that would occur if we divided S by K. This is a crucial and complex issue, very well addressed in [4, 1], and which reasons of space force us to overlook.

- As a consequence of having divided by a row proper form of K, it can now be shown that the remainder matrix P has the property that, given a real vector $\ell \in \mathbb{R}^{1\times s}$, there exists a polynomial vector $b \in \mathbb{R}^{1\times p}[\xi]$ such that $\ell S = bK$ if and only if $\ell P = 0$; in this case we have $b = \ell B$, where B is defined above. The set CH_P of *constant annihilators* of the rows of P is defined as:

$$CH_P = \{\ell \in \mathbb{R}^{1\times s} \ : \ \ell P = 0\},$$

which is a linear subspace of $\mathbb{R}^s$. Therefore standard linear algebra techniques allow us to build a matrix L whose rows span CH_P. In our pseudocode we assume the availability of a procedure **ANNIHIL** that constructs such a matrix. Thus, remembering $b = \ell B$, the checking of condition 2 in the above theorem is reduced to a linear algebra problem, namely the computation of L and the verification of $La = (LB(\frac{d}{dt})J\left(\frac{d}{dt}\right) f)(0)$.

With these remarks at hand, we can now sketch the desired algorithm:

Solv=**SOLVABILITY**(K, J, S, f, a);

> Solv=0;
>
> N=**SYZYGY**(K);
>
> if $N(\frac{d}{dt})\, J(\frac{d}{dt})\, f == 0$ then
>
> Solv=1;
> $[K', U]$=**ROWPROP**(K);
> $[B', P]$=**DIVIDE**(S, K');
> L=**ANNIHIL**(P);
> $B = B'U$;
> if $La \neq (LB(\frac{d}{dt})\, J(\frac{d}{dt})\, f)(0)$ then Solv=0 endif;
>
> endif.

The procedure returns Solv=0 if the problem has no solution, Solv=1 otherwise.

We close this section by showing how the classical Cauchy problem fits into our framework.

Example 2 : Consider the first order system:

$$\left.\begin{array}{l} \dot{x} = Ax + Bf \\ x(0) = a \end{array}\right\},$$

which corresponds, in our notation, to $K = \xi I - A$, $S = I$ and $J = B$. Because K is square and non-singular, the set $\mathfrak{H}_K$ of annihilators of the rows of K contains only the zero vector, so $N = 0$. We see that K is row proper and that the division of S by K gives $P = S = I$ and $B = 0$, so CH_P also contains only the zero vector, i.e. $L = 0$. Therefore, by applying Theorem 1, we conclude the existence of a solution x for arbitrary f and a.

6. OBSERVABILITY

As announced in the introduction, the purpose of the final two sections of this paper is recasting the classical concepts of observability and controllability in the language we have used to describe dynamical systems; for the case of linear differential systems, we shall present algorithms which allow us to verify their properties in terms of the coefficients of the defining equations.

Traditionally, in the context of input/state/output systems of the form $\frac{d}{dt}x = f(x, u)$, $y = h(x, u)$, observability is defined as the possibility of deducing, knowing the laws of the system, the state trajectory $x(\cdot)$ from observation of the input and output trajectories $u(\cdot)$ and $y(\cdot)$. In our context, however, no special variables such as the state show up in the definition of observability, which is now seen as the possibility of deducing the trajectories of a subset of the system variables w, given the laws of the system and observations of the remaining variables, which form a second subset. The first subset is often referred to as to-be-observed variables and denoted by w_2, while the latter is often referred to as observed variables and denoted by w_1.

Formally, let $(\mathbb{T}, \mathbb{W}_1 \times \mathbb{W}_2, \mathfrak{B})$ be a dynamical system for which the splitting of the signal space $\mathbb{W}$ into $\mathbb{W}_1 \times \mathbb{W}_2$ corresponds to the separation of variables into observed and to-be-observed variables. We denote a typical element of the behavior by $w = (w_1, w_2) \in \mathfrak{B}$, and we define w_2 to be *observable* from w_1 if

$$(w_1, w_2) \in \mathfrak{B} \quad \text{and} \quad (w_1, w_2') \in \mathfrak{B} \quad \Rightarrow \quad w_2 = w_2'.$$

In the case when $\mathfrak{B}$ is a linear behavior, this is equivalent to the condition

$$(0, w_2) \in \mathfrak{B} \quad \Rightarrow \quad w_2 = 0.$$

When discussing observability for $\mathfrak{B} = \ker(R\left(\frac{d}{dt}\right))$, it is convenient to partition R into two parts that correspond to its action on the w_1 and w_2 subsets. Therefore we rewrite the behavioral equations $R\left(\frac{d}{dt}\right) w = 0$ as

$$N\left(\frac{d}{dt}\right) w_1 = M\left(\frac{d}{dt}\right) w_2.$$

Before giving conditions for observability we recall two definitions. A matrix $G \in \mathbb{R}^{p\times p}[\xi]$ is said to be *unimodular* if it admits a polynomial inverse, in other words if there exists a matrix F such that $FG = GF = I$; this is equivalent to $\det(G) = g \in \mathbb{R}\backslash\{0\}$. A polynomial matrix $R \in \mathbb{R}^{p\times q}[\xi]$ is said to be *left prime* if $R = GR'$, with $G \in \mathbb{R}^{p\times p}[\xi]$ and $R' \in \mathbb{R}^{p\times q}[\xi]$, implies that G is unimodular; the definition of *right primeness* is analogous.

We can now state a theorem (see [3]) that presents equivalent conditions for observability.

Theorem 3 : *Let $\mathfrak{B}$ be the set*

$$\mathfrak{B} = \left\{w = (w_1, w_2) \in \mathfrak{C}^{\infty}(\mathbb{R}, \mathbb{R}^q) : N\left(\tfrac{d}{dt}\right) w_1 = M\left(\tfrac{d}{dt}\right) w_2\right\},$$

and assume $M \in \mathbb{R}^{p\times \ell}[\xi]$. Then the following properties are equivalent:

1 w_2 is observable from w_1,

2 There exist unimodular matrices U and V such that

$$UMV = \begin{bmatrix} I \\ 0 \end{bmatrix},$$

3 $\operatorname{rank}(M(\lambda)) = \ell$ *for any* $\lambda \in \mathbb{C}$,

4 M is right prime,

5 The rows of M span the full module $\mathbb{R}^{\ell}[\xi]$.

Condition 5 of Theorem 3 can be restated as follows: consider the set of row vectors of degree 0 contained in the module spanned by the rows of M; such a set is obviously an $\mathbb{R}$-vector space contained in $\mathbb{R}^{\ell}$, and, in order for M to be right prime, it must be $\mathbb{R}^{\ell}$ itself. We now sketch an algorithm for investigating such a condition. It recursively builds a generating set for the above vector space, and checks whether the dimension of the space is ℓ. We employ the following notation.

- Standard MATLAB notation will be used to indicate the rows and columns of a matrix, e.g. $M(i,:)$ is the i-th row of M, and $M(i:j,:)$ are rows i to j of M.

- M^0 denotes the set of rows of the matrix M that have degree 0. For example

$$M = \begin{bmatrix} \xi^2+\xi+1 & 2 \\ 1 & 1 \end{bmatrix} \qquad \Rightarrow \qquad M^0 = [1\ 1].$$

- d_i is the degree of the i-th row of M, so in the above example we have $d_1 = 2$ and $d_2 = 0$.

- $M_{hc} \in \mathbb{R}^{p \times \ell}$ is the highest row coefficient matrix of M as defined in the preceding section. Further, $M_{hp} \in \mathbb{R}^{p \times \ell}[\xi]$ is the highest row power matrix of M, meaning that, for $i = 1, 2, \ldots, p$, the row $M_{hp}(i,:)$ contains just the highest power of ξ in $M(i,:)$ with its coefficients. Therefore $M_{hp}(i,:)$ is the product $\xi^{d_i} M_{hc}(i,:)$, and in the above example we have

$$M_{hc} = \begin{bmatrix} 1 & 0 \\ 1 & 1 \end{bmatrix} \quad \text{and} \quad M_{hp} = \begin{bmatrix} \xi^2 & 0 \\ 1 & 1 \end{bmatrix}.$$

- M=**Order**(M) is a procedure that permutes the rows of M into decreasing row degree order. For example

$$M = \begin{bmatrix} \xi & 1 \\ \xi^2 + \xi + 1 & 2 \\ 1 & 1 \end{bmatrix} \quad \Rightarrow \quad \mathbf{Order}(M) = \begin{bmatrix} \xi^2 + \xi + 1 & 2 \\ \xi & 1 \\ 1 & 1 \end{bmatrix}.$$

- Let the $p \times \ell$ matrix M be ordered by **Order**(M), and then let M_{hc} be formed. For any integer i between 1 and $p - 1$, standard linear algebra allows us to find, if it exists, a non-zero real vector n such that $M_{hc}(i,:) = nM_{hc}(i+1:p,:)$. For example,

$$M = \begin{bmatrix} \xi^2 + \xi + 1 & \xi^2 + 2 \\ \xi & 1 \\ 0 & 1 \end{bmatrix} \quad \Rightarrow \quad M_{hc}(1,:) = nM_{hc}(2:3,:)$$

for $n = [1 \ 1]$. Given such an n, the function h=**polann**(n, M) returns a polynomial vector h such that $M_{hp}(i,:) = hM_{hp}(i+1:p,:)$. In the above case, for example, **polann**(n, M) returns $h = [\xi \ \ \xi^2]$. To build h from n is straightforward, because, for $k = i+1, \ldots, p$, the element of h that multiplies the k-th row of M_{hp} is the corresponding element of n times $\xi^{d_i - d_k}$.

- M=**Eliminate**(M, i) is a procedure that removes the i-th row of a matrix M. For example, if

$$M = \begin{bmatrix} \xi^2 + \xi + 1 & \xi^2 + 2 \\ \xi & 1 \\ 0 & 1 \end{bmatrix},$$

then **Eliminate**$(M, 2)$ returns

$$M = \begin{bmatrix} \xi^2 + \xi + 1 & \xi^2 + 2 \\ 0 & 1 \end{bmatrix}.$$

- Let m be a polynomial row vector of degree d_m and length p, and let $M \in \mathbb{R}^{p\times l}[\xi]$ be a matrix ordered as by **Order**; moreover, let j be such that the degree of $M(i,:)$ is less than or equal to d_m for $i \geq j$ and greater than d_m for $i \leq j-1$. Then $[M,j]$=**Insert**(M,m) is a procedure that replaces the matrix M by the matrix with the $p+1$ rows $(M(1:j-1,:),\ m,\ M(j:p,:))$, where, as usual, p is the number of rows of the original M. The procedure also returns j, which is the row index of m in the new matrix. For example, if

$$M = \begin{bmatrix} \xi^2+\xi+1 & \xi^2+2 \\ 1 & 1 \end{bmatrix} \quad \text{and} \quad m = [\xi+1 \;\; \xi],$$

then **Insert**(M,m) returns

$$M = \begin{bmatrix} \xi^2+\xi+1 & \xi^2+2 \\ \xi+1 & \xi \\ 1 & 1 \end{bmatrix} \quad \text{and} \quad j=2.$$

We can now sketch (in MATLAB pseudo-code) our procedure for checking right primeness (RPR) of a matrix M, and hence observability of a differential system.

```
[M,obs]=RPR(M);

    M=Order(M);
    obs=(rank(M^0) == ℓ);
    p=rowdim(M);
    i=p-rowdim(M^0);

    while ((not obs) and (i ≥ 1)) do

        if (∃ real n ≠ 0 such that
         M_hc(i,:) = n M_hc(i+1:p,:)) then

            h=polann(n,M);
            m = M(i,:) - hM(i+1:p,:);
            M=Eliminate(M,i);

            if (m ≠ 0) then
                [M,j]=Insert(M,m);

                if (degree(m) == 0) then
                    obs=(rank(M^0) == ℓ);
                    i = j - 1;
```

```
                else i = j endif;

            else p = p - 1   i = i - 1 endif;

        else i = i - 1;
        endif;

    endwhile.
```

In the above algorithm obs is a boolean variable that tells us whether the matrix we are considering is right prime or not; we call this variable obs because of the relation between right primeness and observability explained in Theorem 3. As already discussed, right primeness is checked by verifying if a generating set for the vectors of degree 0 contained in the module spanned by the rows of M has rank equal to ℓ. After reordering M, we immediately perform a check to see whether the vectors of degree 0 in the original matrix are already sufficient to meet the requirement. If this is not the case, after setting $i = p - \text{rowdim}(M^0)$, which gives the index of the first row from the bottom with degree higher than 0, we enter the main while loop in which we try to generate additional constant vectors by taking combinations of rows of M. This is done by replacing a row $M(i,:)$ of degree higher than 0 by a polynomial combination of $M(i,:)$ and later rows of M, provided that the combination is of lower degree than $M(i,:)$ itself. Such a degree lowering is possible only if the highest coefficient vector of $M(i,:)$ is linearly dependent on the highest coefficient vectors of the rows of equal or lower degree. This condition is tested in the "if" statement by looking for a real vector $n \neq 0$ of suitable dimension such that $M_{hc}(i,:) = nM_{hc}(i+1:p,:)$. If such a dependence is found, then starting from n we build the polynomial vector h such that $M_{hp}(i,:) = hM_{hp}(i+1:p,:)$. Thus the polynomial vector $m = M(i,:) - hM(i+1:p,:)$ has degree lower than $M(i,:)$, as desired.

If such a degree lowering is found to be possible, we then eliminate $M(i,:)$. Further, when the new row vector m is not zero, it is included in the matrix M in the earliest position that maintains the degree ordering of **Order**. If the new vector has degree 0, we check again if the right primeness condition is fulfilled; alternatively, the next iteration of the while loop will check whether the degree of this newly generated vector can itself be lowered (this is the reason for having $i = j$).

The algorithm ends if the condition for right primeness is verified (obs is true) or if no more lowering of degree is possible. When the condition $i < 1$ occurs in the while statement, this means that we have considered all the rows in M, so there is no possibility of further lowering. Because at each step we are replacing a vector by one of lower degree, or are reducing the number of rows of M, the $i < 1$ condition will always hold eventually. Therefore the stopping rule for the algorithm is well defined.

We see that the final matrix M is an output argument of our procedure. This is not needed when the only output of interest is obs; it will, however, be handy in the next section, where **RPR** is a subprocedure of a procedure that checks controllability.

As an example of the above algorithm, we wish to show its application to the observability matrix for state space systems.

Example 4 : We consider the classical problem of deducing the state trajectory $x(\cdot)$, starting from observations of the input and output trajectories $u(\cdot)$ and $y(\cdot)$ in the state space system

$$\left.\begin{array}{rcl} \dot{x} & = & Ax + Bu \\ y & = & Cx + Du \end{array}\right\}.$$

Such a problem corresponds, in our formalism, to

$$w_1 = \begin{pmatrix} u \\ y \end{pmatrix}, \quad w_2 = x \quad \text{and} \quad M = \begin{bmatrix} -\xi I + A \\ C \end{bmatrix},$$

where $M \in \mathbb{R}^{p \times \ell}[\xi]$ and I is the $\ell \times \ell$ identity matrix.

The algorithm provides an efficient way of checking the classical rank conditions on the observability matrix $\begin{pmatrix} C \\ CA \\ \vdots \\ CA^{\ell-1} \end{pmatrix}$. It can be shown that the rows of this matrix are vectors of degree 0 contained in the module spanned by the rows of M. Specifically, the rows of CA can be expressed as $(\xi I)C + C(-\xi I + A)$, and are therefore polynomial linear combinations of degree 0 of the rows of M. By induction, one finds that the rows of $CA^k = (\xi I)CA^{k-1} + CA^{k-1}(-\xi I + A)$ are also 0 degree vectors in the module spanned by the rows of M. Further, the rows of the observability matrix are actually a generating set for the space of all 0 degree vectors spanned by the rows of M; this remark follows easily from the fact that A^ℓ is linearly dependent on $I, A, \ldots A^{\ell-1}$, so no independent row vectors would be added by considering CA^k for $k \geq \ell$.

Checking that the observability matrix has rank ℓ is therefore equivalent to verifying that a generating set for the space of vectors of degree 0 contained in the module spanned by the rows of M has rank ℓ. As discussed above, this task is done by our algorithm in an efficient way. For example, consider the case when C is a non-singular matrix. Then observability is found immediately without computing the rest of the observability matrix; in this case, in fact, our algorithm stops without even entering the main "while" loop.

7. CONTROLLABILITY

In the context of state space systems, the concept of controllability addresses the possibility of reaching any final state starting from any initial state in finite time. The state system $\frac{d}{dt}x = f(x,u)$ is said to be controllable if, for any initial state x_0 and any final state x_f, there exists an input function u and a time T such that the solution to $\frac{d}{dt}x = f(x,u)$ with initial condition $x(0) = x_0$ yields $x(T) = x_f$.

As in the case of observability, we now give a definition of controllability which relies only on properties of the system's trajectories, and not on specific properties of special variables chosen to represent it, namely state variables in the above example.

In particular, if $\mathfrak{B}$ is a continuous-time, time invariant behavior, we define it to be *controllable* if, for any two trajectories w_1, $w_2 \in \mathfrak{B}$, there exists $t_1 \geq 0$ and a third trajectory $w \in \mathfrak{B}$ such that

$$w(t) = \begin{cases} w_1(t), & t \leq 0, \\ w_2(t - t_1), & t \geq t_1. \end{cases}$$

The intuition behind this definition is that, for a behavior to be controllable, one must be able to connect any admissible "undesired" past trajectory to any admissible "desired" future one, through suitable steering.

The concepts of controllability and observability are of central importance in systems theory, in particular in controller design and stabilization [10, 5], and in observer design and filtering [5], respectively. Very roughly speaking, controllability implies that a system can be stabilized by control. More precisely, the system, viewed as a plant, can be interconnected with a controller such that all solutions that satisfy the equations of both the plant and the controller go to zero (at an arbitrarily fast rate) as time goes to infinity. Observability in turn implies the existence of a signal processor that deduces, in a suitable way, the to-be-estimated variables (the w_2's in the definition of observability) from the observed variables (the w_1's in the definition of observability).

Again, as for observability, we now concentrate on the case of linear differential systems $\mathfrak{B} = \ker(R\left(\frac{d}{dt}\right))$. We seek conditions that are equivalent to controllability in terms of the polynomial matrix R.

Unfortunately, we have to examine two situations separately: the case in which R is of full row rank (equivalently, it provides a minimal kernel representation of $\mathfrak{B}$) and the case in which it is not. Notice that in the observability case any attempt to observe "too many" variables (i.e. M does not have full column rank) is excluded by Theorem 3, whereas using "too many" equations to describe $\mathfrak{B}$ (i.e. R does not have full row rank) may allow controllability of $\mathfrak{B}$, even though, as we shall see, the conditions on R become less elegant.

The above remark may seem strange to anyone used to the old principle that controllability and observability are dual concepts; in the setting we are working in we abandon this adage. Indeed, our definitions show that, while controllability is essentially a property of the behavior, observability also depends on the choice of observed and to-be-observed variables. The fact that technical conditions for checking observability and controllability often turn out to be "dual" (in some sense) should not make us forget the fundamental difference just mentioned.

We now state two theorems (see [3]) which present conditions for controllability in terms of the polynomial matrix R. The first result applies to the general case.

Theorem 5 : *The following are equivalent:*

1 $\mathfrak{B} = \ker(R\left(\frac{d}{dt}\right))$ *is controllable,*

2 There exist unimodular matrices U and V such that

$$URV = \begin{bmatrix} I & 0 \\ 0 & 0 \end{bmatrix},$$

3 $\mathrm{rank}(R(\lambda))$ *is independent of λ for any $\lambda \in \mathbb{C}$,*

4 If $N(\xi)$is a minimal generating set for the module spanned by the columns of R, then $N(\xi)$ is a right prime matrix.

In the case when R has full row rank, much more can be said, this time yielding a result which can be regarded as the "dual" of Theorem 3.

Theorem 6 : *Let $R(\xi) \in \mathbb{R}^{p\times q}[\xi]$ be a full row rank polynomial matrix. The following are equivalent:*

1 $\mathfrak{B} = \ker(R\left(\frac{d}{dt}\right))$ *is controllable,*

2 There exist unimodular matrices U and V such that $URV = [I\ 0]$,

3 $\mathrm{rank}(R(\lambda)) = p$ *for any $\lambda \in \mathbb{C}$,*

4 R is left prime,

5 The columns of R span the full module $\mathbb{R}^p[\xi]$.

We now sketch an algorithm that assesses controllability for a given $\mathfrak{B} = \ker(R\left(\frac{d}{dt}\right))$.

To begin with notice that a matrix is left prime if and only if its transpose is right prime. Therefore, given the algorithm of the preceding section, it is easy to build a procedure **LPR** which checks whether a matrix R is left prime (equivalently whether $\ker(R\left(\frac{d}{dt}\right))$ is controllable under the assumptions of Theorem 6). We can use

[R,ctr]=**LPR**(R);

> $M = R^T$;
> [M,ctr]=**RPR**(M);
> $R = M^T$.

We know, however, that left primeness is equivalent to controllability only when R is a full rank matrix; to check controllability in the general case we have to verify the conditions of Theorem 5. In order to sketch an algorithm that does so, we make two remarks.

1 A column proper form of a matrix R is obtained just by transposing the row proper form of R^T, with row proper defined in Section 5. If **LPR** returns ctr=false, then the matrix R which is returned is very close to a column proper form of the original R. Indeed, if left primeness is not verified, than the algorithm stops when the highest column coefficient vectors of all columns with degree higher than 0 are linearly independent of the highest column coefficient vectors of the subsequent columns. For vectors of degree 0, however, we always check their rank but not their independence, so they need not be linearly independent.

   The columns with degree higher than 0, therefore, already satisfy the property that defines the column proper form of a matrix R, so to get a column proper form it is sufficient to replace the degree 0 columns of the returned R by a basis of the $\mathbb{R}$-vector space they generate. The procedure R=**COLPRP**(R) brings the returned R into column proper form in this way.

2 If R is in column proper form, then the number of its columns is equal to the rank of the original matrix R. Therefore, if this number is equal to the row dimension of R, then R is of full row rank, so, by Theorem 6, the test performed by the call of **LPR** is necessary and sufficient for controllability.

   Alternatively, if the rank is smaller than the row dimension of R, we apply Condition 4 of Theorem 5 to check controllability. In this case the column proper form is a minimal generating set for the module spanned by the columns of R. Therefore we need to verify that the column proper form is right prime in order to conclude controllability.

The above remarks provide the following algorithm for checking controllability.

```
[R,ctr]=CTRB(R);

    [R,ctr]=LPR(R);
    if (not ctr) then

        R=COLPRP(R);
        if (rowdim(R)>coldim(R)) then

            [R,ctr]=RPR(R);

        endif;

    endif.
```

As for observability, it can be shown that our algorithm corresponds to the usual test on the controllability matrix, in the case of checking controllability of a state space system, where R has the form $R = (\xi I - A \quad -B)$. We are going to show how it applies a known test for controllability of systems described by a single differential equation.

Example 7 : Assume $R = (r_1 \; r_2)$, where r_1 and r_2 are polynomials and $d_1 = \text{degree}\,(r_1) \geq d_2 = \text{degree}\,(r_2)$. By applying a division algorithm for polynomials, we can write $r_1 = q_2 r_2 + r_3$ with $d_2 = \text{degree}\,(r_2) > d_3 = \text{degree}\,(r_3)$. Going through our algorithm we see that, after at most $d_1 - d_2 + 1$ steps, it will yield $R = (r_2 \; r_3)$. Similarly, we write $r_2 = q_3 r_3 + r_4$, and after at most $d_2 - d_3 + 1$ more steps it will provide $R = (r_3 \; r_4)$. Thus our algorithm applies the classical Euclidean algorithm for computing the greatest common divisor of two polynomials. Therefore at the end of the calculation we will have $R = r$ with $r = \text{GCD}(r_1, r_2)$, so the condition for controllability is equivalent to asking if r is a constant, which is equivalent to r_1 and r_2 being coprime polynomials.

8. CONCLUSIONS

In this paper we have discussed some basic concepts of the behavioral approach to dynamical modelling. We have treated in particular the simulation question, i.e. the problem of adding the specification of externally given signals and internally given initial data, so that the system has a (unique) solution. Further, we elaborated on the notions of controllability and observability in this setting.

Our presentation has concentrated on lumped linear differential systems. Present work aims at extending these ideas to distributed and nonlinear systems, and to extending the range of applications, particularly in the area of $\mathcal{H}_\infty$ control and filtering.

References

[1] W. Adams and P. Loustanau (1994), *An Introduction to Gröbner Bases*, AMS.

[2] T. Cotroneo and J.C. Willems (1999), The initial value problem for high order linear differential systems, *Proceedings 1999 European Control Conference,* Karlsruhe, Germany.

[3] T. Cotroneo and J.C. Willems (1999), Controllability and observability of linear differential behaviors, *Proceedings 38th Conference on Decision Control,* Phoenix, USA.

[4] D. Cox, J. Little, and D. O'Shea (1997), *Ideals, Varieties and Algorithms:An Introduction to Computational Algebraic Geometry and Commutative Algebra*, 2nd edition, Springer Verlag.

[5] T. Kailath (1980), *Linear Systems*, Prentice Hall.

[6] H.K. Pillai and S. Shankar (1999), A behavioral approach to control of distributed systems, *SIAM Journal on Control and Optimization*, 37, pp. 388–408.

[7] J.W. Polderman and J.C. Willems (1998), *Introduction to Mathematical Systems Theory: A Behavioral Approach*, Springer Verlag.

[8] P. Rocha and J.C. Willems (1991), Controllability of 2-D systems, *IEEE Transactions on Automatic Control*, 36, pp. 413–423.

[9] J.C. Willems (1991), Paradigms and puzzles in the theory of dynamical systems, *IEEE Transactions on Automatic Control*, 36, pp. 259–294.

[10] J.C. Willems (1997), On interconnections, control, and feedback, *IEEE Transactions on Automatic Control*, 42, pp. 326–339.

LIPSCHITZIAN STABILITY OF NEWTON'S METHOD FOR VARIATIONAL INCLUSIONS*

Asen L. Dontchev
AMS, Mathematical Reviews
Ann Arbor, MI 48107-8604, USA
ald@ams.org

Abstract We present an overview to Lipschitz-type properties of mappings associated with solutions of optimization problems including variational inequalities and mathematical programs. We show that these properties are inherited in various ways by the mapping acting from parameters of the problem and the starting point to the set of sequences generated by Newton's method. Some new insights into convergence of Newton's/SQP method are also presented.

Keywords: Newton's method, stability, variational inequalities, optimization.

1. INTRODUCTION

In this paper we study continuity properties of solutions to variational problems and associated properties of sequences generated by Newton's method. Our basic model is the following inclusion (generalized equation) with a parameter v,

$$v \in f(x) + F(x), \tag{1}$$

where $v \in \mathbf{R}^m$, $x \in \mathbf{R}^n$ f is a $\mathcal{C}^2$ function from $\mathbf{R}^n$ to $\mathbf{R}^m$ and F is a set-valued map from $\mathbf{R}^n$ to the subsets of $\mathbf{R}^m$ with closed graph. For $F(x) = \{0\}$ we obtain a system of equations, for $F(x) = \mathbf{R}^m_+$, the positive orthant in $\mathbf{R}^m$, we have a system of inequalities. If $F(x) = N_C(x)$, the normal cone mapping to a closed set C in $\mathbf{R}^n$, for $m = n$, then we have a variational inequality; in particular, for $C = \mathbf{R}^n_+$ we obtain a complementarity problem.

*This research was supported by the National Science Foundation.

M.J.D. Powell and S. Scholtes (Eds.), *System Modelling and Optimization: Methods, Theory and Applications.*

The parameter v provides a *shift* perturbation to the inclusion (1). Most of the results presented in this paper can be extended to problems with a *basic* perturbation parameter, u, added to the function f and to the mapping F, thus regarding (1) as embedded in a larger family of inclusions,

$$v \in f(x, u) + F(x, u).$$

The pair (v, u) now represents *canonical* perturbations. In general, shift stability of (1) with respect to v is different from full (canonical) stability with respect to (u, v); in some cases these concepts are equivalent under additional conditions for the dependence of f and F on the "generic" parameter u. In this paper we stay within the format of shift stability of the model (1). We study local Lipschitz-type properties of the map "parameter $v \mapsto$ set of solutions of (1)", denoted $\mathbf{S}$; that is,

$$v \mapsto \mathbf{S}(v) = \{x \in \mathbf{R}^n \mid v \in f(x) + F(x)\}, \tag{2}$$

around a fixed reference point (v^*, x^*) in the graph of $\mathbf{S}$. Of course, $\mathbf{S}$ is the inverse map $(f + F)^{-1}$.

For illustration of our results we consider throughout the following optimization problem:

$$\min_x \, [g(x) - \langle v, x \rangle] \quad \text{subject to} \quad x \in C, \tag{3}$$

where g is a C^3 function from $\mathbf{R}^n$ to $\mathbf{R}$, C is a convex polyhedral set in $\mathbf{R}^n$ and v is a parameter which provides "tilting" perturbations to the cost, in the terminology of [17]. The first-order necessary optimality condition for (3) has the form

$$v \in \nabla g(x) + N_C(x), \tag{4}$$

where ∇g is the derivative of g and $N_C(x)$ denotes the normal cone to the set C at the point x. The variational inequality (4) fits into the general format of (1).

By Newton's method applied to the basic model (1) we mean the procedure which generates a sequence $\{x_1, x_2, \cdots\}$ of points x_n, with a given x_0, according to the rule

$$v \in f(x_n) + \nabla f(x_n)(x_{n+1} - x_n) + F(x_{n+1}) \quad \text{for } n = 0, 1, \cdots. \tag{5}$$

This procedure is the standard Newton for equations when $F = \{0\}$; it is usually called a generalized Newton method or a Newton-type method for solving systems of inequalities when $F = \mathbf{R}^m_+$ or variational inequalities when F is a normal cone mapping, respectively. One may expect that the "true" Newton method for the general model (1) should involve some kind of linearization of the map F, and a number of authors have followed this line of research for various nonsmooth, semismooth, etc. equations, see, e.g., the recent survey in [26]. Here we keep the map F unchanged with the mental picture that F has a "simple" yet set-valued form, e.g. it is a polyhedral map or the normal map to a polyhedral cone, both having already a linear structure. Such a model leads us quite far, also for infinite-dimensional (i.e., optimal control) problems where nonsmooth-analytic tools may become more technical than instrumental. Most of the results in this paper hold in abstract spaces, by changing only the terminology.

Consider the optimization problem (3). Representing the convex polyhedral set C as $C = \{x \in \mathbf{R}^n \mid Bx \leq b\}$ for some matrix B from $\mathbf{R}^n$ to $\mathbf{R}^m$ and a vector $b \in \mathbf{R}^m$, the normal cone to the set C at x has the form

$$N_C(x) = \{v \in \mathbf{R}^n \mid v = B^T y \text{ for some } y \in \mathbf{R}^m_+ \text{ and } x \perp v\}.$$

Then the Newton method (5) applied to the variational inequality (4) becomes

$$\begin{aligned} -v + \nabla g(x_n) + \nabla^2 g(x_n)(x_{n+1} - x_n) + B^T y_{n+1} = 0, \\ Bx_{n+1} \leq b, \quad y_{n+1} \geq 0, \quad (Bx_{n+1} - b)^T y_{n+1} = 0. \end{aligned} \tag{6}$$

This is the best-known form of the sequential quadratic programming (SQP) method for the problem (3); indeed, an iteration in (6) is obtained to solving a quadratic programming problem.

For given x and v, let $\xi = \{x_1, x_2, \cdots, x_n, \cdots\}$ be a Newton sequence, that is, a sequence starting from x and satisfying (5). Denote by $\mathbf{N}(x, v)$ the set of all Newton sequences, starting from the point x for v. Let $\xi^* = \{x^*, x^*, \cdots, x^*, \cdots\}$, that is, ξ^* is the constant sequence with all elements x^*. Note that $\xi^* \in \mathbf{N}(x^*, v^*)$. We equip the set of Newton sequences with a distance induced by the l^∞ norm:

$$\|\xi\|_\infty = \sup_{n \geq 1} \|x_n\|.$$

In this paper we study continuity properties of the map $(x, v) \mapsto \mathbf{N}(x, v)$ in relation to corresponding properties of the map $\mathbf{S}$ defined in (2). We consider three continuity properties which play central roles in quantitative stability analysis of variational problems: the local upper-Lipschitz property, the Aubin continuity, and the Lipschitzian localization property. We also provide some new insights into convergence properties of Newton's method.

In this paper we apply the basic principle of smooth nonlinear analysis, stated by A. Ioffe [10] in the following way: a property of the derivative of a nonlinear map at a given point should hold, in a sufficiently small neighborhood, for the map itself. In the context of the inclusion (1) and the corresponding solution map $\mathbf{S} = (f + F)^{-1}$, this principle means that adding terms of order $o(x)$ to (1) doesn't change certain properties of the map $\mathbf{S}$. Since replacing f by its linearization at x^* corresponds to adding $o(x - x^*)$, we call this specific form of the basic principle *invariance under linearization.*

The basic principle of smooth nonlinear analysis is present already in the classical implicit function theorem and is even more explicit in the Lyusternik and Graves theorems. In his seminal paper [21], Robinson extended this principle to variational inequalities and optimization problems. The main idea in the present paper is to apply the basic principle to Lipschitz-type properties of Newton's map $\mathbf{N}$ defined above.

In Section 2 we study the local upper-Lipschitz continuity, a relatively recently introduced concept which turns out to be quite natural in nonlinear optimization. We present characterizations of the upper-Lipschitz continuity for the map $\mathbf{S}$ and for the Argmin map of the problem (3). In particular, we show that the local upper-Lipschitz continuity is invariant under linearization. We prove that the Argmin map of the problem (3) possesses this property if and only if the standard second-order sufficient optimality condition holds.

In Section 3 we first show that the local upper-Lipschitz continuity of the map $\mathbf{S}$ implies that every Newton sequence within a sufficiently small ball around the solution is quadratically convergent. Then we study the local upper-Lipschitz continuity of the Newton map $\mathbf{N}$. We prove that if the map $\mathbf{S}$ is locally upper-Lipschitz, then the map $\mathbf{N}$ is locally upper-Lipschitz as well (in the l^∞ norm). We also show that the converse implication holds if, in addition, $\mathbf{N}$ is locally nonempty-valued. For the optimization problem (3) this result evolves into the equivalence between the second-order sufficient optimality condition and the property that the Newton (SQP) map is locally upper-Lipschitz and nonempty-valued.

At the end of Section 3 we briefly show that analogous results can be obtained for the proximal point method.

In Section 4 we discuss the Aubin continuity by following the pattern of the previous sections. We give a version of the Lyusternik-Graves theorem (Lemma 4.1) and use it to show that the Aubin continuity of the map $\mathbf{S}$ implies existence of locally quadratically convergent Newton sequences around the reference point. Further, the Newton map $\mathbf{N}$ is Aubin continuous if the map $\mathbf{S}$ is Aubin continuous. As a partial converse, the Aubin continuity of the submap of $\mathbf{N}$ associated with all convergent Newton sequences implies the Aubin continuity of the map $\mathbf{S}$.

Section 5 is devoted to the Lipschitzian localization, a continuity property which appears in implicit function theorems and is best studied in optimization. We show that the Lipschitzian localization of the map $\mathbf{S}$ implies the existence of a unique Newton sequence around the reference point. Also, the Newton map $\mathbf{N}$ has a Lipschitzian localization if and only if the map $\mathbf{S}$ possesses this property. For the problem (3) the Lipschitzian localization of the Argmin map is equivalent to the strong second-order sufficient optimality condition.

There is a vast literature on the continuity concepts discussed in the present paper and their applications to various variational problems. For recent discussions and references see the bibliographical comments in [10, 12, 25].

As a further application of the approach presented in this paper, we target error analysis of discrete approximations to infinite-dimensional variational problems for which the parameter v represents the discretization remainder while $v = v^*$ is identified with the original (continuous) problem. We shall not discuss here discrete approximations; for first results in this direction, see the forthcoming paper [6].

Throughout we denote by $\|\cdot\|$ any norm in $\mathbf{R}^n$, by $B_r(x)$ the closed ball with center x and radius r, and by $\mathbf{B}$ the ball $B_1(0)$. In writing "f maps X into Y" we mean that the domain of f is a (possibly proper) subset of X; thus, a set-valued map F from X to the subsets of Y may have empty values for some points of X. Given a map F from X to the subsets of Y, we define $\operatorname{graph} F = \{(x, y) \in X \times Y \mid y \in F(x)\}$ and $F^{-1}(y) = \{x \in X \mid y \in F(x)\}$. We denote by $\operatorname{dist}(x, A)$ the distance from a point x to a set A and by ^{T} a transposition of a matrix or a vector. Throughout L is a Lipschitz constant of $\nabla f(\cdot)$ in a (sufficiently large) ball centered at x^*.

2. LOCAL UPPER-LIPSCHITZ CONTINUITY

The local upper-Lipschitz continuity is a localized version, for the graph of a map, of the upper-Lipschitz continuity introduced by Robinson [20].

Definition 2.1 *A map $\Gamma : \mathbf{R}^n \to \mathbf{R}^m$ is locally upper-Lipschitz continuous at $(y^*, x^*) \in \operatorname{graph} \Gamma$ with constants a and b for neighborhoods and c for growth if*

$$\Gamma(y) \cap B_a(x^*) \subset \{x^*\} + c\|y - y^*\|\mathbf{B} \quad \textit{for all } \; y \in B_b(y^*).$$

The following properties can be deduced directly from the definition: A locally upper-Lipschitz map Γ at (x^*, y^*) has necessarily x^* as a locally unique (isolated) point of $\Gamma(y^*)$. If a map is locally upper-Lipschitz with constants a, b and c, then for every $0 \leq a' \leq a, 0 \leq b' \leq b$, and $c' \geq c$, it is locally upper-Lipschitz with constants a', b' and c'. If a map Γ is locally upper-Lipschitz at (y^*, x^*) and Σ is a local selection of Γ around (x^*, y^*), that is, for some neighborhoods U of x^* and V of y^* we have $\Sigma(y) \cap U \subset \Gamma(y) \cap U$ for $y \in V$, then Σ is locally upper-Lipschitz at (y^*, x^*).

Robinson [22] proved that in finite dimensions every map whose graph is a polyhedral (possibly nonconvex) set is upper-Lipschitz at every point of its domain. As a consequence, every map $\Gamma : \mathbf{R}^n \to \mathbf{R}^m$, for which x^* is an isolated point of $\Gamma(y^*)$ and $\operatorname{graph} \Gamma$ is a polyhedral set, is locally upper-Lipschitz at (y^*, x^*). Such maps are for instance the solution maps of the linear variational inequalities over convex polyhedral sets when the reference solution is isolated. We note that Robinson used in [20] a different definition of the local upper-Lipschitz continuity where the localization is in the domain of the map.

Rockafellar [23], see also [11, 15], gave a characterization of the upper-Lipschitz property by employing the so-called proto-derivatives (contingent derivatives). We showed in [4] that the local upper-Lipschitz continuity is invariant under linearization in a very abstract setting (see Lemma 2.1 and Corollary 2.1 below). The upper-Lipschitz property was studied in [8] for a general nonlinear programming problem with canonical parameters in both the functional and the constraints. In [8], Theorem 2.6, it was proved that the Karush-Kuhn-Tucker map has this property with the reference point being a local optimal solution exactly when the combination of the strict Mangasarian-Fromowitz condition and the standard second-order sufficient optimality condition holds. Further extensions and generalizations of this result for nonsmooth mathematical programs, as well as extended surveys on the subject, are given in the recent papers [9, 12, 16].

In a related vain, Zolezzi introduced and studied in [27] a condition number for the solutions to abstract optimization problems of the form $\min f(x, y)$, $x \in X$, where y is a parameter. This conditional number is precisely the growth constants c in the definition of the upper-Lipschitz property, where $\Gamma(y)$ is the set of optimal solutions for y.

Note that the local upper-Lipschitz property does not guarantee nonemptiness of the values of Γ near the reference point y^* (as an example, take $\Gamma(y^*) = x^*$ and $\Gamma(y) = \emptyset$ for $y \neq y^*$). Such a requirement leads to a different concept, namely:

Definition 2.2 *A map* $\Gamma : \mathbf{R}^n \to \mathbf{R}^m$ *is locally nonempty-valued at* $(y^*, x^*) \in \operatorname{graph} \Gamma$ *if there exist positive numbers* a *and* b *such that*

$$\Gamma(y) \cap B_a(x^*) \neq \emptyset \quad \textit{for all } y \in B_b(y^*).$$

A map Γ is locally nonempty-valued at (y^*, x^*) if and only if the map Γ^{-1} is open at (x^*, y^*). If $\Gamma(y)$ is the set of solutions of a system of inequalities with data y (matrices, vectors), then the radius of the neighborhood b is a kind of bound on perturbations of the data from y^* such that the system still has a solution x within distance a from the reference solution x^* for y^*. With $a = \infty$ and after normalization, the supremum of such b is exactly the relative condition number introduced by Renegar [19].

The following lemma shows the invariance under linearization of the upper-Lipschitz continuity in a transparent form which is particularly suitable for applications:

Lemma 2.1 *Let G be a map from $\mathbf{R}^n$ to the subsets of $\mathbf{R}^m$ and let G^{-1} be locally upper-Lipschitz at (v^*, x^*) with constants a and b for neighborhoods and c for growth. Let the nonnegative constants α, β and λ satisfy*

$$\alpha \leq a, \quad \lambda c < 1, \quad \beta + \lambda\alpha \leq b. \tag{7}$$

Let $h : \mathbf{R}^n \to \mathbf{R}^m$ be a function which satisfies

$$\|h(x) - h(x^*)\| \leq \lambda \|x - x^*\| \quad \text{for all } x \in B_\alpha(x^*). \tag{8}$$

Then the map $(h + G)^{-1}$ is locally upper-Lipschitz at $(v^* + h(x^*), x^*)$ with constants α and β for neighborhoods and $c/(1 - \lambda c)$ for growth.

Proof. Choose α, β and λ as in (7) and let h satisfy (8). Put $y^* = v^* + h(x^*)$ and let $y \in B_\beta(y^*)$ and $x \in (h + G)^{-1}(y) \cap B_\alpha(x^*)$. Then $x \in G^{-1}(y - h(x)) \cap B_a(x^*)$ and

$$\|y - h(x) - v^*\| \leq \|y - y^*\| + \|h(x) - h(x^*)\| \leq \beta + \lambda\alpha \leq b.$$

Hence, from the upper-Lipschitz continuity of G^{-1} at (v^*, x^*) we obtain

$$\begin{aligned} \|x - x^*\| \leq c\|y - h(x) - v^*\| &\leq c\|y - y^*\| + c\|h(x) - h(x^*)\| \\ &\leq c\|y - y^*\| + c\lambda\|x - x^*\|. \end{aligned}$$

Thus,

$$\|x - x^*\| \leq \frac{c}{1 - \lambda c}\|y - y^*\|$$

and the proof is complete. □

It is clear from the proof that this lemma still holds if we replace $\mathbf{R}^n$ with any metric space and $\mathbf{R}^m$ with a linear space with a shift-invariant metric.

Recall that $\mathbf{S}(v)$ denotes the set of solution to (1) for v. With the solution map $\mathbf{S}$ we associate the map

$$v \mapsto \mathbf{L}(v) = (f(x^*) + \nabla f(x^*)(\cdot - x^*) + F(\cdot))^{-1}(v)$$

obtained by the linearization of f at the point x^*. We have $x^* \in \mathbf{L}(v^*)$ iff $x^* \in \mathbf{S}(v^*)$. From Lemma 2.1 we obtain the following result for the solutions of (1) first stated in [4], Theorem 3.2.

Corollary 2.1 *The following are equivalent:*

(i) The map $\mathbf{L}$ *is locally upper-Lipschitz at* (v^*, x^*).

(ii) The map $\mathbf{S}$ *is locally upper-Lipschitz at* (v^*, x^*).

We note that the local nonempty-valuedness of a map (Definition 2.2) is not invariant under linearization.

In our analysis of Newton's method we use the following map:

$$(v, x) \mapsto \mathbf{P}(v, x) := (f(x) + \nabla f(x)(\cdot - x) + F(\cdot))^{-1}(v). \tag{9}$$

If x^* is a solution to (1) for v^*, then $x^* \in \mathbf{P}(v^*, x^*)$. Also, $\mathbf{L}(v) = \mathbf{P}(v, x^*)$. The following corollary complements Corollary 2.1.

Corollary 2.2 *The following are equivalent:*

(i) The map $\mathbf{L}$ *is locally upper-Lipschitz at* (v^*, x^*).

(ii) The map $\mathbf{P}$ *is locally upper-Lipschitz at* $((v^*, x^*), x^*)$.

Proof. The implication (ii) $\Rightarrow$ (i) is immediate by noting that $\mathbf{L}(v) = \mathbf{P}(v, x^*)$. Let us prove (i) $\Rightarrow$ (ii). Let $\mathbf{L}$ be locally upper-Lipschitz at (v^*, x^*) with constants a, b and c and let the positive numbers α and β satisfy the inequalities (7) with $\lambda = L\beta$. (Recall that, here and later, L is a Lipschitz constant of $\nabla f(\cdot)$ in a (sufficiently large) ball centered at x^*; in this case this is the ball with radius a.)

Let $(v, x) \in B_\beta((v^*, x^*))$ and let $z \in \mathbf{P}(v, x) \cap B_a(x^*)$. We apply Lemma 2.1 with

$$G(z) = f(x^*) + \nabla f(x^*)(z - x^*) + F(z)$$

and

$$h(z) = f(x) + \nabla f(x)(z - x) - f(x^*) - \nabla f(x^*)(z - x^*).$$

For any $z' \in B_\alpha(x^*)$ we have

$$\begin{aligned} \|h(z') - h(x^*)\| &= \|(\nabla f(x) - \nabla f(x^*))(z' - x^*)\| \\ &\leq L\beta\|z' - x^*\| = \lambda\|z' - x^*\|, \end{aligned}$$

hence (8) holds. By assumption, G^{-1} is locally upper-Lipschitz at (v^*, x^*), hence from Lemma 2.1 we obtain that the map $(h+G)^{-1}$ is locally upper-Lipschitz at $(v^* + h(x^*), x^*)$. Note that $z \in (h+G)^{-1}(v) \cap B_\alpha(x^*)$. Then, for $\gamma = c/(1-\lambda c)$, we have

$$\begin{aligned} \|z - x^*\| &\leq \gamma\|v - v^* - f(x) - \nabla f(x)(x^* - x) + f(x^*)\| \\ &\leq \gamma\|v - v^*\| + \frac{1}{2}\gamma L\beta\|x - x^*\|. \end{aligned}$$

Thus the map $\mathbf{P}$ is locally upper-Lipschitz with constants α and β for the radia of the neighborhoods and growth constants γ for v and $\gamma L\beta/2$ for x, respectively. □

Remark 2.1 *Note that the growth constant of* $\mathbf{P}$ *with respect to* x *can be made arbitrarily small, by choosing* β *small.*

Consider the optimization problem (3) and the corresponding first-order necessary optimality condition (4). Let x^* be a solution of (4) for $v = v^*$. Denote by K the critical cone at (v^*, x^*), that is,

$$K = \{x \in T_C(x^*) \mid v^* - \nabla g(x^*) \perp x\},$$

where $T_C(x^*)$ is the tangent cone to C at the point x^*. Then the (standard) second-order sufficient optimality condition at (v^*, x^*) has the form

$$\langle u, \nabla^2 g(x^*)u\rangle > 0 \quad \text{for all nonzero } u \in K. \tag{10}$$

Let Argmin(v) be the set of local solutions for v to the optimization problem (3).

Theorem 2.1 *The following are equivalent:*

(i) The second-order sufficient optimality condition (10) *holds at* (v^*, x^*).

(ii) The Argmin map is locally nonempty-valued and upper-Lipschitz continuous at (v^*, x^*).

Proof. Let $\mathcal{X}$ be the map of the critical points,

$$v \mapsto \mathcal{X}(v) = \{x \in \mathbf{R}^n \mid v \in \nabla g(x) + N_C(x)\}. \tag{11}$$

The solution map of the linearization of (4) at x^* is defined as

$$v \mapsto \mathcal{L}(v) = \{x \in \mathbf{R}^n \mid v \in \nabla g(x^*) + \nabla^2 g(x^*)(x - x^*) + N_C(x)\}. \tag{12}$$

Let (i) hold. For short, denote $a = \nabla g(x^*)$ and $A = \nabla^2 g(x^*)$. First, let us show that x^* is an isolated solution of the linear variational inequality

$$v^* \in a + A(x - x^*) + N_C(x). \tag{13}$$

On the contrary, suppose that there exists a sequence $x_n \to x^*$ such that

$$v^* \in a + A(x_n - x^*) + N_C(x_n) \quad \text{for all } n.$$

Then

$$v^* \in a + \nabla^2 g(x_n)(x_n - x^*) + o(\|x_n - x\|) + N_C(x^* + (x_n - x^*)).$$

By Reduction Lemma in [7],

$$0 \in \nabla^2 g(x_n)(x_n - x^*) + o(\|x_n - x^*\|) + N_K(x_n - x^*);$$

that is,

$$\langle \nabla^2 g(x_n)(x_n - x^*), x_n - x^* \rangle + o(\|x_n - x^*\|^2) = 0. \tag{14}$$

Since

$$b_n := \frac{x_n - x^*}{\|x_n - x^*\|} \in K \quad \text{for all } n \quad \text{and} \quad \|b_n\| = 1,$$

we obtain that a subsequence of b_n is convergent to, say, $u \in K$. From (14) we get

$$\langle \nabla^2 g(x_n) b_n, b_n \rangle + O(\|x_n - x^*\|) = 0.$$

Passing to the limit with (the subsequence of) $n \to \infty$ in the latter equality we come to a contradiction with (i). Thus x^* is an isolated solution of (13).

As noted at the beginning of this section, since $\mathcal{L}$ is a polyhedral map, from a result of Robinson [22] we obtain that the requirement x^* be an isolated point of $\mathcal{L}(v^*)$ is equivalent to the local upper-Lipschitz continuity of $\mathcal{L}$ at (v^*, x^*). Then the local upper-Lipschitz continuity of the critical point map $\mathcal{X}$ at (v^*, x^*) follows from the invariance under linearization (Corollary 2.1).

Consider the problem (3) with the additional constraint $\|x - x^*\| \leq \delta$, for $\delta > 0$. By the Berge theorem, the solution map Argmin_δ of the new problem is upper semicontinuous at $v = v^*$. Clearly, Argmin_δ is nonempty-valued and $\text{Argmin}_\delta(v^*) = \{x^*\}$ for δ sufficiently small. One then sees that for v close to v^* eventually the constraint $\|x - x^*\| \leq \delta$ is not active so that $\text{Argmin}_\delta(v) = \text{Argmin}(v) \cap B_\delta(x^*)$. Thus, Argmin is locally nonempty-valued at (v^*, x^*). Since $\text{Argmin}(v) \subset \mathcal{X}(v)$, we obtain that (ii) holds.

Conversely, let Argmin be locally nonempty-valued and upper-Lipschitz continuous at (v^*, x^*) with a growth constant κ and neighborhoods U and V. Without loss of generality, assume $v^* = 0$. Take any $x \in \mathbf{R}^n$ close x^* such that $v = (x - x^*)/2\kappa \in V$. Let $x_v \in \text{Argmin}(v) \cap U$ be an associated local minimizer, then $\|x_v - x^*\| \leq \kappa \|v\|$. From the optimality we have

$$g(x) - \langle v, x \rangle \geq g(x_v) - \langle v, x_v \rangle,$$

that is

$$g(x) - \frac{1}{2\kappa}\langle x - x^*, x - x^* \rangle \geq g(x_v) - \frac{1}{2\kappa}\langle x - x^*, x_v - x^* \rangle.$$

Since $g(x_v) \geq g(x^*)$ (remember that $v^* = 0$), we continue the chain of inequalities obtaining

$$\begin{aligned} g(x) - \frac{1}{2\kappa}\langle x - x^*, x - x^* \rangle &\geq g(x^*) - \frac{1}{2\kappa}\|x - x^*\|\|x_v - x^*\| \\ &\geq g(x^*) - \frac{1}{2\kappa}\|x - x^*\|\kappa\|v\| \\ &= g(x^*) - \frac{1}{4\kappa}\|x - x^*\|^2. \end{aligned}$$

Thus

$$g(x) \geq g(x^*) + \frac{1}{4\kappa}\|x - x^*\|^2$$

for every x close to x^*. We obtain the so-called growth condition of order two which, as well known, see e.g. [25], p. 606, is equivalent to the second-order sufficient condition (10). □

The above result clarifies the meaning of the second-order sufficient condition: it gives not only optimality at the reference point but also a certain continuity property of the minimizers with respect to certain perturbations and this property is exactly captured by the local upper-Lipschitz continuity.

The approach presented in this section can be applied to other mappings in variational analysis. As an example consider the map

$$(u, v) \mapsto \mathbf{M}_\mu(u, v) = (f(\cdot) + \mu(\cdot - u) + F(\cdot))^{-1}(v), \tag{15}$$

where μ is a positive scalar. In the context of our illustrative problem (3), the map $\mathbf{M}_\mu$ is the stationary point map of the prox-regularization (Moreau-Yosida regularization) of (3):

$$\min_{x \in C}\left[g(x) - \langle v, x \rangle + \frac{1}{2}\mu\|x - u\|^2\right].$$

A straightforward application of Lemma 2.1 gives us

Corollary 2.3 *Let the map* $\mathbf{S}$ *be locally upper-Lipschitz at* (v^*, x^*) *with a growth constant* c. *Then for every positive* $\mu < 1/c$ *the map* $\mathbf{M}_\mu$ *is locally upper-Lipschitz at* $((v^*, x^*), x^*)$ *with growth constants* $c/(1-c\mu)$ *for* v *and* $c\mu/(1-c\mu)$ *for* x.

At the end of the following section we apply this result to show convergence of the proximal-point method.

3. LOCAL UPPER-LIPSCHITZ CONTINUITY OF THE NEWTON MAP

Consider the following perturbed version of the Newton method (5) in which $\nabla f(x_n)$ is replaced by a sequence of matrices H_n and the parameter v may vary from step to step:

$$v_n \in f(x_n) + H_n(x_{n+1} - x_n) + F(x_{n+1}) \quad \text{for } n = 0, 1, \cdots. \tag{16}$$

Our first result relates the local upper-Lipschitz property of the map $\mathbf{L}$ (or, equivalently, $\mathbf{S}$) to convergence of Newton sequences.

Theorem 3.1 *Let* $\mathbf{L}$ *be locally upper-Lipschitz at* (v^*, x^*) *with constants* a *and* b *for neighborhoods and* c *for growth. Let the positive constants* σ *and* κ *satisfy*

$$\sigma \le a, \quad c\kappa < 1, \quad \kappa(1 + 2\sigma) + \frac{1}{2}L\sigma^2 \le b, \tag{17}$$

let $\{v_n\}$ *be a sequence of vectors with* $v_n \in B_\kappa(v^*), n = 1, 2, \cdots$, *and let* H_n *be a sequence of matrices with* $H_n \in B_\kappa(\nabla f(x^*))$, $n = 0, 1, \cdots$. *Then for every initial point* $x_0 \in B_\sigma(x^*)$, *every sequence* $\{x_n\}$ *obtained from (16) and whose elements are all in* $B_\sigma(x^*)$ *for* $n = 1, 2, \cdots$ *satisfies*

$$\|x_{n+1} - x^*\| \le \frac{c}{1 - c\kappa}\Big(\|x_n - x^*\|^2 + \|(H_n - \nabla f(x^*))(x_n - x^*)\| + \|v_n - v^*\|\Big). \tag{18}$$

Proof. Choose κ and σ such that the inequalities (17) hold and let $\{x_n\}$ be a sequence satisfying the conditions of the theorem. Fix $n \ge 1$. We apply Lemma 2.1 with $G = \mathbf{L}^{-1}$, $h(x) = f(x_n) + H_n(x - x_n) - f(x^*) - \nabla f(x^*)(x - x^*)$, $\alpha = \sigma$, $\beta = \kappa(1 + \sigma) + \frac{1}{2}L\sigma^2, \lambda = \kappa$. The so defined

α, β and λ satisfy (7) and h satisfies (8). From Lemma 2.1, the map $(h+G)^{-1}$ is locally upper-Lipschitz at $(v^* + h(x^*), x^*)$ with constants α and β for the neighborhoods and a growth constant $c(1-c\lambda)$. We have

$$x_{n+1} \in (h+G)^{-1}(v_n) \cap B_\sigma(x^*)$$

and

$$\begin{aligned}
&\|v_n - v^* - h(x^*)\| \leq \|v_n - v^*\| + \|f(x_n) + H_n(x^* - x_n) - f(x^*)\| \\
&\leq \|v_n - v^*\| + \|(H_n - \nabla f(x^*))(x_n - x^*)\| + \frac{1}{2}L\|x_n - x^*\|^2 \\
&\leq \kappa + \kappa\sigma + \frac{1}{2}L\sigma^2 = \beta.
\end{aligned}$$

Hence, from the upper-Lipschitz continuity of $(h+G)^{-1}$, we obtain (18). □

Corollary 3.1 *Let the assumptions of Theorem 3.1 hold, let $v_n = v^*$ for all n, and let σ and κ be the associated constants. Choose σ and κ smaller if necessary so that*

$$\frac{c(\sigma+\kappa)}{(1-c\kappa)} \leq 1.$$

Then every Newton sequence $\{x_n\}$ obtained from (16) with $x_0 \in \mathbf{R}^n$ has all elements outside the ball $B_\sigma(x^*)$ or:

(a) x_n is linearly convergent to x^*;

(b) if $H_n \to \nabla f(x^*)$ as $n \to \infty$, then the sequence x_n is superlinearly convergent to x^*;

(c) if $H_n = \nabla f(x_n)$, then x_n is quadratically convergent to x^*.

Proof. Observe that, on our assumptions, if $\{x_n\}$ is a Newton sequence and for some $i \geq 0$ we have $x_i \in B_\sigma(x^*)$, then $x_n \in B_\sigma(x^*)$ for all $n \geq i$. Then apply (18). □

In the remaining part of this section we consider the unperturbed Newton method (5) with a parameter v. Recall that $\mathbf{N}(x, v)$ is the set of all Newton sequences satisfying (5) for v which start from the point x. The following theorem shows the close relation of the map $\mathbf{N}$ with the map $\mathbf{P}$ defined in (9), and hence with the maps $\mathbf{S}$ and $\mathbf{L}$ according to Corollaries 2.1 and 2.2.

Theorem 3.2 *The following are equivalent:*

(i) The map $\mathbf{P}$ *is locally upper-Lipschitz and nonempty-valued at* $((v^*, x^*), x^*)$*;*

(ii) The map $\mathbf{N}$ *is locally upper-Lipschitz and nonempty-valued at* $((x^*, v^*), \xi^*)$.

Proof. (i) $\Rightarrow$ (ii). If $\mathbf{P}$ is locally nonempty-valued, then for each (v, x) close to (v^*, x^*) there exists a Newton step, hence $\mathbf{N}$ is locally nonempty-valued as well. Let $\mathbf{P}$ be locally upper-Lipschitz with constants a and b for the neighborhoods and c and μ for the growth of v and x, respectively. According to Remark 2.1, the constant μ can be made arbitrary small by choosing smaller b. Thus, without loss of generality, suppose that $\mu < 1$. Let $\xi = \{x_1, x_2, \cdots, x_n, \cdots\} \in \mathbf{N}(x, v) \cap B_a(\xi^*)$ for some $(x, v) \in B_b(x^*, v^*)$. Then $x_1 \in \mathbf{P}(v, x) \cap B_a(x^*)$ and from the local upper-Lipschitz continuity of $\mathbf{P}$ we have

$$\|x_1 - x^*\| \le c\|v - v^*\| + \mu\|x - x^*\|.$$

Proceeding by induction, from $x_{n+1} \in \mathbf{P}(v, x_n) \cap B_a(x^*)$ and from the local upper-Lipschitz continuity of $\mathbf{P}$, we obtain

$$\|x_{n+1} - x^*\| \le c(1 + \mu + \mu^2 + \cdots + \mu^n)\|v - v^*\| + \mu^{n+1}\|x - x^*\|.$$

Then,

$$\sup_{n \ge 1} \|x_{n+1} - x^*\| \le \frac{c}{1 - \mu}\|v - v^*\| + \mu\|x - x^*\|.$$

We obtain that the map $\mathbf{N}$ is locally upper-Lipschitz at $((x^*, v^*), \xi^*)$ with constants a, b for the neighborhoods, and growth constants μ for x and $c/(1 - \mu)$ for v.

(ii) $\Rightarrow$ (i). Let $\mathbf{N}$ be locally upper-Lipschitz and locally nonempty-valued at $((x^*, v^*), \xi^*)$ with constants a, b and c. Let $(x, v) \in B_b(x^*, v^*)$ and $z \in \mathbf{P}(v, x) \cap B_a(x^*)$. Then z is the first element of a Newton sequence starting at x for v. From the nonemptiness assumption for $\mathbf{N}$, there exists a Newton sequence $\bar{\xi} = \{\bar{x}_1, \cdots, \bar{x}_n, \cdots\} \in \mathbf{N}(z, v) \cap B_a(\xi^*)$. Observe that the sequence $\eta = \{z, \bar{x}_1, \bar{x}_2, \cdots, \bar{x}_n, \cdots\}$ is an element of $\mathbf{N}(x, v) \cap B_a(\xi^*)$. From the local upper-Lipschitz continuity of $\mathbf{N}$ (in the supremum norm), the first component of η satisfies

$$\|z - x^*\| \le c(\|x - x^*\| + \|v - v^*\|).$$

This means exactly that $\mathbf{P}$ is locally upper-Lipschitz at $((v^*, x^*), x^*)$. The local nonemptiness of $\mathbf{N}(x, v) \cap B_a(\xi^*)$ implies that $\mathbf{P}$ is locally nonempty-valued. □

Remark 3.1 *Observe that, from the above proof, the local upper-Lipschitz continuity of* **P** *implies the local upper-Lipschitz continuity of* **N** *without the requirement* **P** *be locally nonempty-valued.*

Theorem 3.3 *If the map* **S** *is locally upper-Lipschitz at* (v^*, x^*), *then* **N** *is locally upper-Lipschitz at* $((x^*, v^*), \xi^*)$. *Conversely, if the map* **N** *is locally upper-Lipschitz and nonempty-valued at* $((x^*, v^*), \xi^*)$, *then* **S** *is locally upper-Lipschitz at* (v^*, x^*).

Proof. If **S** is locally upper-Lipschitz, then, from Corollaries 2.1 and 2.2, the map **P** is locally upper-Lipschitz at $((v^*, x^*), x^*)$; then Theorem 3.2 (with Remark 3.1) completes the proof. Conversely, if **N** is locally upper-Lipschitz and nonempty-valued at $((x^*, v^*), \xi^*)$, then, from Theorem 3.2, **P** is locally upper-Lipschitz at $((v^*, x^*), x^*)$, hence, from Corollaries 2.1 and 2.2, **S** is locally upper-Lipschitz at (v^*, x^*). □

As an illustration, consider the optimization problem (3). The Newton (SQP) iterate x_{n+1} from x_n defined in (6) is a stationary point of the following quadratic program:

$$\min_{z \in C} \left[.5\langle \nabla^2 g(x_n)(z - x_n), z - x_n \rangle + \langle \nabla g(x_n) - v, z - x_n \rangle \right]. \tag{19}$$

Under the second-order sufficient optimality condition, by using the nonemptiness argument in the proof of Theorem 2.1, we conclude that the intersection of the solution set of this problem for (v, x_n) with any ball around x^* is nonempty provided that (v, x_n) is sufficiently close to (v^*, x^*). Thus, the corresponding map **P** is nonempty-valued locally around $((v^*, x^*), x^*)$. The converse follows from Theorems 2.2 and 3.2. Thus we obtain

Corollary 3.2 *The following are equivalent:*

(i) The second-order sufficient optimality condition (10) *holds at* (v^*, x^*).

(ii) The Newton map **N** *defined as in* (19) *is locally upper-Lipschitz and nonempty-valued at* $((x^*, v^*), \xi^*)$ *and* x^* *is a local solution of (3) for* $v = v^*$.

Further, from Theorems 2.2, 3.1 and from Corollary 3.2 we obtain that under the second-order sufficient optimality condition, every Newton (SQP) sequence x_n within a sufficiently small neighborhood of the solution x^* is quadratically convergent to x^*; moreover, the map of SQP sequences has the local upper-Lipschitz property. Note that we do not impose any conditions on regularity of the constraints. Also note that, because of the polyhedrality and Robinson's theorem [22], the map from v to the set of all sequences of Lagrange multipliers y_n associated with a Newton (SQP) sequence x_n satisfying (6) has the (standard) upper-Lipschitz property.

For problems with nonlinear smooth constraints, the SQP method is obtained when the Newton method is applied to the variational inequality of the first-order optimality conditions (the Karush-Kuhn-Tucker system). To be specific, consider the problem

$$\min g(z) \quad \text{subject to} \quad \varphi_j(z) \le 0, \quad j = 1, 2, \cdots, m, \quad z \in \mathbf{R}^n, \tag{20}$$

with three times continuously differentiable g and $\varphi = (\varphi_1, \cdots, \varphi_m)$ and with a solution z^* at which the Mangasarian-Fromovitz condition holds (to guarantee the existence of Lagrange multipliers, say y^*). The Karush-Kuhn-Tucker optimality system has the form

$$\begin{aligned} \nabla g(z) + \nabla \varphi(z)^T y = 0 \\ \varphi(z) \in N_{\mathbf{R}^n_+}(y). \end{aligned}$$

The SQP method is then the Newton method (5) with $x = (z, y)$,

$$f(x) = \begin{pmatrix} \nabla g(z) + \nabla \varphi^T y \\ \varphi(z) \end{pmatrix} \quad \text{and} \quad F(x) = \begin{pmatrix} 0 \\ N_{\mathbf{R}^n_+}(y) \end{pmatrix}.$$

Note that Corollary 3.1 now claims quadratic convergence of Newton sequences for both the primal variables z and the Lagrange multipliers y. From [8], Theorem 2.6, we obtain that this convergence is guaranteed when the combination of the strict Mangasarian-Fromovitz condition and the standard second-order sufficient optimality condition holds.

As another application of our approach, consider the well-known proximal point method

$$v \in f(x_{n+1}) + \mu(x_{n+1} - x_n) + F(x_{n+1}), \tag{21}$$

where μ is a positive constant. Now the role of the map $\mathbf{P}$ is played by the map $\mathbf{M}_\mu$ defined in (15). Applying Corollary 2.3 we obtain that if the map $\mathbf{S}$ is locally upper-Lipschitz with a growth constant c, then every sequence generated by (21) for v^* and within a sufficiently small ball around x^* is linearly convergent to x^*. Further, let $\mathbf{T}_\mu$ be the map from the pair (x, v) to the set of all sequences obtained by (21) for v and μ and starting from x. The constant sequence $\xi^* = \{x^*, x^*, \cdots\} \in \mathbf{T}_\mu(x^*, v^*)$.

By noting that the growth constant of the map $\mathbf{M}_\mu$ defined in (15) can be < 1 for μ sufficiently small and repeating the argument in Theorem 3.2, we obtain:

Theorem 3.4 *If the mapping* $\mathbf{S}$ *is locally upper-Lipschitz at* (v^*, x^*), *then for a sufficiently small positive* μ *the map* $\mathbf{T}_\mu$ *is locally upper-Lipschitz at* $((x^*, v^*), \xi^*)$.

There is a broad field of applications of our approach to other models in optimization; in particular, to the extended nonlinear programming problem introduced by Rockafellar [24] which covers the conventional nonlinear programming models as well as penalty-function and augmented-Lagrangian type models.

4. AUBIN CONTINUITY

The concept of Aubin continuity can be traced back to the original proofs of the Lyusternik and Graves theorems, see [10, 25] for discussions. In mathematical programming it has appeared as "metric regularity"; in a topological framework, it was called "openness with linear rate". J.-P. Aubin was the first to define this concept as a continuity property of set-valued maps, calling it "pseudo-Lipschitz continuity" [1]. Following [7] we use the name "Aubin continuity".

Definition 4.1 *Let* Γ *map* $\mathbf{R}^n$ *to the subsets of* $\mathbf{R}^m$ *and let* $(y^*, x^*) \in \operatorname{graph}\Gamma$. *We say that* Γ *is Aubin continuous at* (y^*, x^*) *with constants* a, b *for neighborhoods and* c *for growth if for every* $y', y'' \in B_b(y^*)$ *and every* $x' \in \Gamma(y') \cap B_a(x^*)$, *there exists* $x'' \in \Gamma(y'')$ *such that*

$$\|x' - x''\| \leq c\|y' - y''\|.$$

Directly from the definition one can extract the following properties of Aubin continuous maps. If a map Γ is Aubin continuous at (y^*, x^*) with constants a, b and c, then for every $0 < a' \leq a$ and $0 < b' \leq b$ the map Γ is Aubin continuous at (y^*, x^*) with constants a', b' and c. If, in addition, $b' \leq a'/c$, then $\Gamma(y) \cap B_{a'}(x^*) \neq \emptyset$ for all $y \in B_{b'}(y^*)$. If Γ is Aubin continuous at (y^*, x^*) with constants a, b and c, there exist constants a', b' and δ such that for every $(y, x) \in \operatorname{graph}\Gamma \cap B_\delta((y^*, x^*))$ the map Γ is Aubin continuous at (y, x) with constants a', b' and c. Finally, for a set-valued map G, if G^{-1} be Aubin continuous at (y^*, x^*) with constants a, b, and c, then for every $\varepsilon < b$ the map

$$y \mapsto \{x \in \mathbf{R}^n \mid y \in G(x) + B_\varepsilon(0)\}$$

is Aubin continuous at (y^*, x^*) with constants a, $b - \varepsilon$, and c. An infinitesimal characterization of the Aubin property is given by Mordukhovich's coderivative criterion, see [25], Chapter 9.

We study the Aubin continuity by following the pattern established in the previous two sections. Our first and basic observation is that, similar to the upper-Lipschitz property, the Aubin continuity is invariant under linearization. This fact is contained in the classical Lyusternik and Graves theorems and their numerous generalizations, see, e.g., [2, 5, 14]. (It is perhaps less known that both Lyusternik and Graves used in their proof a version of the Newton method, the so-called modified Newton method, where, in terms of our notation in (5), $\nabla f(x_n)$ is replaced by $\nabla f(x^*)$. In some of the Russian literature on the subject, see e.g. [3], the modified Newton method is called "Lyusternik process".) This result is presented below in a form very instrumental for applications.

Lemma 4.1 *Let G maps $\mathbf{R}^n$ to the subsets of $\mathbf{R}^m$ and let G^{-1} be Aubin continuous at (v^*, x^*) with constants a and b for neighborhoods and c for growth. Let the nonnegative constants α, β and λ satisfy*

$$\alpha \leq a, \qquad \lambda c < 1, \quad \alpha + \frac{2\beta c}{1-\lambda c} \leq a, \quad \beta + \lambda\left(\alpha + \frac{2\beta c}{1-\lambda c}\right) \leq b. \quad (22)$$

Then for every function $h : \mathbf{R}^n \to \mathbf{R}^m$ which is Lipschitz continuous on $B_a(x^*)$ with a Lipschitz constant λ, the map $(h+G)^{-1}$ is Aubin continuous at $(v^* + h(x^*), x^*)$ with constants α and β for neighborhoods and $c/(1-\lambda c)$ for growth.

Proof. Let $y', y'' \in B_\beta(y^*)$, where $y^* = v^* + h(x^*)$ and let $x' \in (h+G)^{-1}(y') \cap B_\alpha(x^*)$. Then $x' \in G^{-1}(y' - h(x')) \cap B_\alpha(x^*)$ and for both $y = y'$ and $y = y''$ we have

$$\|y - h(x') - v^*)\| \leq \|y - y^*\| + \|h(x') - h(x^*)\| \leq \beta + \lambda\alpha \leq b.$$

From the Aubin continuity of G^{-1} we obtain that there exists $x_2 \in G^{-1}(y'' - h(x'))$ such that

$$\|x_2 - x'\| \leq c\|y' - y''\|.$$

Set $x_1 = x'$. By induction, suppose that there exists a sequence $\{x_k\}$ such that for all $k = 2, 3, \cdots, n$,

$$\|x_k - x_{k-1}\| \leq (\lambda c)^{k-2}\|x_2 - x_1\|$$

and

$$y'' \in h(x_{k-1}) + G(x_k). \quad (23)$$

Using (22), we have

$$\|x_k - x^*\| \leq \|x_1 - x^*\| + \sum_{j=2}^{k} \|x_j - x_{j-1}\|$$

$$
\begin{aligned}
&\leq \|x_1 - x^*\| + \sum_{j=2}^{\infty} \|x_j - x_{j-1}\| \\
&\leq \alpha + \frac{c}{1-\lambda c}\|y' - y''\| \\
&\leq \alpha + \frac{2\beta c}{1-\lambda c} \leq a
\end{aligned} \tag{24}
$$

and

$$\|y'' - h(x_n) - v^*\| \leq \|y'' - y^*\| + \|h(x_n) - h(x^*)\| \leq \beta + \lambda\left(\alpha + \frac{2\beta c}{1-\lambda c}\right) \leq b.$$

Then there exists $x_{n+1} \in G^{-1}(y'' - h(x_n))$ such that

$$
\begin{aligned}
\|x_{n+1} - x_n\| &\leq c\|h(x_n) - h(x_{n-1})\| \\
&\leq c\lambda(c\lambda)^{n-2}\|x_2 - x_1\| \\
&= (c\lambda)^{n-1}\|x_2 - x_1\|.
\end{aligned}
$$

The induction step is complete. Since $c\lambda < 1$ the sequence $\{x_k\}$ is a Cauchy sequence, hence convergent to, say, x'' which is in $B_a(x^*)$ because of (24). Passing to the limit in (23) we obtain that $x'' \in G^{-1}(y'' - h(x'')) = (h + G)^{-1}(y'')$ and we are done. □

Without changing it, Lemma 4.1 can be stated in abstract spaces, i.e. by replacing $\mathbf{R}^n$ with a complete metric space and $\mathbf{R}^m$ by a linear metric space with a shift-invariant metric.

An application to the maps $\mathbf{S}$ and $\mathbf{L}$ defined in Section 2 gives us the following analog to Corollary 2.1:

Corollary 4.1 *The following are equivalent:*

(i) The map $\mathbf{L}$ *is Aubin continuous at* (v^*, x^*).

(ii) The map $\mathbf{S}$ *is Aubin continuous at* (v^*, x^*).

The next corollary is a generalization of the previous one; it shows that the Aubin property is persistent when passing from $\mathbf{S}$ to the map defined by the linearization at a point close to the reference point (v^*, x^*).

Corollary 4.2 *The following are equivalent:*

(i) The map $\mathbf{S}$ *is Aubin continuous at* (v^*, x^*).

(ii) there exist constants δ, a, b *and* c *such that for every* $(v', x') \in B_\delta((v^*, x^*)) \cap \operatorname{graph} \mathbf{S}$ *the map* $(f(x') + \nabla f(x')(\cdot - x') + F(\cdot))^{-1}$ *is Aubin continuous at* (v', x') *with constants* a, b *and* c.

Proof. Let (i) holds. From the properties of Aubin continuous maps given after Definition 4.1, there exists a $\delta > 0$ such that the map $\mathbf{S}$ is Aubin continuous at any $(v', x') \in B_\delta((v^*, x^*)) \cap \operatorname{graph} \mathbf{S}$ with constants independent of the choice of (v', x'). We now apply Lemma 4.1 with $G = f + F$ and $h(x) = -f(x) + f(x') + \nabla f(x')(x - x')$ at a reference point (v', x'). Let $x', x_1, x_2 \in B_a(x^*)$. Then

$$\|h(x_1) - h(x_2)\| = \| - f(x_1) + f(x_2) - \nabla f(x')(x_1 - x_2)\|$$
$$\leq \int_0^1 \|\nabla f(x_1 + t(x_2 - x_1)) - \nabla f(x')\| dt \|x_1 - x_2\| \leq 2La\|x_1 - x_2\|.$$

Hence, by taking smaller a is necessary, the Lipschitz constant $\lambda = 2La$ of h can be obtained small enough such that $\lambda c < 1$, where c is the growth constant of $\mathbf{S}$. Thus, the map $(h + G)^{-1} = (f(x') + \nabla f(x')(\cdot - x') + F(\cdot))^{-1}$ is Aubin continuous as claimed in (ii).

If (ii) is satisfied, then of course for $(v', x') = (v^*, x^*)$ we obtain that the map $\mathbf{L}$ is Aubin continuous at (v^*, x^*), therefore $\mathbf{S}$ is Aubin continuous at (v^*, x^*), from Corollary 4.1. □

The following is an analog of Corollary 2.2.

Corollary 4.3 *The following are equivalent:*

(i) The map $\mathbf{L}$ *is Aubin continuous at* (v^*, x^*).

(ii) The map $\mathbf{P}$ *is Aubin continuous at* $((v^*, x^*), x^*)$.

Proof. Let (i) hold and let a, b and c be the constants of Aubin continuity of $\mathbf{L}$. Choose α and β such that the inequalities (22) hold with $\lambda = \frac{1}{2}L\beta$. Take α and β smaller if necessary such that

$$L(\beta/4 + 2\alpha) \leq 1. \tag{25}$$

Let (v', x') and (v'', x'') are in $B_{\beta/2}((v^*, x^*))$ and $z' \in \mathbf{P}(x', v') \cap B_\alpha(x^*)$. We apply Lemma 4.1 with

$$G(z) = f(x^*) + \nabla f(x^*)(z - x^*) + F(z)$$

and

$$h(z) = f(x'') + \nabla f(x'')(z - x'') - f(x^*) - \nabla f(x^*)(z - x^*).$$

Let $z_1, z_2 \in B_\alpha(x^*)$. Then

$$\|h(z_1) - h(z_2)\| \leq \|\nabla f(x'') - \nabla f(x^*)\| \|z_1 - z_2\| \leq \frac{1}{2} L\beta \|z_1 - z_2\|,$$

that is, h is Lipschitz continuous with a Lipschitz constant λ. Hence, from Lemma 4.1 the map $(h+G)^{-1}$ is Aubin continuous at $(v^* + h(x^*), x^*)$ with constants α, β and $\gamma = c/(1-\lambda c)$. Note that

$$z' \in (h+G)^{-1}\Big(v' + f(x'') + \nabla f(x'')(z'-x'') - f(x') - \nabla f(x')(z'-x')\Big) \\ \cap B_\alpha(x^*)$$

and

$$\begin{aligned} &\|v' + f(x'') + \nabla f(x'')(z' - x'') - f(x') - \nabla f(x')(z' - x') \\ &\qquad -v^* - f(x'') - \nabla f(x'')(x^* - x'') + f(x^*)\| \\ \leq\ & \|v - v^*\| \\ &+\|f(x^*) - f(x') - \nabla f(x')(x^* - x') + (\nabla f(x') - \nabla f(x''))(x^* - z')\| \\ \leq\ & \frac{\beta}{2} + \frac{1}{2}L(\beta/2)^2 + L\beta\alpha \leq \beta, \end{aligned}$$

because of (25). Hence, there exists $z'' \in (h+G)^{-1}(v'')$ such that

$$\begin{aligned} &\|z'' - z'\| \\ \leq\ & \gamma\|v'' - v' - f(x'') - \nabla f(x'')(z' - x'') + f(x') + \nabla f(x')(z' - x')\| \\ \leq\ & \gamma\|v'' - v'\| + \gamma\|f(x'') - f(x') - \nabla f(x')(x'' - x')\| \\ &\qquad + \gamma\|(\nabla f(x') - \nabla f(x''))(x'' - z')\| \\ \leq\ & \gamma\|v'' - v'\| + L\gamma(3\beta/2 + \alpha)\|x' - x''\|. \end{aligned}$$

Thus $\mathbf{P}$ is Aubin continuous. The converse implication is immediate by noting that $\mathbf{L}(v) = \mathbf{P}(v, x^*)$. □

Remark 4.1 *By choosing sufficiently small constants α and β for the neighborhoods, the growth constant of* $\mathbf{P}$ *with respect to x can be made arbitrary small.*

In the remaining part of this section we study the Newton method for Aubin continuous maps. For simplicity, we consider the unperturbed Newton method (5); the analysis can be carried over without major changes to the perturbed form (16). The following theorem shows that the Aubin property implies the existence of Newton sequences for all v close to v^* that are quadratically convergent to a solution $x(v)$ of (1):

Theorem 4.1 *Suppose that the map* $\mathbf{S}$ *is Aubin continuous at (v^*, x^*). Then there exist constants κ, σ and Θ such that for every $\bar{v} \in B_\kappa(v^*)$*

and every $\bar{x} \in \mathbf{S}(\bar{v}) \cap B_\sigma(x^*)$ *the following holds: for every initial point* $x_0 \in B_\sigma(x^*)$ *there exists a Newton sequence* $\{x_n\}$ *such that for* $n = 0, 1, 2, \cdots,$

$$\|x_{n+1} - \bar{x}\| \leq \Theta \|x_n - \bar{x}\|^2. \qquad (26)$$

Proof. Let δ, a, b and c are the constants the existence of which is claimed in Corollary 4.2; that is, for any $(\bar{v}, \bar{x}) \in B_\delta((v^*, x^*)) \cap \operatorname{graph} \mathbf{S}$ the map $(f(\bar{x}) + \nabla f(\bar{x})(\cdot - \bar{x}) + F(\cdot))^{-1}$ is Aubin continuous at $(\bar{v}, \bar{x})$ with constants a, b and c independent of the choice of $(\bar{v}, \bar{x})$. We repeatedly apply Lemma 4.1 with

$$G(x) = f(\bar{x}) + \nabla f(\bar{x})(x - \bar{x}) + F(x)$$

and

$$h(x) = f(x_n) + \nabla f(x_n)(x - x_n) - f(\bar{x}) - \nabla f(\bar{x})(x - \bar{x}).$$

Choose $\sigma = \delta$ and $\kappa = \delta$ and let $\bar{v} \in B_\kappa(x^*)$ and $\bar{x} \in S(\bar{v}) \cap B_\sigma(x^*)$. We will adjust the constant σ after the first step of Newton's method. Let $x_0 \in B_\sigma(x^*)$ and put $n = 0$. Then $h(x) = f(x_0) + \nabla f(x_0)(x - x_0) - f(\bar{x}) - \nabla f(\bar{x})(x - \bar{x})$ and

$$\|h(x_1) - h(x_2)\| \leq \|\nabla f(x_0) - \nabla f(\bar{x})\| \|x_1 - x_2\| \leq 2L\sigma \|x_1 - x_2\|$$

for every $x_1, x_2 \in \mathbf{R}^n$. Note that

$$(\bar{x}, \bar{v} + f(x_0) + \nabla f(x_0)(\bar{x} - x_0) - f(\bar{x})) \in \operatorname{graph}(h + G)$$

and

$$\|f(x_0) + \nabla f(x_0)(\bar{x} - x_0) - f(\bar{x})\| \leq 2L\sigma^2. \qquad (27)$$

We apply Lemma 4.1 with the so chosen h (for $n = 0$) and G and with

$$\lambda = 2L\sigma, \quad \alpha = \sigma \quad \text{and} \quad \beta = 2L\sigma^2.$$

It is a matter of simple calculation to adjust the constant σ such that the inequalities (22) hold; for example, we choose σ so small that $2L\sigma c < 1$, etc. Then, from Lemma 4.1, the map $(h + G)^{-1}$ is Aubin continuous at $(\bar{v} + f(x_0) + \nabla f(x_0)(\bar{x} - x_0) - f(\bar{x}), \bar{x})$ with constants α, β and $\gamma = c/(1 - \lambda)$. Hence, from (27) there exists

$$x_1 \in (h + G)^{-1}(\bar{v}),$$

that is, a Newton step from x_0 such that

$$\|x_1 - \bar{x}\| \leq \gamma \| - f(x_0) - \nabla f(x_0)(\bar{x} - x_0) + f(\bar{x})\| \leq 2\gamma L\sigma^2. \qquad (28)$$

At this point we take σ smaller if necessary such that

$$\frac{2cL\sigma}{1-2L\sigma} \leq 1.$$

Then, from (28), $x_1 \in B_\sigma(\bar{x})$.

From now on the constants σ is fixed. To obtain a Newton iterate x_2 we apply Lemma 4.1 with the same G and with $h(x) = f(x_1) + \nabla f(x_1)(x - x_1) - f(\bar{x}) - \nabla f(\bar{x})(x - \bar{x})$. Here x_1 will play the role of x_0 in the same way as in the first step obtaining that the map $(h + G)^{-1}$ is Aubin continuous at $(\bar{v} + f(x_1) + \nabla f(x_1)(\bar{x} - x_1) - f(\bar{x}), \bar{x})$ with constants α, β and γ: the same constants as at the first step. Then there exists $x_2 \in (h + G)^{-1}(\bar{v})$, that is, x_2 is a Newton step from x_1 for $\bar{v}$, such that

$$\|x_2 - \bar{x}\| \leq \gamma\| - f(x_1) - \nabla f(x_1)(\bar{x} - x_1) + f(\bar{x})\| \leq \frac{1}{2}\gamma L\|x_1 - \bar{x}\|^2.$$

The Newton iterates x^3, x^4, $\cdots$ are obtained in the same way. Taking $\Theta = \gamma L/2$ the proof is complete. □

Recall that $\mathbf{N}(x, v)$ is the set of all Newton sequences, that is, sequences that satisfy (5) starting from the point x and associated with the value v of the parameter. The constant sequence ξ^* is an element of $\mathbf{N}(x^*, v^*)$.

Theorem 4.2 *Suppose that the map* $\mathbf{S}$ *is Aubin continuous at* (v^*, x^*). *Then the map* $\mathbf{N}$ *is Aubin continuous at* $((x^*, v^*), \xi^*)$.

Proof. From Corollaries 4.1 and 4.3, the map $\mathbf{P}$ is Aubin continuous at $((v^*, x^*), x^*)$; let a, b be the associated constants for neighborhoods, c the growth constant with respect to v and μ the growth constant with respect to x. By taking a and b smaller if necessary, we can have $\mu < 1$. Let $v', v'' \in B_b(v^*)$ and $x', x' \in B_b(x^*)$, and let $\xi' \in \mathbf{N}(x', v') \cap B_a(\xi^*)$, $\xi' = \{x'_1, x'_2, \cdots, x'_n, \cdots\}$. Note that $x'_1 \in \mathbf{P}(v', x') \cap B_a(x^*)$. Then there exists $x''_1 \in \mathbf{P}(v'', x'')$ such that

$$\|x''_1 - x'_1\|| \leq c\|v'' - v'\| + \mu\|x'' - x'\|.$$

Further, $x'_2 \in \mathbf{P}(v'_1, x') \cap B_a(x^*)$, hence there exists $x''_2 \in \mathbf{P}(v'', x''_1)$ such that

$$\|x''_2 - x'_2\|| \leq c\|v'' - v'\| + \mu\|x''_1 - x''_1\| \leq c(1 + \mu)\|v'' - v'\| + \mu^2\|x'' - x'\|.$$

By induction, we obtain that there exists a sequence $\xi'' \in \mathbf{N}(x'', v'')$, $\xi'' = \{x''_1, x''_2, \cdots, x''_n, \cdots\}$. such that

$$\|x''_n - x'_n\|| \leq c(1 + \mu + \mu^2 + \cdots + \mu^{n-1})\|v'' - v'\| + \mu^n\|x'' - x'\|.$$

Taking the supremum norm we obtain that $\mathbf{N}$ is Aubin continuous with constants a and b for the neighborhoods and growth constants μ for x and $c/(1-\mu)$ for v. □

In the above theorem $\mathbf{N}(x,v)$ is the set of *all* Newton sequences starting from x for v. From Theorem 4.1 we know that at least one of these sequences will be convergent provided that x is sufficiently close to x^* and v is sufficiently close to v^*. Denote by $\mathcal{N}(x,v)$ the subset of $\mathbf{N}(x,v)$ containing all *convergent* Newton sequences starting from the point x and associated with v. The following theorem is a kind of converse to Theorem 4.2.

Theorem 4.3 *If the map $\mathcal{N}$ is Aubin continuous at $((x^*,v^*),\xi^*)$, then the map $\mathbf{S}$ is Aubin continuous at (v^*,x^*).*

Proof. Let $\mathcal{N}$ be Aubin continuous with constants a, b and c. Let $v', v'' \in B_b(v^*)$ and let x' be an element of $\mathbf{S}(v') \cap B_a(x^*)$. The constant sequence $\xi' = \{x', x', \cdots, x', \cdots\}$ is an element of $\mathcal{N}(x',v') \cap B_a(\xi^*)$. From the Aubin property of $\mathcal{N}$ there exists $\xi'' = \{x''_1, x''_2, \cdots, x''_n, \cdots\} \in \mathcal{N}(x',v'')$, such that for every $n \geq 1$

$$\|x''_n - x'_n\| \leq c\|v'' - v'\|. \tag{29}$$

Since $\xi'' \in \mathcal{N}(x',v'')$, ξ'' is convergent, say to x''. Clearly, every convergent Newton sequence for v'' is convergent to a solution of (1) for v''. Then x'' is an element of $\mathbf{S}(v'')$. Passing to the limit in (29) we obtain that

$$\|x'' - x'\| \leq c\|v'' - v'\|.$$

Thus $\mathbf{S}$ is Aubin continuous. □

Let us apply the above result to our illustrative optimization problem (3). The basic condition here will be the Aubin continuity of the map $\mathcal{X}$ defined in (11) (or, equivalently, of the map $\mathcal{L}$ defined in (12)). If we assume the Aubin property for $\mathcal{L}$, however, we end up having $\mathcal{L}$ locally singe-valued because of the following result obtained in [7]: *Let A be an $n \times n$ real matrix and let C be a convex polyhedral set. Then the map $(A+N_C)^{-1}$ is Aubin continuous at (v^*,x^*) if and only if it is locally single-valued and Lipschitz continuous around (v^*,x^*).*

Using the terminology of [25], we call the property just described Lipschitzian localization; we consider it in the following section.

5. LIPSCHITZIAN LOCALIZATION

The Lipschitzian localization property simply says that when restricted to a neighborhood of a point in its graph, a possibly set-valued map becomes a Lipschitz continuous (single-valued) function.

Definition 5.1 *Let Γ maps $\mathbf{R}^n$ to the subsets of $\mathbf{R}^m$ and let $(y^*, x^*) \in$ graph Γ. We say that Γ has a Lipschitzian localization at (y^*, x^*) with constants a, b and c if the map $y \mapsto \Gamma(y) \cap B_a(x^*)$ is single-valued (a function) and Lipschitz continuous in $B_b(y^*)$ with a Lipschitz constant c.*

If a map Γ has a Lipschitzian localization at (y^*, x^*), then it is Aubin continuous at (y^*, x^*) with the same constants. Conversely, if Γ is Aubin continuous at (y^*, x^*) with constants a, b and c and, in addition, for some positive constants α and β, $\Gamma(y) \cap B_\alpha(x^*)$ consists of at most one point for every $y \in B_\beta(y^*)$, then Γ has a Lipschitzian localization at (y^*, x^*) with constants a', b', c' provided that

$$0 < a' < \min\{a, \alpha\} \quad \text{and} \quad 0 < b' \leq \min\{b, \beta, a'/c', (\alpha - a')/c'\}.$$

The following lemma is an inverse function theorem which can be proved in various ways. Its statement is parallel to Lemmas 2.1 and 4.1.

Lemma 5.1 *Let G maps $\mathbf{R}^n$ to the subsets of $\mathbf{R}^m$ and let G^{-1} has a Lipschitzian localization at (v^*, x^*) with constants a and b for neighborhoods and c for growth. Let the nonnegative constants α, β and λ satisfy*

$$\alpha \leq a, \qquad \lambda c < 1, \quad c(\beta + \lambda\alpha) \leq \alpha, \quad \beta + \lambda\alpha \leq b. \tag{30}$$

Then for every function $h : \mathbf{R}^n \to \mathbf{R}^m$ which is Lipschitz continuous on $B_\alpha(x^*)$ with a Lipschitz constant λ, the map $(h+G)^{-1}$ has a Lipschitzian localization at $(v^* + h(x^*), x^*)$ with constants α and β for neighborhoods and $c/(1 - \lambda c)$ for growth.

Sketch of Proof. Let α, β, λ satisfy (30) and let h be Lipschitz continuous on $B_a(x^*)$ with a Lipschitz constant λ. Let $v \in B_\beta(v^*)$ and let $y = v + h(x^*)$, $y^* = v^* + h(x^*)$. Then $\|y - h(x) - y^*\| \leq \|v - v^*\| + \|h(x) - h(x^*)\| \leq \beta + \lambda\alpha \leq b$. It is easy to see that, for each $v \in B_\beta(v^*)$, the map

$$x \mapsto \Phi(x) := G^{-1}(y - h(x)) \cap B_\alpha(x^*)$$

is a function from $B_a(x^*)$ to $B_a(x^*)$ and is a contraction, hence it has a unique fixed point, say, $x(y)$, in $B_\alpha(x^*)$. Since $x(y) \in (G + h)^{-1}(y) \cap$

$B_\alpha(x^*)$, the map $(G+h)^{-1} \cap B_\alpha(x^*)$ is a function defined (with nonempty values) on $B_\beta(y^*)$. Moreover, for any $y', y'' \in B_\beta(y^*)$ we have

$$\begin{aligned} &\|x(y') - x(y'')\| \\ &= \|G^{-1}(y' - h(x(y'))) \cap B_\alpha(x^*) - G^{-1}(y'' - h(x(y''))) \cap B_\alpha(x^*)\| \\ &\leq c(\|y' - y''\| + \lambda \|x(y') - x(y'')\|), \end{aligned}$$

hence

$$\|x(y') - x(y'')\| \leq \frac{c}{1 - \lambda c} \|y' - y''\|.$$

Thus $y \mapsto x(y)$ is Lipschitz continuous with a Lipschitz constants $c/(1 - \lambda c)$. □

Applying Lemma 5.1 to the maps **S**, **L** and **P** defined in Section 2 we obtain:

Corollary 5.1 *The following are equivalent:*

(i) The map **L** *has a Lipschitzian localization at* (v^*, x^*).

(ii) The map **S** *has a Lipschitzian localization at* (v^*, x^*)

(iii) The map **P** *has a Lipschitzian localization at* $((v^*, x^*), x^*)$.

In optimization, the existence of Lipschitzian localization of the linearization map **L** is often called "strong regularity", a concept introduced by Robinson; actually, the implication (i) ⇒ (ii) is essentially the contents of Robinson's implicit function theorem in [21]. A characterization of the Lipschitzian localization property of a locally continuous set-valued map in finite dimensions is provided by the so-called strict graphical derivative, see [25], Theorem 9.54. A far reaching recent study of optimization problems with solution maps having this property is given in [18].

For our illustrative problem (3) the Lipschitzian localization property of the solution map is equivalent to the so-called *strong second-order sufficient optimality condition* which has the form:

$$\langle u, \nabla^2 g(x^*) u \rangle > 0 \quad \text{for all nonzero} \quad u \in K - K, \tag{31}$$

where K the critical cone at (v^*, x^*) defined in Section 2. Specifically, we have the following result which can be extracted from [7] or from more recent papers [13, 17, 18]:

Theorem 5.1 *The following are equivalent for the problem (3):*

(i) The strong second-order sufficient optimality condition (31) holds at (v^*, x^*).

(ii) The Argmin map has a Lipschitzian localization at (v^*, x^*).

By combining Theorem 5.1 with the last property of Aubin continuous maps given after Definition 4.1 and the result in [7] given at the end of Section 4, we obtain that the strong second-order sufficient optimality condition is equivalent to the property that for every sufficiently small $\varepsilon > 0$ the ε-stationary map

$$v \mapsto \mathcal{X}_\varepsilon(v) = \{x \in \mathbf{R}^n \mid \text{dist}\,(v - \nabla g(x), N_C(x)) \leq \varepsilon\}$$

is Aubin continuous at (v^*, x^*).

Let us consider the Newton method (5) applied to the general model (1). The analog of Corollary 3.1 and Theorem 4.1 has the following form:

Theorem 5.2 *Suppose that the map* $\mathbf{S}$ *has a Lipschitzian localization at* (v^*, x^*). *Then there exist constants* κ, σ *and* Θ *such that for every* $\bar{v} \in B_\kappa(v^*)$ *and every initial point* $x_0 \in B_\sigma(x^*)$ *there exists a unique Newton sequence* $\{x_n\}$ *and this sequence is quadratically convergent to* $\bar{x} := \mathbf{S}(\bar{v}) \cap B_\sigma(x^*)$ *with a constant* Θ, *that is,*

$$\|x_{n+1} - \bar{x}\| \leq \Theta \|x_n - \bar{x}\|^2, \quad n = 0, 1, 2, \cdots.$$

Proof. The proof is completely analogous to the proofs of Theorem 4.1 with Lemma 5.1 replacing Lemma 4.1. □

The next result is the analog of Theorems 3.3, 4.3 and 4.4.

Theorem 5.3 *The following are equivalent:*

(i) The map $\mathbf{S}$ *has a Lipschitzian localization at* (v^*, x^*).

(ii) The Newton map $\mathbf{N}$ *has a Lipschitzian localization at* $((x^*, v^*), \xi^*)$.

Applied to the illustrative optimization problem (3) and the SQP method (6), we obtain that the strong second-order sufficient optimality condition (31) at (v^*, x^*) is equivalent to the following: for every v near v^* and x near x^* there exists a unique sequence of primal variables x_n starting from x and which is quadratically convergent to the unique solution $x(v)$ of (3) for v, for some sequence of Lagrange multipliers y_n. Moreover the map from (x, v) to the set of all sequences x_n staring from x for v has a Lipschitzian localization at $((x^*, v^*), \xi^*)$.

For the nonlinear program (20), the Newton (SQP) step is performed for the pair (x, y) with x being the primal variable and y the Lagrange multiplier. Theorem 5.4 extends the characterization of the Lipschitzian localization property of the Karush-Kuhn-Tucker map to the SQP map. Specifically, if we assume that the reference solution is isolated, then the Lipschitzian localization property of the SQP method is equivalent to the combination of the linear independence condition for the active constraints and the strong second-order sufficient optimality condition at the reference point.

References

[1] J.-P. Aubin (1984), Lipschitz behavior of solutions to convex minimization problems, *Math. Oper. Res.*, 9, pp. 87–111.

[2] R. Cominetti (1990), Metric regularity, tangent sets, and second-order optimality conditions, *Appl. Math. Optim.*, 21, pp. 265–287.

[3] A.V. Dmitruk, A.A. Milyutin and N.P. Osmolovskiĭ (1980), The Lyusternik theorem and the theory of extremum, *Uspekhi Math. Nauk*, 35, pp. 11–46.

[4] A.L. Dontchev (1995), Characterizations of Lipschitz stability in optimization, in *Recent developments in well-posed variational problems,*, Math. Appl., 331, Kluwer, Dordrecht, pp. 95–115.

[5] A.L. Dontchev (1996), The Graves theorem revisited, *J. Convex Anal.*, 3, pp. 45–53.

[6] A.L. Dontchev, W.W. Hager and V. Veliov, Uniform convergence and mesh independence of Newton's method for discretized variational problems, *SIAM J. Control and Optim.,* to appear.

[7] A.L. Dontchev and R.T. Rockafellar (1996), Characterizations of strong regularity for variational inequalities over polyhedral convex sets, *SIAM J. Optim.*, 6, pp. 1087–1105.

[8] A.L. Dontchev and R.T. Rockafellar (1998), Characterizations of Lipschitzian stability in nonlinear programming, in *Mathematical programming with data perturbations,* Lecture Notes in Pure and Appl. Math., 195, Dekker, New York, pp. 65–82.

[9] W.W. Hager and M.S. Gowda (1999), Stability in the presence of degeneracy and error estimation, *Math. Programming*, 85, pp. 181–192.

[10] A.D. Ioffe, Metric regularity and subdifferential calculus, to appear.

[11] A.J. King and R.T. Rockafellar (1992), Sensitivity analysis for nonsmooth generalized equations, *Math. Programming*, 5, Ser. A, pp. 193–212.

[12] D. Klatte, Upper Lipschitz behavior of solutions to perturbed $C^{1,1}$ programs, preprint.

[13] D. Klatte and B. Kummer (1999), Strong stability in nonlinear programming revisited, *J. Austral. Math. Soc.*, 40, Ser. B, pp. 336–352.

[14] B. Kummer (1998), Lipschitzian and pseudo-Lipschitzian inverse functions and applications to nonlinear optimization, in *Mathematical programming with data perturbations,* Lecture Notes in Pure and Appl. Math., 195, Dekker, New York, pp. 201–222.

[15] A.B. Levy (1996), Implicit multifunction theorems for the sensitivity analysis of variational conditions, *Math. Programming*, 74, Ser. A, pp. 333–350.

[16] A. Levy, Solution sensitivity from general principles, preprint.

[17] R.A. Poliquin and R.T. Rockafellar (1998), Tilt stability of local minima, *SIAM J. Optim.*, 2, pp. 287–299.

[18] A. Levy, R.A. Poliquin and R.T. Rockafellar, Stability of locally optimal solutions, *SIAM J. Optim.,* to appear.

[19] J. Renegar (1995), Linear programming, complexity theory and elementary functional analysis, *Math. Programming*, 70, Ser. A, pp. 279–351.

[20] S.M. Robinson (1979), Generalized equations and their solution, Part I: Basic Theory, *Math. Programming Study,* 10, pp. 128–141.

[21] S.M. Robinson (1980), Strongly regular generalized equations, *Math. Oper. Res.*, 5, pp. 43–62.

[22] S.M. Robinson (1981), Some continuity properties of polyhedral multifunctions, *Math. Programming Study,* 14, pp. 206–214.

[23] R.T. Rockafellar (1989), Proto-differentiability of set-valued mappings and its applications in optimization, *Analyse non linéaire (Perpignan, 1987), Ann. Inst. H. Poincaré Anal. Non Linéaire* 6, suppl., pp. 449–482.

[24] R.T. Rockafellar, Extended nonlinear programming, in *Nonlinear Optimization and Applications 2*, Kluwer, Dordrecht, to appear.

[25] R.T. Rockafellar and R.J.-B. Wets (1997), Variational analysis, Springer-Verlag, Berlin.

[26] H. Xu (1999), Set-valued approximations and Newton's methods, *Math. Programming*, 84, Ser. A, pp. 401–420.

[27] T. Zolezzi (1998), Well-posedness and conditioning of optimization problems, *Proc. 4th International Conf. Math. Methods Oper. Research and 6th Workshop on Well-posedness and Stability of Optimization Problems, Pliska Stud. Math. Bulgar.*, 12, pp. 267–280.

SQP METHODS FOR LARGE-SCALE NONLINEAR PROGRAMMING

Nicholas I. M. Gould
Computational Science and Engineering Department,
Rutherford Appleton Laboratory,
Chilton, Oxfordshire, OX11 0QX, England, EU.
n.gould@rl.ac.uk

Philippe L. Toint
Department of Mathematics,
Facultés Universitaires ND de la Paix,
61, rue de Bruxelles, B-5000 Namur, Belgium, EU.
pht@math.fundp.ac.be

Abstract We compare and contrast a number of recent sequential quadratic programming (SQP) methods that have been proposed for the solution of large-scale nonlinear programming problems. Both line-search and trust-region approaches are studied, as are the implications of interior-point and quadratic programming methods.

1. INTRODUCTION

1.1. PERSPECTIVES

By the start of the 1980s, it was generally accepted that sequential quadratic programming (SQP) algorithms for solving nonlinear programming problems were the methods of choice. Such a view was based on strong convergence properties of such algorithms, and reinforced in the comparative testing experiments of [41], in which SQP methods clearly outperformed their competitors. Although such claims of superiority were made for implementations specifically aimed at small-scale problems,—that is, those problems for which problem derivatives can be stored and manipulated as dense matrices—there was little reason to believe that similar methods would not be equally appropriate when the problem matrices were too large to be stored as dense matrices, but

M.J.D. Powell and S. Scholtes (Eds.), *System Modelling and Optimization: Methods, Theory and Applications.*

rather required sparse storage formats. Remarkably then, it is only in the latter part of the 1990s that SQP methods for sparse problems have started to appear in published software packages, while sparse variants of the methods that SQP was supposed to have superseded (for instance MINOS, see [50], and LANCELOT, see [16]) have been used routinely and successfully during the intervening years.

In our opinion, this curious divergence between what logically should have happened in the 1980s, and what actually came to pass may be attributed almost entirely to a single factor: quadratic programming (QP) methods (and their underlying sparse matrix technology) were not then capable of solving large problems. Witness the almost complete lack of software for solving large-scale (non-convex) quadratic programs even today, especially in view of the large number of available codes for the superficially similar linear programming problem.

The purpose of this paper is to survey modern SQP methods, and to suggest why it is now reasonable to accept the widely-held view that SQP methods really are best. There have been a number of surveys of SQP methods over the past 20 years, and we refer the reader to [1, 17, 56, 57, 61]. Much of the material in this paper is covered in full detail in our forthcoming book on trust-region methods [19], which also contains a large number of additional references.

1.2. THE PROBLEM

We consider the problem of minimizing a (linear or nonlinear) function f of n real variables x, for which the variables are required to satisfy a set of (linear or nonlinear) constraints $c_i(x) \geq 0$, $i = 1, \ldots, m$. For simplicity, we ignore the possibility that some of the constraints might be equations, since these are easily incorporated in what follows, nor shall we consider any special savings that can be made if some or all of the constraints have useful structure (e.g., might be linear). We remind the reader that if x_* is a local solution to the problem, and so long as a so-called constraint qualification holds to exclude pathological cases, it follows that the first-order criticality conditions

$$g(x_*) = A^T(x_*)y_*, \quad c(x_*) \geq 0, \quad y_* \geq 0 \text{ and } y_*^T c(x_*) = 0 \tag{1}$$

will hold. Here $c(x)$ is the vector whose components are the $c_i(x)$, $g(x) = \nabla_x f(x)$ is the gradient of f, $A(x) = \nabla_x c(x)$ is the Jacobian of c, and y_* are appropriate Lagrange multipliers. Notice that the first requirement in (1) is that the gradient $\nabla_x \ell(x_*, y_*)$ of the Lagrangian function $\ell(x, y) = f(x) - y^T c(x)$ should vanish. For future reference, we also denote the Hessian of the Lagrangian function by $H(x, y) = \nabla_{xx} \ell(x, y)$, and will let c_- be the vector whose i-th component is $\min(c_i(x), 0)$,

Throughout this paper, we shall make a blanket assumption that

A1. f and the c_i have Lipschitz-continuous second derivatives (in the region of interest).

Throughout $\|\cdot\|$ will denote a generic norm. While there may be good practical reasons for choosing a specific norm, and while some of the given results have only been established in such a case, we suspect there are very few places where general results in arbitrary norms are not possible.

1.3. GENERIC SQP METHODS

For a most transparent derivation of the basic SQP method, we note that the final requirement in (1) (the complementarity condition) implies that Lagrange multipliers corresponding to inactive constraints (those for which $c_i(x_*) > 0$) must be zero. Thus, so long as the given inequalities hold, (1) may equivalently be written as

$$g(x_*) - A_{\mathcal{A}_*}^T(x_*)(y_*)_{\mathcal{A}_*} = 0 \text{ and } c_{\mathcal{A}_*}(x_*) = 0, \tag{2}$$

where the subscript $\mathcal{A}_*$ indicates the components corresponding to the active set $\mathcal{A}_* = \{i \mid c_i(x_*) = 0\}$. Of course $\mathcal{A}_*$ depends on x_*, but suppose for the time being that we know $\mathcal{A}_*$. We then note that, if $\mathcal{A}_*$ has $m_{\mathcal{A}_*}$ elements, (2) is a set of $n + m_{\mathcal{A}_*}$ nonlinear equations in the $n + m_{\mathcal{A}_*}$ unknowns x and $y_{\mathcal{A}_*}$.

The best-known method for solving such systems (when it works) is Newton's method, and the basic SQP method is simply Newton's iteration applied to (2). This leads to an iteration of the form

$$\begin{pmatrix} x_{k+1} \\ (y_{k+1})_{\mathcal{A}_*} \end{pmatrix} = \begin{pmatrix} x_k + s_k \\ (y_k)_{\mathcal{A}_*} + (v_k)_{\mathcal{A}_*} \end{pmatrix},$$

where

$$\begin{pmatrix} H_k & A_{\mathcal{A}_*}^T(x_k) \\ A_{\mathcal{A}_*}(x_k) & 0 \end{pmatrix} \begin{pmatrix} s_k \\ -(v_k)_{\mathcal{A}_*} \end{pmatrix} = - \begin{pmatrix} g(x_k) - A_{\mathcal{A}_*}^T(x_k)(y_k)_{\mathcal{A}_*} \\ c_{\mathcal{A}_*}(x_k) \end{pmatrix}, \tag{3}$$

to correct the guess $(x_k,\ (y_k)_{\mathcal{A}_*})$. Here H_k is a "suitable" approximation of $H(x_k, y_k)$, where the nonzero components of y_k are those of $(y_k)_{\mathcal{A}_*}$. Since this is a Newton iteration, we expect a fast asymptotic convergence rate in many cases, so long as H_k is chosen appropriately. Interestingly, fast convergence does not require H_k to converge to $H(x_*, y_*)$, and considerable effort over the past 25 years has been devoted to obtaining minimal conditions, along with practical choices of H_k, which permit satisfactory convergence rates. We refer the interested reader to any of the previously mentioned surveys, and the papers cited therein, for more details.

Most revealingly, we may rewrite (3) as

$$\begin{pmatrix} H_k & A^T_{\mathcal{A}_*}(x_k) \\ A_{\mathcal{A}_*}(x_k) & 0 \end{pmatrix} \begin{pmatrix} s_k \\ -(y_{k+1})_{\mathcal{A}_*} \end{pmatrix} = - \begin{pmatrix} g(x_k) \\ c_{\mathcal{A}_*}(x_k) \end{pmatrix},$$

which are the first-order criticality conditions for the (equality constrained) quadratic programming problem

$$\underset{s \in \mathbb{R}^n}{\text{minimize}} \;\; s^T g(x_k) + \frac{1}{2} s^T H_k s \;\; \text{subject to} \;\; c_{\mathcal{A}_*}(x_k) + A_{\mathcal{A}_*}(x_k) s = 0,$$

with $(y_{k+1})_{\mathcal{A}_*}$ being its Lagrange multipliers. Notice that the constraints here are simply linearizations of the active constraints about the current estimate of the solution. Further, this suggests that to avoid having to estimate $\mathcal{A}_*$ in advance, it suffices to consider linearizations of all of the constraints, and to solve the (inequality constrained) quadratic programming problem

$$\underset{s \in \mathbb{R}^n}{\text{minimize}} \;\; s^T g(x_k) + \frac{1}{2} s^T H_k s \;\; \text{subject to} \;\; c(x_k) + A(x_k) s \geq 0. \quad (4)$$

This provides the basic SQP method: given an estimate $(x_k,\ y_k)$ and a suitable H_k, solve (4) to find s_k, update $x_{k+1} = x_k + s_k$, and (if necessary) adjust y_{k+1} to provide convergence to y_*. Remarkably, [60] showed that, so long as x_0 is sufficiently close to x_*, H_0 is sufficiently close to $H(x_*, y_*)$, and $H_k = H(x_k, y_k)$ for $k \geq 1$, as well as

A2. the Jacobian of active constraints $A_{\mathcal{A}_*}(x_*)$ is of full rank,

A3. second-order necessary optimality conditions hold at $(x_*,\ y_*)$, and

A4. strict complementarity slackness occurs (i.e., $[y_*]_i > 0$ if $c_i(x_*) = 0$),

the SQP iteration converges Q-superlinearly, and the set of constraints which are active in (4) is precisely the set $\mathcal{A}_*$ for all sufficiently large k. If y_{k+1} are chosen to be the Lagrange multipliers for (4), the rate is actually Q-quadratic.

The important assumption here is A2, since this ensures that the Lagrange multipliers at x_*, as well as those for (4) for sufficiently large k, are unique. If $A_{\mathcal{A}_*}(x_*)$ is not of full rank, the limiting multipliers may not be unique, and the SQP method using the estimates obtained from (4) may not converge Q-quadratically. Of course A2 is a relatively strong first-order constraint qualification, and [67] shows that it is possible to replace this assumption by a weaker one due to [45] while still obtaining Q-quadratic convergence. To do so, the subproblem (4) must be modified slightly to ensure that its Lagrange multipliers are (locally) unique. In fact, Wright's subproblem is equivalent to minimizing an augmented Lagrangian function for (4) with respect to x and simultaneously maximizing with respect to y while ensuring that $y \geq 0$. To ensure a Q-quadratic rate, the penalty parameter for the augmented Lagrangian must approach zero as $O\left(\max[\|x_k - x_*\|, \|y_k - y_*\|]\right)$. Bonnans and Launay [4] and Hager [38] show that it is also possible to remove A4 if A3 is strengthened.

Since the above iteration is essentially Newton's method, we must, of course, be cautious since in general such methods are not globally convergent. There have been traditionally two types of globalization schemes, linesearch and trust-region methods, and it is these that we now consider.

2. LINESEARCH METHODS

A traditional linesearch SQP method computes s_k by solving (4), and then obtains $x_{k+1} = x_k + \alpha_k s_k$ for some appropriately chosen stepsize α_k. The stepsize is selected so that x_{k+1} is closer in some way to a critical point than its predecessor, and linesearch methods achieve this by requiring that $\varphi(x_{k+1})$ is significantly smaller than $\varphi(x_k)$ for some so-called merit function φ. A highly desirable property of any merit function is that critical points of the merit function correspond to critical points for the underlying nonlinear programming problem. The most widely-used merit functions are non-smooth penalty functions of the form

$$\varphi(x, \sigma) = f(x) + \sigma \|c(x)_-\|, \tag{5}$$

which depends on a positive penalty parameter σ, and also smooth exact penalty functions of the form

$$\begin{aligned}\varphi(x, z, \sigma) = {} & f(x) - y^T(x)(c(x) - z) \\ & + \sigma \left(c(x) - z\right)^T \left(A(x)A^T(x) + Z\right)^{-1} (c(x) - z),\end{aligned} \tag{6}$$

where Z is a diagonal matrix with entries $z_i \geq 0$, and where $y(x)$ is defined by

$$\left(A(x)A^T(x) + Z\right) y(x) = A(x)\nabla_x f(x). \tag{7}$$

Relevant references include [2, 21, 23, 44, 59]. Note that none of these functions is actually ideal, since they may sometimes have critical points at values which do not correspond to those for the underlying nonlinear programming problem—these rogue values usually occur at points which are locally least infeasible. However, it can be shown that critical points for the two problems coincide so long as those for the merit function are feasible, and so long as the penalty parameter is larger than a problem-dependent critical value—for smooth exact penalty functions, a further requirement like assumption A2 may also be required.

It is crucial that the SQP step s_k and the merit function $\varphi(x)$ be compatible, in the sense that the directional derivative (slope in the smooth case) must be negative, for otherwise the linesearch may fail. In many cases, this condition is guaranteed when the penalty parameter is sufficiently large, and when $s_k^T H_k s_k \geq 0$. While the latter condition is likely to hold asymptotically, there is little reason why it should be true far from the solution, unless H_k is itself positive definite. For this reason, most active-set SQP methods work under the blanket assumption that H_k is positive definite, which is of course a far stronger assumption than A3.

When the function (5) is used, the penalty parameter may have to satisfy $\sigma > \|y_{k+1}\|_D$, where y_{k+1} are the Lagrange multipliers for (4) and $\|\cdot\|_D$ is the norm dual to $\|\cdot\|$. Such a condition is consistent with the problem-dependent critical value alluded to earlier, namely that $\sigma > \|y_*\|_D$. An *a priori* bound on the size of the penalty parameter for (6) is harder to obtain, since it depends on the eigenvalues of H_k.

2.1. SECOND-ORDER CORRECTION

The main disadvantage of functions like (5)—indeed, of any merit function which simply tries to balance f against constraint infeasibility—is that there is no guarantee that the SQP step together with a unit stepsize $\alpha_k = 1$ will lead to a reduction of the merit function, however close the iterates are to a critical point. Thus the full Newton (SQP) step may not be taken, and the iterates fail to converge at the anticipated Q-superlinear rate. Indeed, a famous example due to [46] shows that this defect can actually occur. The Maratos effect happens because the linearization of the constraints fails to take adequate account of their nonlinear behaviour.

The idea of using a second-order correction to cope with the Maratos effect first appeared in a number of contemporary papers (see [12, 25, 47]). The idea is to aim to replace the update $x_{k+1} = x_k + s_k$ by a corrected update

$$x_{k+1} = x_k + s_k + s_k^{\mathrm{C}}, \tag{8}$$

where $s_k^{\rm C}$ corrects for the "second-order" effects due to the constraint curvature. Let $\mathcal{A}_k$ be the set of active constraints at the solution to (4). Then a general second-order correction is the solution $s_k^{\rm C}$ to the system

$$\begin{pmatrix} H_k^{\rm C} & A_{\mathcal{A}_k}^T(x_k+p_k) \\ A_{\mathcal{A}_k}(x_k+p_k) & 0 \end{pmatrix} \begin{pmatrix} s_k^{\rm C} \\ -y_k^{CS} \end{pmatrix} = - \begin{pmatrix} g_k^{\rm C} \\ c_{\mathcal{A}_k}(x_k+s_k) \end{pmatrix} \tag{9}$$

for some appropriate $H_k^{\rm C}$, p_k and $g_k^{\rm C}$. In order that the resulting step is suitable, we require that $g_k^{\rm C}$ and p_k are both small, indeed that

$$g_k^{\rm C} = O\left(\|x_k - x_*\| \max[\|x_k - x_*\|, \|y_k - y_*\|]\right) \;\; \text{and} \;\; p_k = O(\|x_k - x_*\|). \tag{10}$$

Moreover, we also require that $H_k^{\rm C}$ is uniformly positive definite on the null-space of $A_{\mathcal{A}_k}(x_k+p_k)$ for all (x_k, y_k) close to (x_*, y_*). Provided that these conditions are satisfied, and so long as A2–A4 hold, it is possible to show that the corrected update (8) is guaranteed to reduce the merit function (5) close to (x_*, y_*). Two popular choices are

$$p_k = 0, \;\; g_k^{\rm C} = 0, \;\; \text{and} \;\; H_k^{\rm C} = H_k,$$

which gives the traditional second-order correction championed by [12, 25, 47], and

$$p_k = s_k, \;\; g_k^{\rm C} = \nabla_x \ell(x_k + s_k, y_{k+1}), \;\; \text{and} \;\; H_k^{\rm C} = H(x_k + s_k, y_{k+1}),$$

which corresponds to a second SQP step, and provides the basis for the "watchdog technique" suggested by [11]. Note that other authors (for example, [53]) have also shown that a small number (> 1) of SQP steps ensure that (5) decreases, but couch their proposal in the language of the non-monotone descent methods made famous for unconstrained minimization by [37]. In the linesearch context, a search should be made along the arc $x_k + \alpha s_k + \alpha^2 s_k^{\rm C}$, with the expectation that ultimately $\alpha_k = 1$ and (8) will occur.

2.2. BOGGS, KEARSLEY AND TOLLE'S APPROACH

The Maratos effect does not occur for (6), and herein lies the popularity of methods based on this function. Traditionally such functions have been viewed somewhat unfavourably by most researchers since at a first glance they require a Jacobian value and the solution of the linear system (7) each time a function value is required—this may be very expensive if a number of different trial steps are required during the linesearch. This difficulty may be avoided by replacing (6) locally by a surrogate (approximation) merit function in which $y(x)$ and $A(x)$ are replaced by an appropriate y and A; it can be shown that such an approximation is valid, and that with care global convergence properties are retained. See [2] for details.

Boggs, Kearsley and Tolle [3] provide an implementation and positive practical experience with such a method. Of particular note is that instead of trying to solve (4) directly, they pick three promising estimates of the required solution, and subsequently find the best approximation to this solution in the subspace spanned by these three vectors, at least one of the three directions being chosen to be a descent direction for the merit function. A trust-region (see Section 3) is used to limit the steps in the tests performed, and updated appropriately, but as yet this enhancement has no theoretical underpinning—the current theory for the linesearch version requires that H_k be positive definite.

2.3. SNOPT

SNOPT is a state-of-the-art linesearch-based SQP method for large-scale nonlinear programming due to [34]. At present, SNOPT uses a positive-definite approximation H_k to the Hessian of the Lagrangian, which exploits the fact that frequently many variables only appear linearly in the problem formulation, and retains information about previously encountered curvature via a limited-memory secant update formula — we understand that a new version capable of using the exact Hessian of the Lagrangian is being tested. Special techniques are used to ensure that subsequent updates to H_k maintain positive definiteness. Feasibility with respect to linear constraints is attained from the outset. An augmented Lagrangian merit function is used to assess steps in both x and the Lagrange multiplier estimates y. The method is designed to be flexible, in that in theory it can use any quadratic programming algorithm, although by default it uses a null-space based active set method, which slightly limits the size and type of problems which can be handled. In practice, numerical tests have shown SNOPT to be most effective.

2.4. FEASIBLE POINT APPROACHES

A particularly appealing idea is to ensure that all iterates remain feasible, since then the objective function is itself a suitable merit function, and additionally the linearized constraints are sure to be consistent as $s = 0$ lies in the set $\{s \mid c(x_k) + A(x_k)s \geq 0\}$. In a sequence of papers, [5, 40, 52, 54] show that this is possible provided precautions are taken. It is easy to show that the SQP direction s_k from (4) at a feasible point x_k is a descent direction for $f(x)$ provided that H_k is positive definite.

However, it may not be a feasible descent direction, i.e., it may not lie in the set

$$\{s \mid s^T g(x_k) < 0 \text{ and } s^T a_i(x_k) < 0 \text{ for all } i \in \mathcal{A}(x_k)\}, \qquad (11)$$

since $s_k^T a_i(x_k)$ may be zero for one or more $i \in \mathcal{A}(x_k)$. Thus an arbitrarily small step along s_k may violate one or more of the (nonlinear) constraints active at x_k. To avoid this difficulty, any feasible descent direction s_k^F, and the "tilted" direction $s_k^T = (1 - \rho_k)s_k + \rho_k s_k^F$, for some $\rho_k \in (0, 1)$, are determined. The tilted direction is itself a feasible descent direction, and, so long as ρ_k converges to zero sufficiently fast, retains the fast asymptotic convergence properties of the original SQP direction. However, since these properties only arise if an asymptotic unit step is taken, a second-order correction rather like s_k^C from (9) may be necessary. The FSQP algorithm of [54] is based on these ideas, and can be shown to be globally convergent under suitable assumptions on H_k, and the requirement that (11) is non-empty, which amounts to a constraint qualification. Further, Q-superlinear convergence is achieved under the additional assumptions A3 and A4. Although this method has only been considered for small problems, it is not difficult to imagine how to generalize it by taking approximate solutions to the various subproblems. As with most linesearch methods, the requirement that H_k be positive definite is its major weakness.

3. TRUST-REGION METHODS

The second important class of methods designed to ensure global convergence of locally convergent minimization algorithms are trust-region methods. Rather than controlling the step taken along the SQP direction (having computed the direction), trust-region methods aim to control the step at the same time as computing the search direction. Such methods hold a distinct advantage over linesearch methods, in that H_k is not required to be positive definite.

To simplify our discussion, consider first the unconstrained minimization of a nonlinear smooth function f. At the k-th iteration, a model $m_k(x_k + s)$ of $f(x_k + s)$ is used. This model is merely required to resemble f increasingly accurately as s approaches zero, and is believed to be a good approximation for all s within a trust region $\|s\|_k \le \Delta_k$ for some appropriate, possibly iteration-dependent, norm $\|\cdot\|_k$ and radius $\Delta_k > 0$. If this is the case, an approximate minimizer of m_k should provide a good estimate of the minimizer of f within the same region. The first stage of a trust-region method is thus to compute a suitable approximate minimizer s_k of m_k within the trust region. If our hypothesis is correct, we would then expect $m_k(x_k) - m_k(x_k + s_k)$ to be a good approximation to $f(x_k) - f(x_k + s_k)$; if this confidence is repaid, we set $x_{k+1} = x_k + s_k$, and possibly increase the radius. On the other hand, when $m_k(x_k) - m_k(x_k + s_k)$ and $f(x_k) - f(x_k + s_k)$ are very different, our hypothesis is invalid, that is to say Δ_k is too large. In this case, we set $x_{k+1} = x_k$, and ensure that Δ_{k+1} significantly less than Δ_k. This extremely simple framework is imbued with very powerful global convergence properties under extremely modest assumptions (see, for example, [19]). In practice, all that is required of the step is that it reduces the model by at least a fixed fraction of the reduction that can be obtained by approximately minimizing the model within the trust-region along a gradient-related direction (such as $-g(x_k)$). This one-dimensional minimization problem is often trivial; the resulting point, the Cauchy point, plays a key role in the convergence theory for trust-region methods. A most important result is that if a Newton model—the first three terms of a Taylor expansion—is used, and if the model is minimized sufficiently accurately, the trust region constraint will become inactive asymptotically, and the resulting full Newton step will provide a Q-superlinear convergence rate.

Turning now to the constrained case, it is reasonable to expect to replace the objective function by a suitable merit function, and to build a model of this merit function. However, if we try to impose a trust-region constraint $\|s\|_k \leq \Delta_k$ on top of the linearized constraints $c(x_k) + A(x_k)s \geq 0$, we immediately see a difficulty. Simply, if $c(x_k)$ is nonzero, the intersection of linearized constraints with the trust region will be empty if Δ_k is too small. Thus, the strategy outlined in the previous paragraph, in which the radius is reduced until the model of the merit function proves to be adequate, is flawed in the constrained case.

In this section, we consider a number of ways of avoiding potential devastation from this discovery.

3.1. Sℓ_1QP-LIKE APPROACHES

This approach avoids the incompatibility issue altogether. Simply, rather than considering an SQP method directly, we instead aim to minimize the unconstrained, non-smooth penalty function (5). Since (5) is non-smooth, we cannot appeal directly to trust-region theory for smooth unconstrained minimization. However, the basic idea remains valid. We model $\varphi(x_k + s, \sigma)$ as

$$m_k(x_k+s) = f(x_k)+s^T g(x_k)+\frac{1}{2}s^T H_k s+\sigma\| \left(c(x_k) + A(x_k)s\right)_- \|, \quad (12)$$

where H_k reflects the curvature in both f and c, and aim to approximately minimize this model within the trust region. All that is really required is to change the definition of the Cauchy point, since the gradient may not exist at x_k. Instead of the negative gradient, it suffices to consider the steepest descent direction

$$d(x_k) = -\arg \min_{g \in \partial\varphi(x_k,\sigma)} \|g\|,$$

where $\partial\varphi(x_k, \sigma)$ is the generalized gradient of $\varphi(x, \sigma)$ at x_k. Because of the polyhedral convex structure of the non-differentiable term $\min(c, 0)$ in the definition of φ, this generalized gradient may be found by solving a linear or convex quadratic program in the commonly occurring cases that use the ℓ_1, ℓ_2 or ℓ_∞ norm. While computing the Cauchy point is thus undoubtedly more expensive than in the smooth case, the alternative of exactly minimizing (12) within the trust region is usually even more expensive, since the latter problem may be non-convex (depending on H_k). The original idea here is due to [24] and [26], Section 14.4, who proposed minimizing (12) using the ℓ_1 norm within an ℓ_∞-norm trust region. The resulting so-called ℓ_1QP subproblem can be converted to an ordinary QP (with a given initial feasible point) by adding additional variables, but is probably best solved as is. A significant advantage of this method over almost all of its competitors is that an independence assumption like A2 is not required to assure global convergence to a critical point of the merit function.

Asymptotically, so long as the penalty parameter is large enough, the SQP and ℓ_1QP directions coincide, and thus we might expect a fast asymptotic convergence rate provided that the trust-region constraint is inactive. However, as we noted in Section 2, the SQP direction may suffer from the Maratos effect, and the same is true of the Sℓ_1QP direction. Thus, the Sℓ_1QP direction may not be acceptable, and consequently the trust-region radius will be reduced to exclude this step. The cure is exactly as before, namely a second-order correction should be added to correct for constraint curvature. In view of (9), an appropriate correction is obtained by minimizing

$$s^{\mathrm{c}T} g_k^{\mathrm{c}} + \frac{1}{2} s^{\mathrm{c}T} H_k^{\mathrm{c}} s^{\mathrm{c}} + \sigma \| \left(c(x_k + s_k) + A(x_k + p_k) s^{\mathrm{c}} \right)_- \|$$

within the trust region $\|s_k + s^{\mathrm{c}}\|_k \leq \Delta_k$. A fairly intricate algorithm, based on such a correction and proposed by [25], was shown by [72] to ensure that the trust-region radius is asymptotically inactive, and thus the iterates can converge Q-superlinearly under assumptions A2–A4. Perhaps more simply, all that is required is that the trust region radius is reset to at least a fixed positive value whenever a successful step is taken, for then the trust-region will not ultimately interfere with the next step s_k and, if needed, the correction s_k^c. Thus either s_k or $s_k + s_k^c$ will ultimately be accepted, and the radius is subsequently bounded away from zero. Convergence to a second-order critical point—one for which (weak) second-order necessary conditions hold—may also be guaranteed, so long as significant negative curvature is exploited in the model, and that this negative curvature is reflected in the true problem—this is the case, for example, if H_k converges to $H(x_*, y_*)$.

To date, it is unclear whether it is better to update σ as the iteration proceeds, or to wait until a critical point of $\varphi(x,\sigma)$ has been found before doing so. The advantage of the former is that a sequence of problems will not be solved, while the disadvantage is that any automatic value which aims to predict the correct value may also over-estimate it, leading to a poorer conditioned penalty function.

3.2. VARDI-LIKE APPROACHES

The remaining approaches to be considered may be termed composite-step methods. A composite step s_k is computed as the sum of two components q_k and t_k, each of which has different aims. The (quasi-) normal component q_k is simply intended to improve the linearized infeasibility as much as possible while satisfying the trust region constraint. Thus the merit function is ignored in this part of the computation. By contrast, the tangential component t_k aims not to degrade the improved infeasibility obtained in the normal step, while now concentrating on reducing a model of the merit function.

For simplicity, we shall suppose in this and the next two sections, that our constraints are equations, $c(x) = 0$—we shall return to the inequality case in Section 3.5. Recognising that the set

$$\mathcal{F}_k = \{q \mid c(x_k) + A(x_k)q = 0 \text{ and } \|q\|_k \le \Delta_k\}$$

may be empty, Vardi [66] and Byrd, Schnabel and Shultz [8] instead replace the linearized constraints by $\alpha_k c(x_k) + A(x_k)q = 0$ for some $0 < \alpha_k \le 1$, where α_k is chosen so that

$$\mathcal{F}_k(\alpha_k) = \{q \mid \alpha_k c(x_k) + A(x_k)q = 0 \text{ and } \|q\|_k \le \Delta_k\}$$

is non-empty. Clearly $\mathcal{F}_k(0)$ is non-empty, and any value $\alpha_k \le \alpha_{\max}$ is also suitable, where $\alpha_{\max}$ is the greatest α in $(0,1]$ such that

$$\min_{\|q\|_k \le \Delta_k} \|\alpha c(x_k) + A(x_k)q\| = 0.$$

As finding $\alpha_{\max}$ may be expensive,—it may require the computation of the projection $q^c(x_k)$ of the origin onto the set $\{q \mid c(x_k) + A(x_k)q =$

$0\}$ — in practice an approximation q_k^{c} to $q^{\mathrm{c}}(x_k)$ that satisfies $c(x_k) + A(x_k)q_k^{\mathrm{c}} = 0$ may be computed instead, and α_k subsequently found so that $q_k = \alpha_k q_k^{\mathrm{c}}$ lies within the trust region. In fact, this last condition is strengthened so that q_k lies strictly within the trust region, allowing some "elbow room" for the subsequent tangential step. Notice, however, the implicit requirement that the linearized constraints be compatible, which is the weakest point of the whole approach.

Having found the normal step, the tangential step is chosen to reduce a model of the merit function. Specifically, if we consider a merit function of the form

$$\varphi(x, \sigma) = f(x) + \sigma \|c(x)\|,$$

and model $\varphi(x_k + s, \sigma)$ by $m_k(x_k + s) = m_k^{\mathrm{o}}(x_k + s) + \sigma m_k^{\mathrm{c}}(x_k + s)$, where

$$m_k^{\mathrm{o}}(x_k + s) = f(x_k) + s^T g(x_k) + \tfrac{1}{2} s^T H_k s$$
$$\text{and} \quad m_k^{\mathrm{c}}(x_k + s) = \|c(x_k) + A(x_k)s\|,$$

we see that following the normal step, $m_k^{\mathrm{c}}(x_k + q_k)$ will have decreased, but $m_k^{\mathrm{o}}(x_k + q_k)$ may have increased. To cope with this, we pick the tangential step so that

$$m_k^{\mathrm{c}}(x_k + q_k + t_k) = m_k^{\mathrm{c}}(x_k + q_k) \;\text{ and }\; m_k^{\mathrm{o}}(x_k + q_k + t_k) \leq m_k^{\mathrm{o}}(x_k + q_k)$$

(the latter inequality being strict unless $g(x_k) + H_k q_k = 0$) by approximately solving the problem

$$\begin{aligned} &\underset{t \in \mathbb{R}^n}{\text{minimize}} \quad t^T(g(x_k) + H_k q_k) + \tfrac{1}{2} t^T H_k t \\ &\text{subject to} \quad A(x_k)t = 0, \quad \text{and} \quad \|t\| \leq \Delta_k - \|q_k\|. \end{aligned}$$

A suitable Cauchy point for this problem is readily available. So long as the linearized constraints are compatible, both normal and tangential steps satisfying the above requirements may be computed using suitable conjugate-gradient methods, the accuracy required being measured by suitable measures of the violation of the criticality conditions for the underlying problem.

Note that there is no *a priori* guarantee that the separate choices of q_k and t_k provide $m_k(x_k + q_k + t_k) < m_k(x_k)$. However, since $m_k^{\mathrm{c}}(x_k + q_k + t_k) < m_k^{\mathrm{c}}(x_k)$, one way of ensuring that the model of the merit function does decrease is to increase σ if necessary as the iteration proceeds. A simple rule is to increase the parameter to ensure that

$$m_k(x_k) - m_k(x_k + q_k + t_k) \geq \tau\sigma(m_k^{\mathrm{c}}(x_k) - m_k^{\mathrm{c}}(x_k + q_k + t_k)),$$

where the value $\tau \in (0, 1)$ is arbitrary but preferably very small. These, then, are the essential ingredients in the algorithm, which otherwise follows the standard trust-region paradigm. Any limit point of such an algorithm can be shown to be first-order critical. Moreover, the penalty parameter cannot grow arbitrarily large.

Of course, as usual in methods based upon the non-smooth merit function φ, it does not follow automatically that a desirable rate of convergence occurs. The cure, as alway, is to include a second-order correction $s_k^{\rm C}$, satisfying (9), when needed. A suitable rule is to consider a second-order correction only when the normal step q_k lies well within the trust region,—by implication this step is feasible for the linearized constraints—and when the original step $s_k = q_k + t_k$ does not provide a sufficient reduction in the merit function. In fact, the actual form of second-order correction required depends on the form of s_k. If s_k is the standard SQP step (3), then any second-order correction for which (10) holds is permitted. On the other hand, if the standard SQP step lies outside the trust-region, a specific second-order correction for which $g_k^{\rm C} = 0$ is advised—since the first-order criticality conditions for the model are not satisfied, there is little sense in trying to correct for them, but it is still important to try to correct for constraint curvature. It can then be shown that with the usual assumptions A2–A4, the algorithm sketched above converges Q-superlinearly so long as the SQP step is attempted (asymptotically) whenever possible, so long as the second-order correction is discarded if it lies too far outside the trust region, and so long as the trust-region radius is not reduced when the SQP step is acceptable but has a "small" component q_k. The same conditions suffice to ensure convergence to at least one second-order critical point x_* under assumption A2 if $H_k = H(x_k, y_k)$ and y_k converges to the corresponding y_*.

3.3. BYRD–OMOJOKUN-LIKE APPROACHES

A different composite-step approach is due to (Byrd and) Omojokun [6, 51]. It forms the basis of the NITRO, ETR and BECTR algorithms of [7, 43, 55], respectively. This approach has a major advantage over that in the previous section in that there is no requirement that the linearized constraints be compatible, but is otherwise quite similar.

The major, and essentially only, difference is in the computation of the (quasi-) normal step. Rather than shifting the linearized constraint, another possibility is to compute q_k to approximately

$$\underset{q \in \mathbb{R}^n}{\text{minimize}} \ \|c(x_k) + A(x_k)q\| \ \text{subject to} \ \|q\|_k \leq \xi^N \Delta_k \qquad (13)$$

for some $0 < \xi^N < 1$. This problem may have a large number of solutions—the minimum-norm solution will give a component which is normal to t_k. Since computing an exact solution may be expensive, a cheaper option is to find a q that gives a reduction in $\|c(x_k) + A(x_k)q\|$ that is at least a fraction of that achievable at a suitable Cauchy point for this problem, such a point being

$$q_k^{\mathrm{C}} = -\alpha_k^{\mathrm{C}} A^T(x_k)c(x_k), \tag{14}$$

where

$$\alpha_k^{\mathrm{C}} = \arg\min_{0 \le \alpha \le \xi^N \Delta_k / \|A^T(x_k)c(x_k)\|_k} \|c(x_k) - \alpha A(x_k)A^T(x_k)c(x_k)\|.$$

As before, such a requirement is satisfied at the first iteration of a suitable conjugate-gradient method, and subsequent conjugate-gradient steps may be used to further reduce the violation. From a theoretical point of view, the normal step needs to have a non-trivial component in the above-mentioned minimum-norm solution.

The resulting algorithm offers essentially the same guarantees as its predecessor. So long as $A(x_k)$ is of full rank, it follows that $\lim_{k\to\infty} A^T(x_k)c(x_k) = 0$, indicating that, at worst, limit points are locally least infeasible. If the limiting Jacobian is also of full rank, we deduce not only $\lim_{k\to\infty} c(x_k) = 0$, but also that the remaining first-order criticality conditions hold, and the penalty parameter remains finite.

Turning to the issue of fast convergence, essentially the same precautions as before may be used. Since we are ultimately interested in using a full SQP step, we shall require that eventually either the normal step lies on the "shrunken" trust-region boundary, i.e., $\|q_k\| = \xi^N \Delta_k$, or that a step that satisfies the linearized constraints, and lies within the shrunken trust-region, is possible, i.e., $c(x_k) + A(x_k)q_k = 0$ and $\|q_k\| \le \xi^N \Delta_k$. In the latter case, if the standard SQP step (3) does not provide a sufficient reduction in the merit function, a second-order correction is attempted. As before, the exact form depends upon whether the SQP step satisfies the trust-region constraint, in which case a general correction is allowed, or if the SQP step lies outside, in which case a restricted correction in which $g_k^{\mathrm{C}} = 0$ is used. The resulting algorithm then converges at a Q-superlinear rate under exactly the same conditions as its predecessor, and at least one limit point x_* is second-order critical if additionally $H_k = H(x_k, y_k)$ and y_k converges to the corresponding y_*.

3.4. CELIS–DENNIS–TAPIA-LIKE APPROACHES

A third way of dealing with the possibility that the linearized constraints and the trust region have no common feasible point is to replace the former by

$$\|c(x_k) + A(x_k)s\| \le \theta_k, \tag{15}$$

where θ_k is chosen so that (15) and the trust region bound $\|s\|_k \le \Delta_k$ can both be satisfied by some s. Clearly, since we wish to reduce the infeasibility, we should insist at the very least that

$$\min_{\|s\|_k \le \Delta_k} \|c(x_k) + A(x_k)s\| \le \theta_k \le \|c(x_k)\|, \tag{16}$$

while another possibility is to require

$$\min_{\|s\|_k \le \xi_1 \Delta_k} \|c(x_k) + A(x_k)s\| \le \theta_k \le \min_{\|s\|_k \le \xi_2 \Delta_k} \|c(x_k) + A(x_k)s\|, \tag{17}$$

where $0 < \xi_2 \le \xi_1 < 1$. Since solving problems of the form

$$\min_{\|s\|_k \le \xi \Delta_k} \|c(x_k) + A(x_k)s\|$$

for some $0 < \xi \le 1$, which are needed to ensure that θ satisfies (16) or (17), may be expensive, a cheaper possibility is to find *any* step q which lies within the trust region but which also significantly reduces $\|c(x_k) + A(x_k)q\|$. The most popular choice is, of course, the Cauchy step (14), but any step which further decreases $\|c(x_k) + A(x_k)q\|$ is also possible.

Although the computation of a suitable shift q to reduce the infeasibility is reminiscent of the composite step methods considered, q_k is actually only used to find

$$\theta_k = \|c(x_k) + A(x_k)q_k\| \le \|c(x_k)\|, \tag{18}$$

where the inequality in (18) is strict unless $c(x_k) = 0$. The overall step is computed as an approximate solution to the problem

$$\underset{s \in \mathbb{R}^n}{\text{minimize}} \quad s^T g(x_k) + \tfrac{1}{2} s^T H_k s \tag{19a}$$

$$\text{subject to} \quad \|c(x_k) + A(x_k)s\| \le \theta_k \quad \text{and} \quad \|s\| \le \Delta_k, \tag{19b}$$

for some appropriate Lagrange multiplier estimates y_k and approximation, H_k, to the Hessian of the Lagrangian. Methods based on these suggestions have been proposed by Celis, Dennis and Tapia [10] and Powell and Yuan [58]. Notice that by considering the whole feasible region (19b) rather than successive normal and tangential components, there is potential for greater reductions in the objective function (19a) than with the previous two approaches. Unfortunately, this advantage may also be regarded as its Achilles' heel.

The main disadvantage of these approaches is apparent if one considers (19). If polyhedral norms are used, this subproblem reduces to a (possibly non-convex) *inequality*-constrained quadratic program which may prove rather expensive to solve. On the other hand, if we choose the ℓ_2 norm, the subproblem involves *two* quadratic constraints. Thus it is unclear if or how the powerful techniques which have been developed for the simpler subproblem involving a single quadratic constraint (see, for example, [36, 48]) may or can be applied. In particular, it is far from evident how to compute the model minimizer, nor is it obvious how to derive a useful approximation—some results for convex models have been obtained by [39, 73] and others. Indeed, given that global and local convergence theories matching those of the other methods we have considered in this section can be developed, we can only surmise that the lack of any reported implementation based on the approach taken here may be attributed to this disadvantage. We mention in passing that all of the methods we are aware of that use this approach use smooth exact penalty functions like (6) to force global convergence, but methods based on (5) seem to be equally possible.

3.5. INEQUALITY CONSTRAINTS

We now return to the case where the constraints are inequalities, $c(x) \geq 0$. There are two basic approaches. The first is to extend the model problem to include inequalities. As we have already noted, it may be that the set

$$\mathcal{F}_k = \{s \mid c(x_k) + A(x_k)s \geq 0 \text{ and } \|s\|_k \leq \Delta_k\}$$

is empty when Δ_k is small. Thus, we may instead have to be content with a step which moves us towards a solution of the model problem. The methods we have considered in the three previous sections have achieved this by decomposing the step as $s_k = q_k + t_k$, where the (quasi-) normal step q_k is chosen to reduce the (linearized) infeasibility, and the tangential step t_k is then determined to reduce the model without worsening the infeasibility attained during the normal step. For the general problem, much the same approach is valid.

There are obvious variants of all of the three main approaches we have discussed. Consider first the normal step. To extend the Vardi-like methods, sketched in Section 3.2, we need to compute a trial step q_k^{C} which satisfies the linear constraints $c(x_k) + A(x_k)s \geq 0$, and which is not significantly longer than the projection onto the linearized feasible region. We then take a step α_k in this direction as far, or almost as far, as we can within the trust region, and set $q_k = \alpha_k q_k^{\rm C}$. For the Byrd–Omojokun-like approaches of Section 3.3, the normal step should be calculated by finding q_k to approximately

$$\operatorname*{minimize}_{q \in \mathbb{R}^n} \; \| \left(c(x_k) + A(x_k)q \right)_- \| \text{ subject to } \|q\| \leq \xi^N \Delta_k, \tag{20}$$

for some $0 < \xi^N < 1$, essentially as we did in (13). Note that, when the problem is defined in terms of the ℓ_1, ℓ_2 or ℓ_∞ norm, and if a polyhedral trust-region is used, (20) can be reformulated as a linear or convex quadratic program, and thus, in principle, there are effective methods for (approximately) solving it. Finally, to extend the Celis–Dennis–Tapia-like approaches of Section 3.4, we merely need the normal step to give us at least as much reduction in $\| \left(c(x_k) + A(x_k)q \right)_- \|$ as a step to a generalized Cauchy point for this problem.

Turning to the tangential step, extensions to both the Vardi- and Byrd–Omojokun-like approaches require that the step solves approximately

$$\begin{array}{rrcl} & \operatorname*{minimize}_{t \in \mathbb{R}^n} & \multicolumn{2}{l}{t^T(g(x_k) + H_k q_k) + \tfrac{1}{2} t^T H_k t} \\ \text{subject to} & A(x_k)t & \geq & -\max\left[c(x_k) + A(x_k)q_k, 0\right] \text{ and} \\ & \|t\|_k & \leq & \Delta_k - \|q_k\|_k. \end{array}$$

Notice that the linearized infeasibility is made no worse, and attention turns instead to reducing the model value. In theory, all that is required is that the reduction in the model at t_k is a positive fraction of that attainable at a generalized Cauchy point, such as that proposed by [15]. For Celis–Dennis–Tapia-like approaches, the tangential step must be calculated to approximately

$$\begin{array}{rrcl} & \operatorname*{minimize}_{s \in \mathbb{R}^n} & \multicolumn{2}{l}{s^T g(x_k) + \tfrac{1}{2} s^T H_k s} \\ \text{subject to} & \| \left(c(x_k) + A(x_k)s \right)_- \| & \leq & \theta_k \text{ and} \\ & \|s\| & \leq & \Delta_k, \end{array} \tag{21}$$

where $\theta_k = \|(c(x_k) + A(x_k)q_k)_-\|$. As before, this approach is less attractive in practice than its predecessors as effective methods for approximately minimizing (21) are not known. Other details extend in an obvious way. In particular, the same merit functions as before are appropriate, provided we replace every mention of $c(x)$ by $c(x)_-$.

The second way of moving from the equality-constrained to the general problem is to handle inequalities using barrier/interior-point methods (see, for example, [14, 22, 32, 33, 62, 63, 65]). That is to say, we embed the inequality problem within a sequence of barrier problems of the form

$$\underset{x}{\text{minimize}} \;\; f(x) + b(c(x), \mu_k),$$

where $b(c(x), \mu)$ is a barrier term like $-\mu \sum_{i=1}^m \log c_i(x)$, and $\{\mu_k\}$ is a sequence of barrier parameters which converge to zero from above. For this class of methods, we insist on starting from a strictly feasible point for the inequality constraints, that is that $c(x) > 0$, and we require all subsequent iterates to remain strictly feasible for these constraints.

A typical trust-region method for such a problem models the barrier term using either a Newton (primal) or quasi-Newton (primal-dual) approximation. However, since such quadratic models have little influence in dissuading the iterates from violating one or more of inequality constraints, it is crucial to either adjust the shape of the trust region to keep the iterates feasible, or to add explicit extra constraints to the trust-region subproblems to do this (or both). The main difficulty when there are nonlinear inequality constraints present is that any additional constraints imposed on the trust-region subproblem may be nonlinear. For this reason, inequality constraints are often converted to equations by introducing slack variables. That is, we replace $c(x) \geq 0$ by the equivalent conditions

$$c(x) - v = 0 \;\text{ and }\; v \geq 0,$$

and then we solve a sequence of equality constrained minimization problems

$$\underset{x,\,v}{\text{minimize}} \;\; f(x) + b(v, \mu_k) \;\text{ subject to }\; c(x) - v = 0.$$

The advantage of this approach is that we believe that the methods given throughout Section 3.2–3.4 are well-able to deal with nonlinear equality constraints, while any of the barrier/interior-point, affine scaling, or [13] algorithms are especially suited to linear, and particularly simple bound, constraints. Indeed, a careful combination of the Byrd–Omojokun and Coleman–Li approaches forms the basis of the algorithm proposed by [6] and implemented as NITRO by [7], while the method proposed by [71] (see also [68, 69, 70]) is essentially a Vardi-like primal-dual method. Both of these methods are reported to perform most effectively in practice.

There are some disadvantages of adding slack variables. Firstly, we have most definitely increased the dimension of the problem. To counter this, it is important to realize that the dominant cost of most algorithms (at least when function values are inexpensive) tends to be that for the linear algebra. In practice, significant algebraic savings may be made be recognising that slack variables only occur linearly in the problem reformulation, and each slack variable is associated with a single constraint. The second disadvantage is that a suitable scaling of the slack variables is often difficult to find—in practice, it is more usual to pick the slacks so that $c(x) - Dv = 0$ and $v \geq 0$, where the diagonal matrix D is supposed to reflect "typical" values of $c(x)$, but the very fact that c is nonlinear indicates that a uniformly good D may be hard to determine. This has further repercussions for trust-region methods since it is usual to scale the trust-region norm to account for different scalings of the variables. We also note that in practice the trust-region scaling needs to reflect the interaction between the nonlinear constraints and the simple bounds (see [14]).

We conclude this short section on inequality constraints with the remark that blending good methods for coping with equality constraints with good ones for dealing with inequalities is an extremely active area of research. For this reason, we shall say no more here, but await further developments, and particularly comparisons of the numerous possibilities, with interest.

3.6. FILTER METHODS

The last method we shall consider is the youngest, and certainly one of the most promising. The central idea is to dispense with the idea of using a merit function as a means of encouraging global convergence as far as is practically possible, and instead to use a mechanism which is less likely to reject candidate iterates. One such mechanism is a so-called filter.

Suppose $\theta(x)$ is some measure of the infeasibility of the constraints at x, for example $\theta(x) = \|c(x)_-\|$. A filter is a list of pairs $\{(f(x_i), \theta(x_i))\}$, with the property that no member of the filter is dominated by another, that is there are no two $(f(x_i), \theta(x_i))$ and $(f(x_j), \theta(x_j))$ $(i \neq j)$ for which

$$f(x_i) \leq f(x_j) \text{ and } \theta(x_i) \leq \theta(x_j).$$

The key point is that the filter may be used as a mechanism to accept or reject candidate iterates: a candidate will only be rejected if it gives "larger" values of both the function value and constraint violation than have been observed before. Contrast this to a merit function, which tries to combine these two (conflicting) requirements in a somewhat arbitrary way. An SQP-filter method aims to use the filter as a means of assessing iterates $x_k + s_k$, where s_k is a suitable approximation to the solution of the trust-region SQP subproblem

$$\underset{s\in\mathbb{R}^n}{\text{minimize}}\; m_k(x_k+s) \;\text{ subject to }\; c(x_k)+A(x_k)s \geq 0 \;\text{ and }\; \|s\|_k \leq \Delta_k, \tag{22}$$

where $m_k(x_k+s) = f(x_k) + s^T g(x_k) + \frac{1}{2} s^T H_k s$. The filter evolves as new iterates are accepted; the new iterate (or rather its (f,θ) pair) may be added to the filter, while the act of adding a new pair can result in the removal of previous members which are now dominated by the newcomer.

Of course, the reader may object immediately that such a simple-minded approach has obvious flaws. The first is, as always, that (22) may not have a solution because either the trust-region radius is too small, or because the linearized constraints are inconsistent. The cure is simply to abandon temporarily the objective function, and to enter a restoration phase, whose sole purpose is to reduce the infeasibility $\theta(x)$. The end of the restoration phase is reached at $x_k + r_k$, at which either the set $\{s \mid c(x_k+r_k) + A(x_k+r_k)s \geq 0 \text{ and } \|s\|_{k+1} \leq \Delta_{k+1}\}$ is non-empty for some $\Delta_{k+1} > 0$ and for which $\big(f(x_k+r_k), \theta(x_k+r_k)\big)$ is acceptable for the filter, or $x_k + r_k$ is a critical point for $\theta(x)$—in either case, such a point may be achieved by (approximately) minimizing $\theta(x)$. The second flaw is that it is easy to imagine a sequence of iterates each of which is barely acceptable to the (current) filter, but whose limit point is not critical—such a potential difficulty arises in most minimization methods, and the cure as always is to require that the iterates provide a "sufficient" improvement in the filter. A suitable rule is that an acceptable iterate must satisfy

$$f(x_k+s_k) < f(x_j) - \gamma\theta(x_j) \;\text{ or }\; \theta(x_k+s_k) < (1-\gamma)\theta(x_j)$$

for all x_j in the filter, where $\gamma \in (0,1)$.

Fletcher and Leyffer [30] demonstrate that an SQP-filter method based on the above, and including a number of other heuristics, is most effective in practice. The stated goal of requiring minimal interference from the filter is vindicated, and evidence is provided to show that other SQP methods (specifically the Sℓ_1QP method discussed in Section 3.1) frequently suffer more interference from their merit functions. In order to prove convergence of an SQP-filter method, specific rules for when to include an iterate in the filter, what sort of approximate step may be tolerated, and how to adjust the trust-region radius are required. The first-such convergence result (for an SLP-filter) was provided by [31], and this has now been extended to the SQP case by [29]. The step is computed as the composite $s_k = q_k + t_k$, essentially as we have considered in the previous four sections—if an infeasible subproblem is detected during the normal-step calculation, then the restoration phase is started straight away. (A variation in which the step s_k is computed as a whole is also possible, although one may have to retreat to the composite step under unfavourable circumstances.) Once s_k has been computed, it is rejected if either it is unacceptable to the filter or if $m_k(x_k + s_k)$ offers a "sufficient" improvement over $m_k(x_k)$ but this predicted improvement does not translate into an actual improvement in $f(x)$. The trust-region radius is reduced whenever a step is rejected. The iterate is added to the filter if either it leads to a restoration phase, or if it has been accepted despite $m_k(x_k + s_k)$ not giving a "sufficient" improvement over $m_k(x_k)$. Second-order convergence issues are still open, and are under investigation.

4. QP METHODS

Without a doubt, in our opinion, the primary reason SQP methods are back in the ascendant is that large-scale quadratic programming (QP) methods have matured considerably over the past few years. There are a number of reasons for this. At the start of the 1980s, the vast majority of QP methods (see the surveys by [26], Chapter 10, and [27], and the bibliography in [20]) were of the active set variety, most were specifically designed for convex (H positive semi-definite) or even strictly convex (H positive definite) problems, and few (if any) were capable of solving even medium size problems (for exceptions, see [28, 35]). The latter defect was due to two factors. Firstly, the dominant linear algebraic requirements usually treated all relevant matrices and associated factorizations as dense—while it was easy to anticipate using sparse factorizations, this ruled out some of the most successful (orthogonal transformation) methods developed for the dense case. Secondly, as problem size increased, the number of iterations rose quite rapidly—in the worst case, an exponential number of changes in the active set was possible, and while the expected and observed behaviour did not get close to such dire predictions, it was a cause for concern.

By the turn of the decade, the theoretical (polynomially bounded) promise of [42] interior-point linear programming (LP) approach, and its successors, had been shown to be realized in practice, and theoretical extensions to convex QP were immediate—we note that, as in the LP case, only a small fraction of the methods proposed and analysed have ever been implemented (for exceptions, see [9, 64]). Most of the implementations differ from their theoretical counterparts in order to obtain good practical performance, and all of them appear to perform considerably better than their worst-case polynomial bound.

It is now accepted that interior-point and active-set methods are useful alternatives, but frequently the former are the methods of choice when the number of variables is very large. Of course, modern SQP methods often require the (approximate) solution of non-convex QPs, for which the above-mentioned interior-point methods cannot offer the same theoretical guarantees, since non-convex QP is known to be an NP-hard problem. Nonetheless, it is possible to construct interior-point-like methods, which are both globally convergent, and whose asymptotic convergence behaviour is similar to the locally convex case (see, for example, [14, 18, 65]). Early computational experience indicates considerable promise for large (say $n \sim 10^5$) problems.

One of the means by which methods for unconstrained minimization made the transition from small to large problems was the recognition that it is not necessary to solve the relevant model problem very accurately, at least when far from the solution. As we have indicated in Section 3, the same is true for SQP methods. However, at present, this either requires that the step is computed as a composite, in which two Cauchy points are determined (see Sections 3.2, 3.3, and 3.6), or as a single step in which an auxiliary computation may be necessary (see Section 3.1). As yet, the only method we are aware of that allows a direct truncation of the QP subproblem is the active set method of [49]. The subproblems in both active-set and interior-point methods may be solved by iterative (conjugate gradient-like) methods, although it is crucial, especially for the latter, to use suitable preconditioners.

Finally, as to which of the two QP alternatives we suggest is appropriate for SQP methods, our answer is both! To justify this, we believe that interior-point methods probably hold the advantage for early SQP iterations when the active set has far from settled down. By contrast, when the active set is essentially known, a few active-set iterations are often cheaper than applying an interior-point method, since the latter is difficult to "warm start", i.e., start from a known near optimal (but possibly un-centred) vector of variables. Thus we contend that any new SQP method for large-scale nonlinear programming should have access to both interior-point and active-set non-convex QP algorithms.

5. CONCLUSIONS

In this paper, we have surveyed many of the most recent SQP algorithms for nonlinear programming. The majority of them are well-suited to large-scale problems, and recent numerical results (see, for example, [30, 71]) indicate that such methods are often considerably better than state-of-the-art implementations of other nonlinear programming algorithms (such as MINOS and, it hurts us to say, LANCELOT). We should add that the tests performed are all on medium-sized problems (say $n \sim 10^3$–10^4), but we expect this trend to continue for large ones (say $n \sim 10^5$–10^6) in the near future, provided that options for solving core linear systems by iterative (say, preconditioned conjugate-gradient) methods are incorporated. It still remains to be seen if the major way in which modern SQP algorithms will benefit from interior-point technology is in the improvements these give to quadratic programming algorithms, or in the ways these suggest for handling inequality constraints. It also remains to be seen if new ideas, such as the SQP-filter method described in Section 3.6, fulfil their early promise.

Acknowledgments

The authors are extremely grateful to Michael Powell for his numerous comments and suggestions, many of which have led to significant improvements in this paper.

References

[1] P. T. Boggs and J. W. Tolle (1995), Sequential quadratic programming, *Acta Numerica*, 4, pp. 1–51.

[2] P. T. Boggs, A. J. Kearsley and J. W. Tolle (1999), A global convergence analysis of an algorithm for large scale nonlinear programming problems, *SIAM Journal on Optimization*, 9(4), pp. 833–862.

[3] P. T. Boggs, A. J. Kearsley and J. W. Tolle (1999), A practical algorithm for general large scale nonlinear optimization problems, *SIAM Journal on Optimization*, 9(3), pp. 755–778.

[4] J. F. Bonnans and G. Launay (1992), Implicit trust region algorithm for constrained optimization, Technical report, Institute Nationale Recherche Informatique et Automation, Le Chesnay, France.

[5] J. F. Bonnans, E. R. Panier, A. L. Tits and J. L. Zhou (1992) Avoiding the Maratos effect by means of a nonmonotone line search II. inequality constrained problems—feasible iterates. *SIAM Journal on Numerical Analysis*, **29**(4), pp. 1187–1208.

[6] R. H. Byrd, J. Ch. Gilbert and J. Nocedal (1996), A trust region method based on interior point techniques for nonlinear programming, Technical Report 2896, INRIA, Rocquencourt, France.

[7] R. H. Byrd, M. E. Hribar and J. Nocedal (2000), An interior point algorithm for large scale nonlinear programming, *SIAM Journal on Optimization*, 9(4), pp. 877–900.

[8] R. H. Byrd, R. B. Schnabel and G. A. Shultz (1987), A trust region algorithm for nonlinearly constrained optimization, *SIAM Journal on Numerical Analysis*, 24, pp. 1152–1170.

[9] T. J. Carpenter, I. J. Lustig, J. M. Mulvey and D. F. Shanno (1993), Higher-order predictor-corrector interior point methods with application to quadratic objectives, *SIAM Journal on Optimization*, 3(4), pp. 696–725.

[10] M. R. Celis, J. E. Dennis and R. A. Tapia (1985), A trust region strategy for nonlinear equality constrained optimization, in P. T. Boggs, R. H. Byrd and R. B. Schnabel, eds, *Numerical Optimization 1984*, SIAM, Philadelphia, USA, pp. 71–82.

[11] R. M. Chamberlain, M. J. D. Powell, C. Lemaréchal and H. C. Pedersen (1982), The watchdog technique for forcing convergence in algorithms for constrained optimization, *Mathematical Programming Studies*, 16, pp. 1–17.

[12] T. F. Coleman and A. R. Conn (1982), Non-linear programming via an exact penalty-function: global analysis, *Mathematical Programming*, 24(2), pp. 137–161.

[13] T. F. Coleman and Y. Li (1996), An interior trust region approach for nonlinear minimization subject to bounds, *SIAM Journal on Optimization*, 6(2), pp. 418–445.

[14] A. R. Conn, N. I. M. Gould, D. Orban and Ph. L. Toint (1999), A primal-dual trust-region algorithm for minimizing a non-convex function subject to bound and linear equality constraints, Technical Report TR99-04, Department of Mathematics, University of Namur, Namur, Belgium.

[15] A. R. Conn, N. I. M. Gould, A. Sartenaer and Ph. L. Toint (1993), Global convergence of a class of trust region algorithms for optimization using inexact projections on convex constraints, *SIAM Journal on Optimization*, 3(1), pp. 164–221.

[16] A. R. Conn, N. I. M. Gould and Ph. L. Toint (1992), LANCELOT*: a Fortran package for Large-scale Nonlinear Optimization (Release A)*, Springer Series in Computational Mathematicsm Springer Verlag, Heidelberg, Berlin, New York.

[17] A. R. Conn, N. I. M. Gould and Ph. L. Toint (1997), Methods for nonlinear constraints in optimization calculations, in I. Duff and A. Watson, eds, *The State of the Art in Numerical Analysis*, Oxford University Press, pp. 363–390.

[18] A. R. Conn, N. I. M. Gould and Ph. L. Toint (1999), A primal-dual algorithm for minimizing a nonconvex function subject to bound

and linear equality constraints, in G. Di Pillo and F. Gianessi, eds, *Nonlinear Optimization and Applications 2*, Kluwer Academic Publishers, Dordrecht, pp. 15–50.

[19] A. R. Conn, N. I. M. Gould and Ph. L. Toint (2000), *Trust-region methods*, To appear. SIAM, Philadelphia.

[20] R. W. Cottle, J.-S. Pang and R. E. Stone (1992), *The Linear Complementarity Problem*, Academic Press, London.

[21] G. Di Pillo, F. Facchinei and L. Grippo (1992), An RQP algorithm using a differentiable exact penalty function for inequality constrained problems, *Mathematical Programming*, 55(1), pp. 49–68.

[22] A. V. Fiacco and G. P. McCormick (1968), *Nonlinear Programming: Sequential Unconstrained Minimization Techniques*, J. Wiley and Sons, Chichester, England. Reprinted as *Classics in Applied Mathematics 4*, SIAM, Philadelphia, USA, 1990.

[23] R. Fletcher (1973), An exact penalty function for nonlinear programming with inequalities, *Mathematical Programming*, 5(2), pp. 129–150.

[24] R. Fletcher (1982), A model algorithm for composite nondifferentiable optimization problems, *Mathematical Programming Studies*, 17, pp. 67–76.

[25] R. Fletcher (1982), Second order corrections for non-differentiable optimization, in G. A. Watson, ed., *Numerical Analysis, Proceedings Dundee 1981, Lecture Notes in Mathematics 912*, Springer Verlag, Heidelberg, Berlin, New York, pp. 85–114.

[26] R. Fletcher (1987), *Practical Methods of Optimization*, J. Wiley and Sons, Chichester, England, second edn.

[27] R. Fletcher (1987), Recent developments in linear and quadratic programming, in A. Iserles and M. J. D. Powell, eds, *The State of the Art in Numerical Analysis*, Oxford University Press, pp. 213–243.

[28] R. Fletcher (1993), Resolving degeneracy in quadratic programming, *Annals of Operations Research*, 46-47(1-4), pp. 307–334.

[29] R. Fletcher, N. I. M. Gould, S. Leyffer and Ph. L. Toint (1999), Global convergence of trust-region SQP-filter algorithms for nonlinear programming, Technical Report 99/03, Department of Mathematics, University of Namur, Namur, Belgium.

[30] R. Fletcher and S. Leyffer (1997), Nonlinear programming without a penalty function, Numerical Analysis Report NA/171, Department of Mathematics, University of Dundee, Dundee, Scotland.

[31] R. Fletcher, S. Leyffer and Ph. L. Toint (1998), On the global convergence of an SLP-filter algorithm, Technical Report 98/13, Department of Mathematics, University of Namur, Namur, Belgium.

[32] A. Forsgren and P. E. Gill (1998), Primal-dual interior methods for nonconvex nonlinear programming. *SIAM Journal on Optimization*, 8(4), pp. 1132–1152.

[33] D. M. Gay, M. L. Overton and M. H. Wright (1998), A primal-dual interior method for nonconvex nonlinear programming, in Y. Yuan, ed., *Advances in Nonlinear Programming*, Kluwer Academic Publishers, Dordrecht, pp. 31–56.

[34] P. E. Gill, W. Murray and M. A. Saunders (1997), SNOPT: an SQP algorithm for large-scale constrained optimization, Technical Report SOL97-3, Department of EESOR, Stanford University, Stanford, California 94305, USA.

[35] N. I. M. Gould (1991), An algorithm for large-scale quadratic programming, *IMA Journal of Numerical Analysis*, 11(3), pp. 299–324.

[36] N. I. M. Gould, S. Lucidi, M. Roma and Ph. L. Toint (1999), Solving the trust-region subproblem using the Lanczos method, *SIAM Journal on Optimization*, 9(2), pp. 504–525.

[37] L. Grippo, F. Lampariello and S. Lucidi (1986), A nonmonotone line search technique for Newton's method, *SIAM Journal on Numerical Analysis*, 23(4), pp. 707–716.

[38] W. W. Hager (1999), Stabilized sequential quadratic programming, *Computational Optimization and Applications*, 12(1–2), pp. 253–273.

[39] M. Heinkenschloss (1994), On the solution of a two ball trust region subproblem, *Mathematical Programming*, 64(3), pp. 249–276.

[40] J. Herskovits (1986), A 2-stage feasible directions algorithm for nonlinear constrained optimization, *Mathematical Programming*, 36(1), pp. 19–38.

[41] W. Hock and K. Schittkowski (1981), *Test Examples for Nonlinear Programming Codes*, Number 187 in *Lecture Notes in Economics and Mathematical Systems*, Springer Verlag, Heidelberg, Berlin, New York.

[42] N. Karmarkar (1984), A new polynomial-time algorithm for linear programming, *Combinatorica*, 4, pp. 373–395.

[43] M. Lalee, J. Nocedal and T. D. Plantenga (1998), On the implementation of an algorithm for large-scale equality constrained optimization, *SIAM Journal on Optimization*, 8(3), pp. 682–706.

[44] S. Lucidi (1992), New results on a continuously differentiable exact penalty function, *SIAM Journal on Optimization*, 2(4), pp. 558–574.

[45] O. L. Mangasarian and S. Fromovitz (1967), The Fritz John necessary optimality conditions in the presence of equality and inequality constraints, *Journal of Mathematical Analysis and Applications*, 17, pp. 37–47.

[46] N. Maratos (1978), *Exact penalty function algorithms for finite-dimensional and control optimization problems*, PhD thesis, University of London, England.

[47] D. Q. Mayne and E. Polak (1982), A superlinearly convergent algorithm for constrained optimization problems, *Mathematical Programming Studies*, 16, pp. 45–61.

[48] J. J. Moré and D. C. Sorensen (1983), Computing a trust region step, *SIAM Journal on Scientific and Statistical Computing*, 4(3), pp. 553–572.

[49] W. Murray and F. J. Prieto (1995), A sequential quadratic programming algorithm using an incomplete solution of the subproblem, *SIAM Journal on Optimization*, 5(3), pp. 590–640.

[50] B. A. Murtagh and M. A. Saunders (1982), A projected Lagrangian algorithm and its implementation for sparse non-linear constraints, *Mathematical Programming Studies*, 16, pp. 84–117.

[51] E. O. Omojokun (1989), *Trust region algorithms for optimization with nonlinear equality and inequality constraints*, PhD thesis, University of Colorado, Boulder, Colorado, USA.

[52] E. R. Panier and A. L. Tits (1987), A superlinearly convergence feasible method for the solution of inequality constrained optimization problems, *SIAM Journal on Control and Optimization*, 25(4), pp. 934–950.

[53] E. Panier and A. L. Tits (1991), Avoiding the Maratos effect by means of a nonmonotone linesearch I. general constrained problems, *SIAM Journal on Numerical Analysis*, 28, pp. 1183–1195.

[54] E. R. Panier and A. L. Tits (1993), On combining feasibility, descent and superlinear convergence in inequality constrained optimization, *Mathematical Programming*, 59(2), pp. 261–276.

[55] T. D. Plantenga (1999), A trust-region method for nonlinear programming based on primal interior point techniques, *SIAM Journal on Scientific Computing*, 20(1), pp. 282–305.

[56] M. J. D. Powell (1983), Variable metric methods for constrained optimization, in A. Bachem, M. Grötschel and B. Korte, eds, *Math-*

ematical Programming: The State of the Art, Springer Verlag, Heidelberg, Berlin, New York, pp. 288–311.

[57] M. J. D. Powell (1987), Methods for nonlinear constraints in optimization calculations, in A. Iserles and M. J. D. Powell, eds, *The State of the Art in Numerical Analysis*, Oxford University Press, pp. 325–357.

[58] M. J. D. Powell and Y. Yuan (1990), A trust region algorithm for equality constrained optimization, *Mathematical Programming*, 49(2), pp. 189–213.

[59] B. N. Pschenichny (1970), Algorithms for general problems of mathematical programming, *Kibernetica*, 6, pp. 120–125.

[60] S. M. Robinson (1974), Perturbed Kuhn-Tucker points and rates of convergence for a class of nonlinear programming algorithms, *Mathematical Programming*, 7(1), pp. 1–16.

[61] R. W. H. Sargent (1995), The development of the SQP algorithm for nonlinear programming, Technical Report C95-35, Centre for Process Systems Engineering, Imperial College, London, England.

[62] R. W. H. Sargent and M. Ding (1998), A new SQP algorithm for large-scale nonlinear programming, Technical Report C98-5, Centre for Process Systems Engineering, Imperial College, London, England.

[63] D. F. Shanno and R. J. Vanderbei (1999), Interior-point methods for nonconvex nonlinear programming: orderings and higher-order methods, Technical Report SOR 99-05, Program in Statistics and Operations Research, Princeton University, New Jersey, USA.

[64] R. J. Vanderbei (1994), LOQO: an interior point code for quadratic programming, Technical Report SOR 94-15, Program in Statistics and Operations,Research, Princeton University, New Jersey, USA.

[65] R. J. Vanderbei and D. F. Shanno (1997), An interior point algorithm for nonconvex nonlinear programming, Technical Report SOR 97-21, Program in Statistics and Operations Research, Princeton University, New Jersey, USA.

[66] A. Vardi (1985), A trust region algorithm for equality constrained minimization: convergence properties and implementation, *SIAM Journal on Numerical Analysis*, 22(3), pp. 575–591.

[67] S. J. Wright (1999), Superlinear convergence of a stabilized SQP method to a degenerate solution, *Computational Optimization and Applications*, 11(3), pp. 253–275.

[68] H. Yabe and H. Yamashita (1997), Q-superlinear convergence of primal-dual interior point quasi-Newton methods for constrained

optimization, *Journal of the Operations Research Society of Japan*, 40(3), pp. 415–436.

[69] H. Yamashita and H. Yabe (1996), Nonmonotone SQP methods with global and superlinear convergence properties, Technical report, Mathematical Systems, Inc., Sinjuku-ku, Tokyo, Japan.

[70] H. Yamashita and H. Yabe (1996), Superlinear and quadratic convergence of some primal-dual interior point methods for constrained optimization, *Mathematical Programming, Series A*, 75(3), pp. 377–397.

[71] H. Yamashita, H. Yabe and T. Tanabe (1997), A globally and superlinearly convergent primal-dual point trust region method for large scale constrained optimization, Technical report, Mathematical Systems, Inc., Sinjuku-ku, Tokyo, Japan.

[72] Y. Yuan (1985), On the superlinear convergence of a trust region algorithm for nonsmooth optimization, *Mathematical Programming*, 31(3), pp. 269–285.

[73] Y. Yuan (1991), A dual algorithm for minimizing a quadratic function with two quadratic constraints, *Journal of Computational Mathematics*, 9(4), pp. 348–359.

ALGORITHMIC IMPLICATIONS OF DUALITY IN STOCHASTIC PROGRAMS

Julia L. Higle
Department of Systems and Industrial Engineering,
The University of Arizona,
Tucson, AZ 85721, USA
julie@sie.arizona.edu

Suvrajeet Sen
As above

Abstract In this paper, we study primal and dual formulations of multistage stochastic programs (SP). Using a dual formulation, we discuss a decomposition/cutting plane algorithm that can be used to solve such problems. The algorithm, which is based on a scenario decomposition derived from the dual statement of the problem, is best viewed as a conceptual algorithm. Nevertheless, it lends itself to the use of sampled data, and enhancements necessary to produce a computationally viable method are discussed.

1. INTRODUCTION

Stochastic programming (SP) is a powerful modeling tool that allows decision-making models to incorporate uncertain parameters. One of the main strengths of the SP methodology is its ability to consider the impact of a variety of scenarios when evaluating a proposed solution, in contrast to the more restrictive approach of deterministic optimization models, in which only a single scenario is considered. Also, despite the large-scale nature of stochastic optimization models, several successful applications of SP models have been reported in the literature (see, e.g., Cariño et al [1994], Sen, Doverspike, and Cosares [1994]).

As in other areas of optimization, duality has implications for both SP modeling as well as the development of SP algorithms (see, e.g., Rockafellar and Wets [1991], Higle and Sen [1996a]). In this paper, we present an easily accessible development of a dual problem for a multi-

M.J.D. Powell and S. Scholtes (Eds.), *System Modelling and Optimization: Methods, Theory and Applications.*

stage SP. This accessiblity is due, in large part, to the use of stochastic analogs of deterministic constructs that are already well understood in the mathematical programming literature. Based on this setting, we present a decomposition/cutting plane algorithm that may be used to solve a multistage stochastic program. The problem is decomposed by "scenario", as in Rockafellar and Wets [1991], rather than by time stage, as in Birge [1985] and Gassmann [1990]. Within our framework, we make no distinctions regarding the nature of the random variables involved; discrete and continuous random variables are considered under a common umbrella.

This paper is organized as follows. In §2, we present a generic formulation of a multistage stochastic program, and suggest alternate representations of the so-called "nonanticipativity constraints". These constraints model the time-staged evolution of information, and are characteristic components of a stochastic program. In §3, we present a dual representation of a stochastic program. In §4, we propose a decomposition/cutting plane algorithm for the solution of stochastic programs. It may be best to think of this as a "conceptual algorithm", from which a computationally viable method can be obtained. As such, some of the enhancements necessary to obtain a viable algorithm are also discussed in §4. Concluding remarks may be found in §5. While this paper is largely expository in nature, with most of the mathematical details omitted, the reader is referred to Higle and Sen [1997] for a fully detailed presentation of the duality results in §3, and to Sen, Higle and Rayco [1999] for a full presentation of a viable algorithm, along with preliminary computational results.

2. PRIMAL PROBLEM

In what follows, we consider a problem in which "decisions", which we denote as x, and random variables are interwoven over time. An initial decision is made, after which an outcome of a random variable is observed. In response to the observation, a subsequent decision is made, after which another outcome is observed, etc. As a result of the multistage nature of the problems that we consider, our model is one in which both randomness and decisions evolve over time. In stage 1, we have the current (certain) information, denoted ω_1. Information beyond the first stage is uncertain, and is modeled through a sequence of random variables $\tilde{\omega}_2, \ldots, \tilde{\omega}_T$. We use the index t to denote a stage in the decision problem, $t = 1, \ldots, T$, whereas $x = \{x_t\}$ and $\omega = \{\omega_t\}$ are associated with decisions and outcomes, respectively. In this sense, x_t indicates a decision made in stage t and ω_t indicates an observation made in stage t. The sequence of random variables $\tilde{\omega} = \{\tilde{\omega}_t\}_{t=1}^T$ is defined on a probability space $(\Omega, \mathcal{A}, \mathcal{P})$, and in the stochastic programming literature, any realization ω is termed as a *scenario*. Although we consider "randomness" as exogenous to the problem, so that a particular choice of $x = \{x_t\}_{t=1}^T$ does not have a distributional impact upon $\tilde{\omega}$, a feasible choice of x is nonetheless dependent upon $\tilde{\omega}$. Thus, for each possible scenario $\omega \in \Omega$, there is a set of feasible solutions, $X(\omega)$, and an objective function $g(x, \omega)$ which influences the choice of x. When a particular variable, such as x is explicitly defined via the consideration of all possible scenarios in Ω, we will denote it as $\{x(\omega)\}_{\omega\in\Omega}$. Otherwise, we will simply denote it as x. Finally, throughout our development, we will assume that all vectors are appropriately dimensioned. Consequently, for each possible scenario $\omega \in \Omega$, we have the following "scenario problem", also known as the "wait and see problem":

$$\begin{aligned} \text{Min} \quad & g(x, \omega) \\ \text{s.t.} \quad & x \in X(\omega). \end{aligned} \tag{P_ω}$$

Let $x(\omega)$ denote the solution to the scenario problem (P_ω). Note that the "scenario solutions", $\{x(\omega)\}_{\omega\in\Omega}$, represent the result of posterior optimization. That is, $x(\omega)$ would be an optimal response if one knew for certain that scenario ω would occur. The difficulty, of course, is that one is rarely (if ever) graced with such knowledge, and while $x(\omega^1)$ is an optimal response to ω^1, it may be a disasterous response to ω^2. Thus, a model that leads to a more balanced response, one whose cost under all scenarios has been explicitly considered, is necessary.

To develop such a model, it is important to first note that scenarios that share a common history up to some point in time must implement the same decision at that time. That is, if we want to be able to implement a decision at a particular time, it cannot depend on information that will only become available at a later time. This requirement leads to constraints known as the nonanticipativity constraints. If we let $\mathcal{N}$ denote the set of nonanticipative solutions, we may formulate a stochastic programming model as follows:

$$\begin{aligned} \text{Min} \quad & E[g(x(\tilde{\omega}), \tilde{\omega})] \\ \text{s.t.} \quad & x(\tilde{\omega}) \in X(\tilde{\omega}) \qquad a.s. \\ & x(\tilde{\omega}) \in \mathcal{N}. \end{aligned}$$

We will assume throughout that the constraints, $x(\tilde{\omega}) \in X(\tilde{\omega})$ have a property known as <u>relatively complete recourse</u>. That is, we assume that if $x_t(\tilde{\omega})$ appears to be feasible on the basis of all decisions and observations made through time t, it cannot be rendered infeasible as a result of some event that cannot be observed until some later time. With this assumption, the scenarios are only coupled through the nonanticipativity constraints. (This constraint qualification is common in the stochastic programming literature.) When formulating a stochastic program, relatively complete recourse can be incorporated through the judicious use of penalties for infeasibilities.

To model the nonanticipativity requirements, let $\mathcal{H}_t$ denote the operator that projects any multistage entity onto the space corresponding to the first t stages. Then

$$\mathcal{H}_t\omega = (\omega_1, \omega_2, \dots, \omega_t)$$

reflects the evolution of scenario ω through the first t periods. Let Ω_t denote the set of possible "truncations" in period t, or

$$\Omega_t = \{\underline{\omega}_t \mid \mathcal{H}_t\omega = \underline{\omega}_t \text{ for some } \omega \in \Omega\}.$$

Note that $\Omega_T = \Omega$. Next, define the point-to-set map $\mathcal{H}_t^{-1}$, which identifies sets of scenarios with common truncations in period t. That is,

$$\mathcal{H}_t^{-1}\underline{\omega}_t = \{\omega \in \Omega \mid \underline{\omega}_t = \mathcal{H}_t\omega\}.$$

Note that for each t, $\{\mathcal{H}_t^{-1}\underline{\omega}_t \mid \underline{\omega}_t \in \Omega_t\}$ forms a partition of Ω. Clearly, if the first t components of ω^1 and ω^2 are identical, then $\omega^2 \in \mathcal{H}_t^{-1}(\mathcal{H}_t\omega^1)$. In this case, ω^1 and ω^2 are indistinguishable in period t, and nonanticipativity requires that $\mathcal{H}_t x(\omega^1) = \mathcal{H}_t x(\omega^2)$. We may use this notion to express the nonanticipativity constraints in terms of state variables $\{z_t\}_{t=1}^T$, where $z_t : \Omega_t \to \Re^n$. Then the state variable representation of the nonanticipativity constraints may be represented as:

$$x_t(\tilde{\omega}) - z_t(\mathcal{H}_t\tilde{\omega}) = 0 \qquad wp1, \quad \forall\, t. \tag{1a}$$

Note that these constraints will ensure that all scenarios that share a common history at period t yield a common solution in period t as well. Alternatively, we may express the nonanticipativity constraints using conditional expectations. That is, we note that if $\tilde{\omega}'$ is defined on the same probability space as $\tilde{\omega}$, then

$$x_t(\tilde{\omega}) - E[x_t(\tilde{\omega}') \mid \tilde{\omega}' \in \mathcal{H}_t^{-1}(\mathcal{H}_t\tilde{\omega})] = 0 \qquad wp1, \qquad \forall\, t \tag{1b}$$

will ensure that $x(\tilde{\omega})$ is nonanticipative. Either (1a) or (1b) may be used to represent the nonanticipativity requirements. There are additional representations as well, and the choice of representation will typically be guided by the solution method adopted. For the purposes of our algorithmic presentation, we will adopt the state variable representation,

(1a). Finally, for ease of exposition, we define the following extended valued function

$$\varphi(x,\omega) = \begin{cases} g(x,\omega) & \text{if } x \in X(\omega) \\ \infty & \text{otherwise.} \end{cases} \tag{2}$$

With this function, we may now specify a multistage stochastic program as follows:

$$\begin{array}{lllll} \text{Min}_x & E[\varphi(x(\tilde{\omega}),\tilde{\omega})] & & & \\ \text{s.t.} & x(\tilde{\omega}) - z_t(\mathcal{H}_t\tilde{\omega}) & = & 0 & wp1, \quad t = 1,\ldots,T. \end{array} \tag{P}$$

3. DUAL PROBLEM

As with the primal problem (P), the dual problem that we study is valid for both continuous and discrete random variables. In this section, we propose a stochastic analog of conjugate dual problems. One of the key features that distinguish these duals from their deterministic counterparts is the role played by multipliers associated with the nonanticipativity constraints (Wets [1975]).

Throughout our development, we will introduce a number of "variables" which correspond to measurable functions of random variables. For example, $\{x_t(\omega)\}_{\omega\in\Omega}$, or equivalently, $x(\tilde{\omega})$, is one such measurable function mapping Ω to $\Re^n$. We assume that the functions $x \in \mathcal{L}^\infty$ are essentially bounded and $\mathcal{P}$-measurable. Let $\mathcal{L}^1$ denote the space of functions that are integrable with respect to $\mathcal{P}$. Given $\xi \in \mathcal{L}^1$, it is convenient to define the following linear operation.

$$\xi \circ x = \int_\Omega \xi(\omega)^\top x(\omega)\mathcal{P}(d\omega). \tag{3}$$

Maintaining an eye toward the stochastic properties of the mathematical program under consideration, it is interesting to note that this inner product is equivalently stated as

$$\xi \circ x = E[\xi(\tilde{\omega})^\top x(\tilde{\omega})], \tag{4}$$

which we recognize as the expected value of the traditional counterpart from deterministic mathematical programming. Throughout our development, we use the representations in (3) and (4) interchangably.

Consider the function φ defined in (2), and its conjugate function

$$\varphi^*(\xi,\omega) = \sup_{x\in\Re^n}\{\xi^\top x - \varphi(x,\omega)\}. \tag{5}$$

Relying on the state variable representation of the nonanticipativity constraints, the following problem

$$\begin{array}{llll} \sup\limits_{\xi \in \mathcal{L}^1} & -E[\varphi^*(\xi(\tilde{\omega}),\tilde{\omega})] & & (D) \\ \text{s.t.} & E[\xi_t(\tilde{\omega}) \mid \tilde{\omega} \in \mathcal{H}_t^{-1}(\mathcal{H}_t\tilde{\omega}')] = 0 & wp1, \quad t=1,\ldots,T & \end{array}$$

is dual to (P), as indicated in the following result from Higle and Sen [1997].

Theorem 1 *Let $\varphi(\cdot,\tilde{\omega})$, as defined in (2), be a convex normal integrand, and assume that (P) has relatively complete recourse. Let v_p and v_d denote the optimal values of (P) and (D), respectively. Then*

(a) $v_p \geq v_d$.

(b) If (P) posseses an optimal solution denoted $(\hat{x},\hat{z})$, and $\partial\varphi(\hat{x}(\tilde{\omega}),\tilde{\omega})$ is nonempty (wp1), then there exists $\hat{\xi}(\tilde{\omega}) \in \partial\varphi(\hat{x}(\tilde{\omega}),\tilde{\omega})$, wp1, such that

$$E[\hat{\xi}_t(\tilde{\omega}) \mid \tilde{\omega} \in \mathcal{H}_t^{-1}(\mathcal{H}_t\tilde{\omega}')] = 0 \qquad \text{(wp1)},$$

where $\tilde{\omega}$ and $\tilde{\omega}'$ are defined on the same sample space. Furthermore, $-E[\varphi^*(\hat{\xi}(\tilde{\omega}),\tilde{\omega})] = v_d = v_p$.

Proof (See Higle and Sen [1997]).

It is interesting to note that (D) may be construed as a *stochastic* conjugate dual. Dual feasibility requires only that the conditional means of the dual vectors be zero at each decision point. For notational convenience, we let Ξ denote the set of dual feasible solutions. In the following section, we use this dual statement of the multistage stochastic programming problem to develop a cutting plane approximation of the problem.

4. A CUTTING PLANE ALGORITHM

Our algorithmic development depends on the dual representation, (D). Assuming the existence of an optimal solution, we restate the problem as

$$\begin{array}{lll} \text{Max}_{\xi \in \mathcal{L}^1} & E[\text{Min}_{x(\tilde{\omega})}\{\varphi(x(\tilde{\omega}),\tilde{\omega}) - \xi^\top(\tilde{\omega})x(\tilde{\omega})\}] & \\ \text{s.t.} & E[\xi_t(\tilde{\omega}) \mid \tilde{\omega} \in \mathcal{H}_t^{-1}(\mathcal{H}_t\tilde{\omega}')] = 0 & wp1, \quad t=1,\ldots,T \end{array}$$

Notice that in this representation, the scenarios are linked via the outer maximization, while the inner minimization is separable by scenario. As such, the inner minimization can be decomposed by scenario, with the outer maximization providing the coordination necessary to identify an optimal solution.

To begin, let

$$v(\xi,\omega) = \operatorname*{Min}_{x}\{\varphi(x,\omega) - \xi^\top x\} \leq \varphi(x,\omega) - \xi^\top x \qquad \forall\ x,\omega$$

with equality holding if $x \in \operatorname{argmin}_x\{\varphi(x,\omega) - \xi^\top x\}$. Thus, given $\xi(\tilde{\omega})$ and $x(\tilde{\omega})$, we have

$$E[v(\xi(\tilde{\omega}),\tilde{\omega})] \leq E[\varphi(x(\tilde{\omega}),\tilde{\omega})] - E[\xi(\tilde{\omega})^\top x(\tilde{\omega})] = E[\varphi(x(\tilde{\omega}),\tilde{\omega})] - \xi \circ x$$

with equality holding if $x(\tilde{\omega}) \in \operatorname{argmin}_{x(\tilde{\omega})}\{\varphi(x(\tilde{\omega}),\tilde{\omega}) - \xi(\tilde{\omega})^\top x(\tilde{\omega})\}$, with probability one. Note that $E[\varphi(x(\tilde{\omega}),\tilde{\omega})]$ is constant with respect to ξ. Thus, this affine bound provides the basis for a cutting plane approximation to the expected value of the inner minimization.

The Basic Scenario Decomposition method may be stated as follows:

Step 0. Initialize $\varepsilon > 0$, an error tolerance, is given.

$k \leftarrow 0$, $\xi_k(\omega) \leftarrow 0$ for all $\omega \in \Omega$, $v^k \equiv +\infty$.

Step 1. Solve the Subproblem $k \leftarrow k+1$.

Let $x^k(\omega) \in \operatorname{argmin}_x\{\varphi(x,\omega) - \xi^{k-1}(\omega)^\top x\}$ wp1.

Step 2. Update the Master Problem Let $\alpha^k = E[\varphi(x^k(\tilde{\omega}),\tilde{\omega})]$, and add the constraint $v \leq \alpha^k - \xi \circ x^k$ to the master problem.

Step 3. Solve the Master Problem Let (v^k,ξ^k) denote a solution to:

$$\begin{array}{llll} \text{Max} & v & & \\ \text{s.t.} & v \leq \alpha^j - \xi \circ x^j & & j = 1,\ldots,k \\ & \xi \in \Xi & & \end{array}$$

If $\alpha^k - \xi^k \circ x^k - v^k < \varepsilon$, then stop.
Otherwise, continue from Step 1.

One recognizes that the Basic Scenario Decomposition algorithm is Kelley's cutting plane method applied to the dual statement of the multistage stochastic program. As such, it follows that the iterates $\{\xi^k(\tilde{\omega})\}_{k=1}^{\infty}$ converge to an optimal solution. We note, however, that there are several impediments to the practical use of this method.

By way of example, note that in Step 1 of the method, we require that $x^k(\tilde{\omega})$ be an optimal solution with probablity one. If $\tilde{\omega}$ is a discrete random variable, this requires the solution of the indicated problem for all possible scenarios. Otherwise, it is necessary to obtain an optimal solution for all scenarios outside a set of probablity measure zero. Clearly, as the number of possible scenarios increases, each execution of Step 1 becomes a substantial task. Related to this is the fact in Step 3 of the method, the columns may be described as: $\{\xi(\omega)\}_{\omega \in \Omega}$. Again, as the number of scenarios increases, the column dimension grows quickly out of hand. Additionally, as with all cutting plane methods, the proliferation of cuts in the master program will become problematic. Because of these observations, as the number of scenarios increases, the ability to actually execute Step 3 diminishes.

Both of these problems are related to the specification of Ω. Of course, when dealing with large sample spaces, there is a natural tendency to look toward sample-based methods. Indeed, these have been successfully used in the solution of large scale two-stage stochastic programs. In particular, Stochastic Decomposition (SD) and its variants, which use a successive sampling scheme (as opposed to a fixed sample) is the only such method which is capable of identifying an optimal solution to the problem with probability one (see, for example Higle and Sen [1991, 1996a]). SD begins with a small sample size, and increases it as iterations progress. In this manner, Steps 1 and 3 may be executed rather quickly in the early interations, when one expects the iterates, (ξ^k) to be far from an optimal solution. As iterations progress, the information obtained via the sample becomes increasingly accurate, as do the resulting cutting planes and solutions. We note that safeguards are required to prevent cutting planes derived in the early iterations (i.e., when the sample-based information is most likely to be inaccurate) from interfering with the search for an improved solution. Within the context of the Scenario Decomposition method, note that the use of sampled observations of Ω will impact the definition of dual feasiblity. Moreover, as iterations progress and larger sample sizes are used, cut proliferation and the corresponding growth in the master problem column dimension becomes problematic.

Sampled Scenario Decomposition (SSD) is an algorithmic methodology based on the Basic Scenario Decomposition scheme described in this paper. SSD is designed specifically to address the issues identified above. As the name indicates, it is based on the use of sampled observations of $\tilde{\omega}$. In addition, as iterations progress it incorporates a row and column aggregation scheme designed to control the size of the master program. Additional features are also included. We refer the interested reader to Sen, Higle and Rayco [1999] for a full description of the method, its properties, and preliminary computational results.

5. CONCLUSIONS

In this paper, we have studied a dual approach to multi-stage stochastic programming problems. This development is closely tied to deterministic conjugate duality, with the only additional requirement in the stochastic case being the inclusion of constraints requiring the conditional expectation of dual multipliers to be zero. By interpreting dual multipliers as subgradients of the cost-to-go function, dual feasibility requires the natural optimality condition. This approach is not only elegant, but leads to algorithms in which we can avoid stage-by-stage recursions. The advantage of doing so is that we can address a larger class of problems, especially those in which the objective function is not separable by stage. This approach is also amenable to the use of sample-based algorithms, although the size of the master problems can become arbitrarily large. By using appropriate aggregation schemes, we are able to maintain approximations of reasonable size. Such computations are reported in a forthcoming paper.

References

[1] J.R. Birge (1985), Decomposition and partitioning methods for multistage stochastic linear programs, *Operations Research*, 22, pp. 989–1007.

[2] D. Cariño, T. Kent, D. Myers, C. Stacy, M. Sylrams, A. Turner, K. Watanabe and W. Ziemba (1994), The Russell–Yasuda Kasai Model: An asset/liablity model of a Japanese insurance company using multi–stage stochastic programming, *Interfaces*, 24, pp. 29–49.

[3] M.A.H. Dempster (1981), The expected value of perfect information in the optimal evolution of stochastic systems, in M. Arato, E. Vermes, and A.V. Balakrishnan (eds.), *Stochastic Differential Systems*, Lecture Notes in Information and Control, 36, Springer, Berlin, pp. 25–40.

[4] M.A.H. Dempster (1988), On stochastic programming II: Dynamic problems under risk, *Stochastics*, 25, pp. 15–42.

[5] H.G. Gassmann (1990), MSLiP: A computer code for the multistage stochastic linear programming problem, *Mathematical Programming*, 47, pp. 407–423.

[6] J.L. Higle and S. Sen (1991), Stochastic Decomposition: An Algorithm for Two Stage Linear Programs with Recourse, *Mathematics of Operations Research*, 16, no. 3, pp. 650–669.

[7] J.L. Higle and S. Sen (1996a), *Stochastic Decomposition: A Statistical Method for Large Scale Stochastic Linear Programming*, Kluwer Academic Publishers, Dordrecht.

[8] J.L. Higle and S. Sen (1996b), Duality and statistical tests of optimality for two stage stochastic programs, *Mathematical Programming*, (Series B), v75, pp. 257–275.

[9] J.L. Higle and S. Sen (1997), Multi–stage Stochastic Programs: Duality and its Implications, submitted to *Mathematical Programming.*

[10] R.T. Rockafellar and R.J-B. Wets (1991), Scenarios and policy aggregation in opimization under uncertainty, *Mathematics of Operations Research*, 16, pp. 119–147.

[11] S. Sen, R.D. Doverspike and S. Cosares (1994), Network planning with random demand, *Telecommunications Systems*, 3, pp. 11–30.

[12] S. Sen, J.L. Higle and B. Rayco (1999), A Scenario Stochastic Decomposition Algorithm for Multistage Stochastic Programming, SIE Department, University of Arizona, Tucson, AZ 85721.

[13] R.J-B. Wets (1975), On the relation between stochastic and deterministic optimization, in A. Bensoussan and J.L. Lions (eds.), *Control Theory, Numerical Methods, and Computer Systems Modeling*, Lecture Notes in Economics and Mathematical Systems, 107, pp. 350–361.

[14] R.J-B. Wets (1989), Stochastic Programming, in: G.L. Nemhauser, A.H.G. Rinooy Kan, and M.J. Todd (eds.) *Handbook of Operations Research: Optimization*, North-Holland, Amsterdam, pp. 573–629.

A MODEL WHICH INTEGRATES STRATEGIC AND TACTICAL ASPECTS OF FOREST HARVESTING

Alastair McNaughton
Division of Science and Technology
Tamaki Campus, University of Auckland,
Private Bag 92019, New Zealand
a.mcnaughton@auckland.ac.nz

Mikael Rönnqvist
Division of Optimisation,
Linkoping University,
58183 Linkoping, Sweden
miron@math.liu.se

David Ryan
Department of Engineering Science,
University of Auckland,
Private Bag 92019, Auckland, New Zealand
d.ryan@auckland.ac.nz

Abstract The forest harvesting problem involves the construction of a schedule for felling the individual blocks of trees which comprise a large commercial plantation. A strategic model sets long-term harvesting goals in terms of total area to be cut each year, but fails to identify individual blocks. A tactical model produces a short-term schedule of actual blocks. Until recently, most planning by forest managers has involved these as two separate models, often resulting in contradictory recommendations. We present an integrated model, which embraces both strategic and tactical decisions, which can be solved by optimisation methods.

This model achieves a detailed formulation by means of a non-standard column generation structure. The solution algorithm solves the resulting relaxed linear program formulation. This is then combined

M.J.D. Powell and S. Scholtes (Eds.), *System Modelling and Optimization: Methods, Theory and Applications.*

with constraint branching techniques to obtain the desired optimal integer solution to the integrated model.

Numerical output from a case study involving Whangapoua Forest in Coromandel, New Zealand, will be used to demonstrate the performance level of this algorithm in terms of the quality of optimisation, the size of the data base and the computational time. However, the main emphasis will be placed on the theoretical aspects of the model and its solution algorithm as the approach may well be transferable to other, quite distinct, applications.

1. INTRODUCTION TO A CASE STUDY

The Coromandel Peninsula is a magnificent bush-clad region of dissected hill country fringed by beautiful sandy coves and estuaries located in Northern New Zealand. Although the climate is generally warm with enough gentle rain to promote ideal growing conditions for trees, the location is prone to intense rainfall events and the associated risk of severe mass movement. Many tourists visit the Coromandel, making the preservation of the natural environment a top priority. The local community is deeply involved in conservation issues. The Whangapoua Forest occupies 7365 hectares on this peninsula. The forest is operated by Ernslaw One Ltd. The management of this company use a rotation of about 28 years. As a consequence, a horizon of about 30 years is needed for the long term strategic harvest planning. The strategic plan contains many constraints. Typical of these are the constraints resulting from the demanding environmental restrictions which limit the proportion of each catchment that may be harvested over any consecutive 5-year period. None of the strategic constraints concern specific blocks of trees. A second type of planning, called tactical planning, is needed for this purpose. Such detailed area-specific planning is required for a much shorter period of 2 to 5 years. At present the forest contains 145 surveyed mature blocks which contain trees representing about 30 crop types. This latter type of planning involves decisions on when each specific block will be felled. Each block must be felled as a single harvesting operation. It is not permitted to partially cut a block and then return later to complete the harvest. Further environmental and operational restrictions apply to this tactical planning. These include the so called *cable logging adjacency* requirements, which identify specific pairs or cliques of blocks and impose restrictions on when each member may be harvested in relation to the other members of the clique.

At present, harvest planning in the Whangapoua Forest is done in the following manner. Linear programming software is used to get an optimal strategic plan, followed by 2 or 3 days of manual activity to convert this into a feasible tactical plan for the next 2 or 3 years. The connection between the strategic plan and the tactical plan is very tenuous, so the optimal aspects of the strategic plan are lost. When the process is repeated 1 or 2 years later a totally different optimal strategic plan appears! Also forest managers are worried about issues of sustainability and long-term block-feasibility. For these reasons the Forest Research Institute was approached for advice, and our team, seeing this as an ideal research case study, became involved from early 1994.

2. SALIENT CHARACTERISTICS

The following distinctive characteristics of this application impact substantially on the model which follows. There are distinct short and long term planning tasks. A great many complex constraints are present in both situations. Mixed integer programming is required with large numbers of integer variables. Many of the management decisions are inter-related, such as road construction and removal of harvested logs. Terrain and the road network provide a partition of the forest into significant smaller units. There are problems associated with the sheer size of this application. The actual harvesting decisions concern a moderate number of blocks, 145, to be harvested over a small number of years, 6. The excessive size of the application results from the combinatorial aspects of this situation. The model and the algorithm which follows may be transferable to other applications with similar characteristics.

3. HISTORICAL SETTING

Mathematical optimisation techniques have been applied to the solution of forest harvesting problems for over 30 years. During this time the nature of the problem has evolved. The growing power of the conservation movement has resulted in the inclusion of more and more constraints of greater severity and greater complexity. The physical size of planning tasks has greatly increased. The recent global drop in timber prices has caused minimal profit margins and a compelling need to attain optimal, or near-optimal solutions if a competitive edge is to be maintained.

It is helpful to consider the literature in three main categories. First consider the optimisation stream. This includes basic papers applying linear programming to the strategic plan, such as that by Manley, Threadgill and Wakelin [9]. Realising that the tactical plan also had to be dealt with, attempts were made to extend a linear programme into a mixed integer programme. One of the best of these was that by Kirby, Hager and Wong [8], who developed the integrated resources planning model, (IRPM). This is a mixed integer formulation of an integrated forest harvesting model. The most singular achievement of IRPM is that it does find the optimal solution. Other researchers recognised the quality of Kirby's work but noted the limitations of IRPM to only small applications. O'Hara, Faaland and Bare [15] report IRPM can only deal with at most 50 integer variables along with a matrix of 250 rows and 250 columns.

One positive development has been to attempt to reduce the number of constraints by constraint aggregation. A number of papers such as those by Meneghin et al. [10], and Murray and Church [12] indicate that currently this is an approach widely regarded as promising. The present model includes constraint aggregation techniques in certain places.

When it became apparent that there were extreme difficulties in constructing an integrated model capable of the optimisation of an operational-sized application, a number of publications appeared advocating a hierarchical solution. These include Hof and Baltic [4], Weintraub and Cholaky [21], and Hof [5]. This approach involves first optimising the strategic plan, and then incorporating this strategic solution as a constraint in a separate optimisation of the tactical plan. Although the optimisation attained in this way will always be inferior to that obtained from an integrated model, the quality of the solution may be quite acceptable provided there are not too many constraints applicable to the tactical plan. Such an approach contains the assumption that, due to the aggregated nature of the variables used in the optimal strategic solution, a feasible solution will always be obtainable to the constrained tactical plan. In practice, this is almost never true, especially as the present trend continues for governments to impose ever more harsh area-sensitive environmental constraints on the tactical plan. As a consequence, the implementation of hierarchical solution algorithms has generally been accompanied by an observable deterioration in objective value between the strategic optimal solution and the final area-sensitive tactical solution. Daust and Nelson [1] have recorded this as a loss of between 4 percent and 20 percent. Such findings question the credibility of the optimisation process in these hierarchical models. An integrated model will avoid this problem.

The second category of forest harvesting research may be termed the heuristic stream. Here attention is focused on the integer variables, with many of these models relating exclusively to the tactical plan. A representative paper here would be that by O'Hara, Faaland and Bare [15]. They use a Monte Carlo method in which a subtle weighting is introduced to facilitate random generation of good quality solutions. The performance achieved is the heuristic solution, not necessarily optimal, of a problem containing 242 blocks, with a tactical horizon of 3 years, in 15 minutes on a VAX 8700.

As one could well anticipate, this stream has indulged in ever more complicated heuristic methods. For example, a recent paper by Yoshimoto, Brodie and Sessions [24] proposes an intricate heuristic procedure in which the forest is partitioned into numerous sub-problems. Each subproblem is solved independently as a regional optimisation problem. A measure of the performance of this method is given by the solution of an application involving 10 time periods in 2 hours on a 486 PC.

Weintraub et al. [23] present a heuristic method they term *heuristic integer planning*, (HIP). This is based on the IRPM model of Kirby [8]. Kirby's optimisation problem has the disadvantage of producing output at the relaxed linear programme stage in which many of the integer variables have fractional values. HIP is a method that assigns some of these variables to value 0 or 1. The problem is then re-optimised and more fractional variables are assigned integer values until an integer solution is obtained. The process is classified as heuristic because the assignment process is heuristic, being the enforcement of a set of rules. These rules may be either the outcome of the user's intuition or the trend observed from numerical trials. As the integer solution is obtained after 7 to 12 iterations it is clear that many variables must be assigned integer values every iteration. This model is capable of solving an application involving 44 road segments, and 28 blocks over a 3-period tactical horizon. A total of 190 integer variables is involved with the complete matrix containing 638 rows and 1071 columns. The solution process takes about 20 minutes.

A comparison such as that by Nelson and Brodie [13] of a heuristic with an optimisation method is very insightful. In this paper a forest harvesting application involving 45 blocks, 52 road segments and 3 time periods, that is 291 integer variables, is first solved to optimality by a mixed integer programme algorithm. Using two computers, a 80386PC and a hyper-LINDO PC this is achieved in 60 hours. The same application is then solved by a Monte-Carlo integer programming (MCIP) heuristic in 9 hours, but the best objective value obtained is 97 percent of the optimal value obtained from the optimisation method.

Sessions and Sessions [17] have developed heuristic software that produces a tactical plan without any reference to the strategic plan. The tactical model is formulated and solved first without including adjacency constraints. Monte Carlo methods are used to generate large numbers of candidate solutions that are then checked against adjacency constraints to determine which, if any, are feasible. The resulting software is named SNAP II. John Sessions has collaborated with Nelson and Brodie [14] to investigate techniques to use SNAP II in association with the linear programming strategic software FORPLAN. Since these two packages are quite distinct an integrated optimisation is extremely difficult.

The third and final division consists of multiple-stage methods. These also deal primarily with the tactical plan and the difficulties relating to the integer variables contained in it. A number of recent papers recommend methods that consist of several stages, with different methods at each stage. One of the best of these would be that by Weintraub, Barahona and Epstein, [22] who suggest a column-generating technique to solve a forest harvesting model. Unfortunately, even after some major simplification assumptions, the formulation leads to a series of very difficult and complex sub-problems for which an optimal solution cannot be attained. Probably the most serious weakness of this type of approach is the extreme difficulty of driving a multi-stage method close to optimality. The task of trying to establish a sound theoretical foundation for such a structure is daunting. Weintraub et al report successful implementation of their algorithm on a number of operational Chilean forests, but unfortunately do not include the computing times involved.

Another significant multiple-stage model is that of Hoganson and Borges [7], who apply dynamic programming techniques to a multitude of subproblems. Other recent formulations include those of Snyder and ReVelle [19, 20]. These contain interesting mathematical insights, but have the weakness that they are presented in the context of hypothetical integerised grid data. Hof, Bevers and Pickens [6] utilise a similar digitised grid, but they show how this is derived from a real map of an actual forest. Naturally their grid cells do not correspond to actual cutting blocks, a factor that must cause significant difficulties when any adjacency constraints are needed. Hochbaum and Pathria [3] also use a grid structure for their model. It is noteworthy that they attempt to negotiate these difficulties involving adjacency constraints, by penalising, rather than prohibiting, infringements of adjacency restrictions.

Sherali, Adams and Driscoll [18] present a helpful generalised theoretical model which stresses the importance of attaining a tight linear programming relaxation, which curiously is a feature of the model to be presented below.

The most salient aspect of this substantial body of research literature is its extreme diversity. This indicates that despite much excellence and insight, not one of the many and varied solution methods has won any degree of wide-spread acceptance. This is a strong indicator that the forest harvesting problem is at present unsolved in an optimisation sense.

One extremely important related problem is that of scheduling the flights of airline crews into tours of duty, along with the related problem of rostering crews to these tours of duty. In simple terms, the models involved are huge mixed integer structures. Each possible roster for a crew member is represented by a column. Binary variables are used to determine which crew member is assigned to which roster. When compared with forest harvesting problems there are fundamental differences. The crew scheduling problem belongs to the class known as set-partitioning problems, whereas forest harvesting problems do not. However, the important factor is that recent major advances have occurred in the technology of solving set-partitioning problems. These involve column generation in association with constraint branching. Desrochers et al [2] and Ryan [16] have developed these methods so as to obtain optimisation solutions to applications previously thought too large for anything other than a heuristic treatment. As a consequence, very definite and significant practical improvements are now available to airline management. The solution algorithm of McNaughton [11] applied constraint branching to the forest harvesting problem for the first time.

4. MAIN FEATURES OF THE PROPOSED MODEL

An integrated model has been chosen so as to get maximum benefits from optimisation, and to avoid conflict between tactical and strategic plans. This means that the strategic part of the plan, which involves continuous variables, will be joined by linking constraints to the tactical part of the plan which contains binary integer variables. An objective function will be used which represents the present net worth of the forest, with revenue and costs from planned future harvests discounted appropriately. It is significant that the revenue part of the objective is associated with the strategic part of the model, while the costs are mostly generated by the tactical part. The solution will then be obtained by a single optimisation process acting on the whole model. We will allow 30 years for the strategic plan and 6 years for the tactical plan, using a time period of 1 year. The strategic part of the model will contain about 1820 continuous variables and about 1120 constraints. The tactical part of the model will contain about 1560 binary integer variables and about 1100 constraints. Although it is not possible to list every one of these constraints in detail, the most significant of them will be displayed and explained in Section 5. Road decision variables will be included in the tactical plan. Many more binary integer variables will be introduced by column generation during the solution process as will be explained below.

5. SOME MODEL DETAIL

The variables in the strategic part of the model, x_{cet}, are continuous, with

x_{cet} = hectares of croptype c established in year e harvested in year t.

These are used to construct constraints in the strategic plan, which has the structure of a linear programme. For example, there may be a requirement that no more than 200 hectares be felled in a year. This would involve the maximum area constraint

$$\sum_{ce} x_{cet} \leq 200. \tag{1}$$

This constraint, as with the others which follow, is representative of a large block of constraints. In this case a separate constraint is required for each establishment year for each croptype in the forest. The other constraints in the strategic part of the model are mostly of equally simple formulation. These concern matters such as acheiving a non-declining yield and ensuring that no more of a given croptype is harvested than is currently available. Since the structure of this part is that of a linear programme, it poses no significant difficulties in the solution process, other than those associated with its size, comprising as it does of many hundreds of constraints.

For the tactical part of the model the formulation is much more complex. The following key concept is of pivotal importance in the model as it provides the definitions for the principle decision variables. A *road harvest plan* is a set of tactical decisions all pertaining to the harvesting of blocks on one given road. These may span the entire tactical horizon. Only one road harvest plan will be permitted to be operating in the solution for each road, although many alternative road harvest plans may be generated in the model. Here is an example of a road harvest plan:

On road 14, *cut block* 2 *in year* 3, *block* 3 *in year* 2, *and block* 5 *in year* 6.

In the linear programming matrix representation of the model, a unique column with an associated integer variable will be associated with each road harvest plan. In this case the variable g_{jtn} is used.

$$g_{jtn} = \begin{cases} 1 & \text{if the } n\text{-th plan on road } j \text{ is chosen,} \\ & \quad \text{harvesting starting in year } t, \\ 0 & \text{if the } n\text{-th plan on road } j \text{ is not chosen.} \end{cases}$$

For each road, one of these variables will represent a null harvest plan in which no harvesting occurs during the planning horizon.

Other columns, also with associated integer variables, represent road construction tasks. In this case the variable r_{jt} is used.

$$r_{jt} = \begin{cases} 1 & \text{if road } j \text{ is constructed in year } t, \\ 0 & \text{if road } j \text{ is not constructed in year } t. \end{cases}$$

The structure of the integrated model ensures that an optimal integer solution to the associated road problem is automatically produced without any explicit manipulation of these parts of the matrix, provided an optimal integer solution is obtained for the road harvest plan variables.

Two of the tactical constraints have a pivotal role in the solution algorithm and need to be studied closely. Firstly a plan constraint is required for each road. This will ensure one and only one road harvest plan is selected for a particular road, j_o, say. These constraints are of the form:

$$\sum_t \sum_n g_{j_o tn} = 1. \tag{2}$$

Secondly, road construction constraints are used to allow harvesting on road j_o only after this road has been constructed. If n^* represents the null harvest plan, then these are of the form:

$$-\sum_{\bar{t}=1}^{t} r_{j_o \bar{t}} + \sum_{n \neq n^*} g_{j_o tn} \leq 0. \tag{3}$$

Other types of tactical constraints are of interest too. Road sequential constraints are required to ensure that a continuous access route is available from the harvesting site to the timber mill. Suppose either road j_1 or j_2 is needed to provide access to road j_3. Then the necessary constraints will be of the form

$$-\sum_{\bar{t}=1}^{t} \left(r_{j_1 \bar{t}} + r_{j_2 \bar{t}} \right) + r_{j_3 t} \leq 0. \tag{4}$$

The construction of the linking constraints is of major significance. Recall that the model consists of two approximately equal parts, one generating revenue, the other cost. They involve quite different variables and are completely disjoint apart from the linking constraints. If these are used to impose a strict equality between the croptype areas, designated by the continuous variables, x_{cet}, in the strategic part, and the areas of the appropriate specific blocks chosen for harvest, represented by expressions involving the integer variables, g_{jtn}, in the tactical part, then the model becomes too stiff. In this case the performance of the solution algorithm is greatly impaired. On the other hand, if slack variables are introduced the solution algorithm runs fast but the integrity of the optimisation is violated. Instead new continuous variables, s_{cet}, called *overlap* variables, are introduced.

$s_{cet} =$ the area of croptype c established in year e harvested in year $t+1$ which is located in blocks where harvesting will commence in year t.

Let the constant a_{cejk} represent the area of croptype c established in year e located in block k on road j. Let G_{jtk} be the set of all g_{jtn} which involve the harvesting of block k on road j in year t. The linking constraints will then be of the form

$$-x_{cet} - s_{cet} + s_{ce[t-1]} + \sum_{jk} a_{cejk} \sum_{G_{jtk}} g_{j\bar{t}n} = 0. \tag{5}$$

The optimisation process, in association with the discounted nature of the objective and the requirement that the tactical variables be integer, will ensure that only a very few of these overlap variables, s_{cet}, are non-zero in the final solution.

Finally there will be many adjacency constraints which record restrictions on the permitted harvesting time for a block in relation to that of a given neighbouring block. As an example suppose that for technical reasons the logging of block a on road j_a required the logging of block b on road j_b within at most 2 years. This occurs if a harvesting technique called *cable logging* is being used and certain local terrain conditions apply. Let the tactical horizon be 3 years. Then the necessary adjacency constraint would be

$$\begin{aligned} -1 \;\leq\; & 3\sum_{G_{j_a 1a}} g_{jtn} - 3\sum_{G_{j_b 1b}} g_{jtn} + 2\sum_{G_{j_a 2a}} g_{jtn} \\ & -2\sum_{G_{j_b 2b}} g_{jtn} + \sum_{G_{j_a 3a}} g_{jtn} - \sum_{G_{j_b 3b}} g_{jtn} \;\leq\; 1. \end{aligned} \tag{6}$$

6. SOLUTION ALGORITHM

During the solution algorithm the difficulties associated with the integer variables are circumvented by considering the relaxed linear programme in which each binary variable is considered as a continuous variable, bounded to the interval [0,1]. This allows extensive use of standard linear programming techniques throughout the solution process. In addition two other processes are involved.

The first of these is column generation. If every possible road harvest plan was to be represented explicitly by a column in the matrix representation of the linear programme, then the size of the matrix would be too large. So at first only a few representative columns are included. In the present model these correspond to elementary road harvest plans, e_{jtk}, which each concern the harvesting of just one single block in a given year.

$$e_{jtk} = \begin{cases} 1 & \text{if block } k \text{ on road } j \text{ is cut in year } t, \\ 0 & \text{otherwise.} \end{cases}$$

The variables e_{jtk} are a subset of the more general g_{jtn}. Each iteration new columns are added to the linear programme matrix, each representing a road harvest plan. Each new column will be a composition of 2 or more of these elementary road harvest plans, which is selected to improve the current objective value, until an optimal relaxed linear programme solution is obtained. In this model a non-standard type of column generation has been used which delivers a high performance level. The role of the e_{jtk} variables in this column generation will be explained in the next section.

The second process is constraint branching. This is a special type of branch and bound. It is used to remove fractional values from the solution to the relaxed linear programme. It works much faster than traditional variable branching, but can only be used if the model has been constructed appropriately.

During the implementation of the solution algorithm these two processes work together. First a phase of column generation is used to solve the relaxed linear programme. Then a suitable constraint branch is implemented. The problem is then re-optimised with a further phase of column generation as the new constraint branch may well have made certain new columns desirable. The rapidity of the column generation process makes this easy. If the new relaxed linear programme has an integer solution then we stop. Otherwise another constraint branch is chosen and the process continues iteratively.

7. TECHNICAL DETAILS OF THE COLUMN GENERATION

When a linear programme is being solved by the revised simplex method, a vector called the reduced cost is obtained. The reduced cost of a basic variable is always 0. So it is only necessary to define the reduced cost for the non-basic variables. Let the linear programme be of the form

$$\text{minimise} \quad c^T x \quad \text{subject to} \quad Ax \leq b,\ 0 \leq x.$$

Let B be the submatrix of basic columns of A, c_{B} be a vector of objective coefficients of the basic variables, A_{N} be the submatrix of non-basic columns of A, and c_{N} be a vector of objective coefficients of the non-basic variables. The reduced cost vector, rc, is defined by

$$rc = c_{\mathrm{N}}^T - c_{\mathrm{B}}^T B^{-1} A_{\mathrm{N}}. \tag{7}$$

In the revised simplex method a negative component of the reduced cost vector indicates that the corresponding variable may be added to the basis without causing a deterioration of the objective value. A reduced cost vector which is entirely non-negative is taken as the test for optimality.

When a column generation method is being used, the matrix A contains only a representative sample of all the many possible columns the application permits. In this case the reduced cost may be evaluated for any of the other columns which are being considered as possible entering columns. Once again, a negative reduced cost indicates that this new column may be added to matrix A, in fact to the basis in A, without causing a deterioration of the objective value. If it can be shown that no entering column with a negative reduced cost exists, then this may be used as a test for optimality. Unfortunately, the definition in Equation 7 is unsuitable for use in a column generation context, as the set of all possible entering columns is not explicitly accessible. Also the task of evaluating the reduced cost for every such column would generally be unacceptably tedious.

Most column generation methods deal with this problem by constructing a subproblem designed to produce an entering column with a negative reduced cost. These subproblems are of varied types, depending on the nature of the application, but are generally complicated and difficult to solve. A good example of this has already been noted in the work of Weintraub et al [22]. In the present model, information obtained from the solution to the current linear programme is used to determine the minimum reduced cost of any entering column. The solution to a linear programme includes both a value of the reduced cost variable, rc, associated with each variable, and also a value of the dual variable, π, associated with each constraint. A vector of dual variable values may be defined by

$$\pi = c_{\mathrm{B}}^T B^{-1}. \tag{8}$$

The dual variables and the reduced cost variables are related by the following equation, where $A = (a_{ij})$, a matrix with n rows.

$$rc_j = c_j - \sum_{i=1}^{n} a_{ij}\pi_i. \tag{9}$$

Equation 9 allows the reduced cost of any entering column to be determined. Unfortunately it requires the new column to be constructed explicitly before the reduced cost can be computed. This would be unsatisfactory in view of the very large number of possible entering columns. So a refinement of Equation 9 will be derived which will allow a reduced cost to be found without having to construct the new column.

Recall that the initial A matrix contains one column for each elementary road harvest plan. Each entering column will be a composite road harvest plan involving the harvesting of 2 or more blocks on the same road. When such a composite column is formed, most entries will be a simple vector addition of the corresponding entries in the appropriate elementary columns. If this were true for all entries, then the reduced cost of the entering column would be merely the sum of the known reduced costs of these elementary columns. However, in the case of the plan constraints, and any trigger constraints present such as the road construction constraints, the composite column is not formed as a sum, but rather as the maximum of the set of corresponding entries.

Consider the column corresponding to a road harvest plan g_{jtn}. From Equation 2 this column will contain an entry of 1 in the plan constraint associated with road j. Let the associated dual variable be π_j. Similarly, from Equation 3 this same column will contain an entry of 1 in each of the road construction constraints associated with road j from year t to t', where t' is the end of the tactical planning horizon. Let the dual variables associated with these road constraints be π_{jt}.

Now consider $\overline{rc}(g_{jtn})$, the reduced cost of a column, corresponding to a road harvest plan g_{jtn}, in which the entries for the plan constraints and the road construction constraints are removed. Since this truncated column is the vector sum of the appropriate truncated elementary columns, $\overline{rc}(g_{jtn})$ is the sum of the reduced costs of these truncated elementary columns.

$$\overline{rc}(g_{jtn}) = \sum_{e_{jtk} \sim g_{jtn}} \overline{rc}(e_{jtk}). \tag{10}$$

Also, it follows from Equation 9 that the truncated reduced cost for an elementary column, $\overline{rc}(e_{jtk})$, is readily computed by

$$\overline{rc}(e_{jtk}) = rc(e_{jtk}) + \pi_j + \sum_{\bar{t}=t}^{t'} \pi_{j\bar{t}}. \tag{11}$$

Let g_{jtn} represent a composite plan with block k_i harvested in year t_i. Observe $t = \min\{t_i\}$. Then

$$\begin{aligned} rc(g_{jtn}) &= \overline{rc}(g_{jtn}) - \pi_j - \sum_{\bar{t}=t}^{t'} \pi_{j\bar{t}} && \text{(as in (11))} \\ &= \sum_{e_{jt_ik} \sim g_{jtn}} \overline{rc}(e_{jt_ik}) - \pi_j - \sum_{\bar{t}=t}^{t'} \pi_{j\bar{t}} && \text{(from (10))} \\ &= \sum_{e_{jt_ik} \sim g_{jtn}} \left(rc(e_{jt_ik}) + \pi_j + \sum_{\bar{t}=t_i}^{t'} \pi_{j\bar{t}} \right) - \pi_j - \sum_{\bar{t}=t}^{t'} \pi_{j\bar{t}}. \end{aligned} \tag{12}$$

Equation 12 allows the minimum reduced cost for any entering column to be determined merely by scanning the known values of the reduced costs of the elementary columns and the dual variables of the appropriate trigger constraints. Any elementary column excluded by the active constraint branches is omitted from this scan. If this minimum value is negative, then the appropriate new column can be easily constructed and added to the matrix. If this minimum value is non-negative, optimality has been attained.

8. TECHNICAL DETAILS OF THE CONSTRAINT BRANCHING

Although r_{jt} and g_{jtn} are all binary integer variables, during the solution process the problem is treated as just one large relaxed linear programme. This is necessary whenever column generation techniques are to be applied. Integer solutions are obtained by a branch and bound process. Until recently, this was generally done by a process called *variable branching* in which a single binary variable was adjusted to a value of 1, on the 1-branch, or 0, on the 0-branch. The use of variable branching results in a slow convergence to an integer solution with a massive binary tree requiring to be searched.

Instead of variable branching, the present algorithm uses constraint branching, similar to that developed by Desrochers et al [2] and Ryan

[16]. In the forest harvesting application this entails using the sets G_{jtk} which have already been introduced in relation to the linking constraints of Equation 5 and the adjacency constraints of Equation 6. For example, a typical branching node consists of a 0-branch

$$\sum_{t \leq T} \left[\sum_{G_{jtk}} g_{j\bar{t}n} \right] = 0,$$

and a 1-branch

$$\sum_{t \leq T} \left[\sum_{G_{jtk}} g_{j\bar{t}n} \right] = 1.$$

Thus the 0-branch prevents the adoption of any road harvest plan in which block k is felled by year T. The 1-branch requires that one of these harvest plans must be chosen, without specifying exactly which of all the many eligible plans this will be. One compelling reason for the use of constraint branching is the dramatic improvement in computational time which results. However, this technique cannot be applied unless the model has been formulated in an appropriate manner. In the present case the necessary sets of road harvest plans, G_{jtk} , have been carefully built into the model for this purpose.

A most helpful enhancement of constraint branching concerns what may be called *allied constraint branches.* These involve the identification and automatic implimentation of any other constraints that are in some way logical consequences of the chosen constraint branch. For example, in forest harvesting an operational requirement may force a pair of adjacent blocks to be felled within two years of each other, as in Equation 6. Let the blocks be a and b on roads j_a, and j_b respectively. In this case if the branch

$$\sum_{t \leq T} \left[\sum_{G_{j_a ta}} g_{j_a \bar{t}n} \right] = 1$$

is applied, then so too should be the allied branch

$$\sum_{t \leq T+1} \left[\sum_{G_{j_b tb}} g_{j_b \bar{t}n} \right] = 1.$$

Effort spent searching out allied constraint branches is amply rewarded by very significant reductions in computational time.

Another good feature of this constraint branching procedure is that it involves the use of only one type of decision node. It is not necessary to impose separate branches using the road variables, as the model has been constructed in such a way that once all the sets of variables designated by G_{jtk} have been forced to integer sums then all other non-continuous components will automatically assume integer values. As a result of this the preferred direction in the branch and bound process is readily apparent, allowing a good quality integer solution to be attained at a relatively shallow depth.

Applications which involve column generation almost always require the use of branch and bound to obtain an integer solution. Consequently, the evaluation of a column generation technique should take into account the associated branch and bound process. In the forest harvesting application, the value of the column generation would be largely wasted without the constraint branching. During the implementation of the solution algorithm both these processes work together with alternating phases of column generation and constraint branching as described in Sections 6 and 7. The quality of the resulting solution will next be presented.

9. PERFORMANCE LEVELS ATTAINED IN CASE STUDY

The main output from this case study includes 5 complete years of a detailed cutting plan, a road construction schedule to match, and a harmonised 30 year strategic plan. An optimal integer solution is obtained at a depth of about 80 nodes, after about 6 minutes computational time on a solaris-2.6 (sparc) computer, with an objective value of 99.95 percent that of the relaxed linear programme.

The robustness of the model has been tested by numerical trials in which various parameters of the problem have been increased. For example, the number of years in the tactical plan has been increased from 3 to 10. Also the number and type of adjacency constraints has been varied. From 0 to 132 cable logging adjacency constraints of the form given in Equation 6 have been imposed. Also, up to 144 alternative *green-up* adjacency constraints have been used. These prevent the harvesting of adjacent blocks within a *green-up* period of 5 years. Such modifications result in changes to the objective value, but in each case the optimal objective value remains within half of one percent of that of the corresponding relaxed linear programme. This indicates that this is a tight linear programme relaxation, as advocated by Sherali et al [18].

In all these trials the computational times lie between 6 and 8 minutes. This computational level should be compared with that of traditional MIP models as already reported in Section 3. Such models can only be implemented for very small applications as noted by O'Hara et al [15]. The present Whangapoua Forest case study is far too large to model as a traditional MIP formulation. Such a model would require explicit representation of all possible road harvest plans which in this case would be many millions. It is significant that the present model not only optimises large r problems than those mentioned in Section 3, but it also requires much less computational time.

10. CONCLUSIONS

The idea of developing an integrated model has been vindicated. The use of column generation in association with constraint branching has provided the technological power to achieve this end. The advantages of an integrated model that addresses both tactical and strategic issues are immediately apparent from the results of the case study presented above. When these are compared with the performance levels attained by other research workers addressing similar problems, as outlined in Section 3, it is seen that significant advances have been made. High quality objective values have been attained. Also the solution time required is very short. Application size appears to present no problem, with the appropriate numerical trials indicating no undue escalation of either elapsed user time, or memory requirements, when the size of the model is increased.

The most significant line of associated research concerns the development of models for other applications with characteristics similar to those noted in Section 2.

Acknowledgments

This research is supported by the Forest Research Institute, Private bag 3020, Rotorua, New Zealand.

References

[1] D.K. Daust and J.D. Nelson (1993), Spatial Reduction Factors for Strata–Based Harvest Schedules, *Forest Science*, 39, pp. 152–165.

[2] M. Desrochers, J. Desrosiers and M. Solomon (1992), A New Optimisation Algorithm for the Vehicle Routing Problem with Time Windows, *Operations Research*, 40, pp. 342–354.

[3] D.S. Hochbaum and A. Pathria (1997), Forest Harvesting and Minimum Cuts: A New Aroach to Handling Spatial Constraints, *Forest Science*, 43, pp. 544–554.

[4] J. Hof and T. Baltic (1991), A Multilevel Analysis of Production Capabilities of the National Forest System, *Operations Research*,

39, pp. 543–552.

[5] J. Hof (1993), *Coactive Forest Management*, Academic Press, San Diego.

[6] J. Hof, M. Bevers and J. Pickens (1996), Chance–Constrained Optimisation with Spatially Autocorrelated Forest Yields, *Forest Science*, 42, pp. 118–123.

[7] H.M. Hoganson and J.G. Borges (1998), Using Dynamic Programming and Overlaing Subproblems to Address Adjacency in Large Harvest Scheduling Problems, *Forest Science*, 44, pp. 526–538.

[8] M. Kirby, W. Hager and P. Wong, (1986), Simultaneous Planning of Wildlife Management and Transportation Alternatives, *TIMS Studies in the Management Sciences*, 21, pp. 371–387.

[9] B.R. Manley, J.A. Threadgill and S.J. Wakelin (1991), *Alication of FOLPI : A Linear Programming–based Forest Estate Model*, Ministry of Forestry, FRI Bulletin 164, New Zealand.

[10] B.J. Meneghin, M.W. Kirby and J.G. Jones (1988), An algorithm for writing adjacency constraints efficiently in Linear Programming Models, in *Proceedings of the 1988 symposium on Systems Analysis in Forest Resources, USDA For. Serv. Gen. Tech. Rep.*, pp. 46–53.

[11] A.J. McNaughton (1998), *Long–term Scheduling of Harvesting with Adjacency and Trigger Constraints*, Ph.D. thesis, Univ. of Auckland, Auckland, New Zealand.

[12] A. Murray and R. Church (1996), Analyzing Cliques for Imposing Adjacency Restrictions in Forest Models, *Forest Science*, 42, pp. 166–175.

[13] J. Nelson and J.D. Brodie, (1990), Comparison of a Random Search Algorithm and Mixed Integer Programming for Solving Area–based Forest Plans, *Canadian Journal of Forestry Research*, 20, pp. 934–942.

[14] J. Nelson, J.D. Brodie and J. Sessions (1991), Integrating Short–Term Area–Based Logging Plans with Long–Term Harvest Schedules, *Forest Science*, 37 (1), pp. 101–122.

[15] A.J. O'Hara, B.H. Faaland and B.B. Bare (1989), Spatially Constrained Timber Harvest Scheduling, *Canadian Journal of Forestry Research*, 19, pp. 715–724.

[16] D.M. Ryan (1992), Solution of Massive Generalised Set Partitioning Problems, *Journal of the Operational Research Society*, 43 (5), pp. 459–467.

[17] J. Sessions and J.B. Sessions (1992), Tactical Harvest Planning Using SNAP 2.03, in *Proceedings of Conference of International Union of Forestry Research*, pp. 112–120.

[18] H.D. Sherali, W.P. Adams and P.J. Driscoll (1998), Exploiting Special Structures in Constructing a Heirachy of Relaxations for 0–1 Mixed Integer Problems, *Operations Research*, 46, pp. 396– 405.

[19] S. Snyder and C. ReVelle (1996), The Grid Packing Problem: Selecting a Harvesting Patter in an Area with Forbidden Regions, *Forest Science*, 42, pp. 27–34.

[20] S. Snyder and C. ReVelle (1997), Dynamic Selection of Harvests with Adjacency Restrictions: the SHARe Model, *Forest Science 43*, pp. 213–222.

[21] A. Weintraub and A. Cholaky (1991), A Hierarchical aroach to Forest Planning, *Forest Science*, 37, pp. 439–460.

[22] A. Weintraub, F. Barahona and R. Epstein (1994), A Column Generation Algorithm for Solving General Forest Planning Problems with Adjacency Constraints, *Forest Science*, 40 (1), pp. 142–161.

[23] A. Weintraub, G. Jones, M. Meacham, A. Magendzo, A. Magendzo and D. Malchuk (1995), Heuristic Procedures for Solving Mixed–Integer Harvest Scheduling–Transportation Planning Models, *Canadian Journal of Forest Research*, 25, pp. 1618–1626.

[24] A. Yoshimoto, J.D. Brodie and J. Sessions (1994), A New Heuristic to Solve Spatially Constrained Long–term Harvest Scheduling Problems, *Forest Science*, 40, pp. 365–396.

OPTIMIZATION OF CHEMICAL PROCESSES UNDER UNCERTAINTY

G. M. Ostrovsky
Department of Chemical Engineering, U-222
University of Connecticut
191 Auditorium Road, Storrs, CT 06269, USA.
ostrovsk@engr.uconn.edu

L. E. K. Achenie
As above
achenie@engr.uconn.edu

Yu. M. Volin
Karpov Institute of Physical Chemistry
Vorontsovo Pole 10
Moscow, 103064, Russia.
volin@cc.nifhi.ac.ru

Abstract The problem of determining optimal equipment sizes and regimes of chemical processes that guarantee flexibility of the processes under uncertainty is considered in this paper. The effect of multiextremality of the solution is investigated. A comparative analysis of deterministic methods available in the open literature has been carried out as a result. All the methods require some convexity (or concavity) assumptions, which are difficult to verify for practical problems. To address the latter, we have developed algorithms with minimal convexity (or concavity) requirements for this problem.

Keywords: Flexibility Analysis, Uncertainty Analysis, Chemical Process Design, Nonlinear Programming

1. INTRODUCTION

In chemical process (CP) design some design specifications such as (a) safety, (b) ecological, and (c) performance specifications must always be

M.J.D. Powell and S. Scholtes (Eds.), *System Modelling and Optimization: Methods, Theory and Applications.*

met. Often such specifications are formulated either as soft or as hard constraints, with the presumption that a violation of *hard* constraints is not allowed. In this paper, we confine ourselves to hard constraints only.

The satisfaction of design specifications is complicated by the presence of uncertainties in design models, such as

- inherent inaccuracies of coefficients in the mathematical models,
- changes in some of the coefficients in the mathematical models during the CP operation (for example rate constants, heat and mass transfer coefficients),
- variations in some of the parameters (e.g. temperature, flow rates, species concentrations) associated with external streams during the CP operation.

It is of critical importance for a safe and economic operation of the process to account for the uncertainties discussed above when the optimal design structure, equipment sizes and regimes of the CP operation subject to design specification are determined. Several authors have addressed aspects of optimal design under uncertainty in the recent literature (Biegler, Grossmann and Westerberg, 1997), (Halemane and Grossmann 1983), (Grossmann and Floudas, 1987), (Pistikopoulos and Grossmann, 1989), (Pistikopoulos and Ieraptritou, 1995). Many contributions in the operation literature, however, rely on the availability of probability distribution functions for characterizing parametric uncertainties. Since practicing engineers very rarely know these distributions (without significant guess work), we believe it is important to consider ways of dealing with uncertainty which do not require probability distributions to be known. Halemane and Grossmann (1983) developed two formulations for analyzing the optimal design problem under uncertainty. The first formulation is an evaluation of the CP flexibility (its ability to satisfy process specifications during the operation stage), which has the form

$$F_1(d) \leq 0 \tag{1}$$

where the flexibility function $F_1(d)$ is of the form

$$F_1(d) = \max_{t \in T} \min_{z \in Z} \max_{j \in J} g_j(d, z, t). \tag{2}$$

Here, $J = \{1, ..., m\}$, d is a vector of design variables, z is a vector of control variables, Z is a region of admissible values of control variables, and

$$T = \{t : \bar{t} \le t \le \bar{\bar{t}}\}$$

is the domain for the uncertain parameters. The reduced process constraints

$$g_j(d, z, t) \le 0, \quad j = 1, ..., m \tag{3}$$

are obtained from the original mathematical model

$$\varphi(d, x, z, t) = 0 \tag{4}$$
$$\bar{g}(d, x, z, t) \le 0 \tag{5}$$

by explicitly solving for the vector of state variables x which has the same dimension as φ. Equations (4) are state equations (i.e. material and heat balance equations), while inequalities (5) are design specifications.

$F_1(d)$ can be represented in the form

$$F_1(d) = \max_{t \in T} h(d, t) \tag{6}$$

where

$$h(d, t) = \min_{z \in Z} \max_{j \in J} g_j(d, z, t) \tag{7}$$

Also

$$h(d, t) = \min_{z \in Z, v} v \tag{8}$$
$$g_j(d, z, t) \le v, \quad j = 1, ..., m$$

Therefore, the calculation of the flexibility function $F_1(d)$ is reduced to the maximization of $h(d, t)$ with respect to t. Halemane and Grossmann (1983) have shown that, in general, $h(d, t)$ is multiextremal and nondifferentiable.

The second formulation (of Grossmann et. al.) is the *two-stage optimization problem under uncertainty* (TSOP), which has the form

$$f_1 = \min_{d \in D} E[f^*(d, t)]$$
$$F_1(d) \le 0$$

where D is a region of admissible values of the vector d, $E[f^*(d, t)]$ is the mathematical expectation of $f^*(d, t)$ with respect to t with $f^*(d, t)$ given by

$$\begin{aligned} f^*(d,t) &= \min_{z \in Z} f(d,z,t) \\ g_j(d,z,t) &\leq 0, \quad j = 1,...,m. \end{aligned}$$

Here $f(d,z,t)$ is the objective function in the original optimization problem.

Using Gaussian quadrature (Carnahan, 1969) to approximate the multiple integral in the objective function, one can reduce the latter problem to the following problem (Halemane and Grossmann, 1983)

$$\begin{aligned} f_1 &= \min_{z^i, d \in D} \sum_{i \in I_1} w_i f(d, z^i, t^i) \qquad (9) \\ g_j(d, z^i, t^i) &\leq 0, \quad j \in I_1, \\ F_1(d) &\leq 0, \end{aligned}$$

where w_i are weights, t^i is an approximation point, z^i is a vector of control variables associated with the point t^i, and I_1 is a set of indices of the approximation points. Notice that the distribution functions which are required for the calculation of the mathematical expectation are often unknown. In this case the weights and approximation points must be selected using engineering insight. Similar problems arise in the design of other technical systems such as electrical circuits. Conceptually, the TSOP (9) determines the optimal design margins that guarantee satisfaction of process specifications under uncertainty, assuming the given uncertainty bounds are correct. This guarantee, however, can only be given if inequality $F_1(d) \leq 0$ holds for the global solution of the flexibility problem (6). Therefore, it is appropriate that we discuss different deterministic methods and the conditions under which the global solution of (6) is achieved.

2. COMPARATIVE ANALYSIS OF CURRENT METHODS

In this section, we compare current methods for estimation of process flexibility and optimization of chemical processes under uncertainty. Specifically, we limit ourselves to the following methods: (i) the method of Halemane and Grossmann (1983) (the HG method), (ii) the active constraint sets method (the ACS method of Grossmann and Floudas, 1987 and Pistikopoulos and Grossmann, 1989), (iii) the SG method of Swaney and Grossmann (1985), and (iv) the upper and lower bounds method (the ULB method of Ostrovsky and Volin, 1994, 1997). These methods are fairly representative of the research effort documented in the open literature, and they all employ local optimization methods.

Let us consider the first method. Swaney and Grossmann (1985) have shown that the global solution of the flexibility problem in (6) is obtained at a vertex of the parameter set T if the following condition is met

Condition 1: The functions $g_j(d, z, t)$ are jointly quasi-convex in z and d. In addition they are one-dimensional quasi-convex in t.

We note that a function $f(x)$ is one-dimensional quasi-convex in x if it is quasi-convex separately with respect to each component of x. The search for the maximum of $h(d, t)$ in T can now be reduced to a search among the vertices of T (Halemane and Grossmann, 1983). The computational effort of the method is, in general, proportional to the number of vertices of T , 2^r, where r is a dimension of the vector t. Swaney and Grossmann (1985) introduced the flexibility index, which characterizes the largest uncertainty region that the design can handle for feasible operation. For the case when each $g_j(d, z, t)$ is monotonic is t, Swaney and Grossmann (1985) proposed an algorithm for the flexibility index problem which uses a branch and bound (BB) strategy to search among the vertices of T. Kabatek and Swaney (1992) suggested a modification, which permits to find non-vertex solutions; however, the procedure does not guarantee a global solution.

We next consider the ACS method. Under certain conditions, the number of active constraints in problem (8) is equal to $q + 1$, where q is the number of control variables. Based on this, Grossmann and Floudas (1987) suggested a method for solving (2) as follows:

1 Identify all potentially active constraint sets $AS(k)$ (k is the index of the set) consisting of $q+1$ constraints. Let us denote the number of such sets as n_{AS}.

2 Solve the problem

$$\begin{aligned} u^k &= \max_{t,z} u \\ g_j(d, z, t) &= u, \qquad j \in AS(k) \end{aligned}$$

for all $k = 1, ..., n_{AS}$.

3 Determine

$$F_1 = \max_k u^k \tag{10}$$

Pistikopoulos and Grossmann (1989) generalized the method for the retrofit design problem. Grossmann and Floudas (1987) showed that the

ACS method gives the global solution of problem (2) under satisfaction of the following:

Condition 2:

1 Functions $g_j(d, z, t)$ are strictly quasi-convex in z for fixed t

2 Functions $g_j(d, z, t)$ are jointly quasi-concave in z and t.

One can show that if *Condition 2.1* is not satisfied, the ACS method does not guarantee a local or global solution. The alternative ULB method (Ostrovsky, Volin et. al., 1994, 1997) uses the branch and bound strategy for the calculation of F_1. It seeks the maximum of $h(d, t)$ over a partitioning of the region T into subregions T_i and is based on the inequality (Ostrovsky, Volin et. al., 1994)

$$F_{2i} \geq h(d, t), \quad \forall t \in T_i, \tag{11}$$

where

$$F_{2i} = \min_{z \in Z} \max_{j \in J} \max_{t \in T_i} g_j(d, z, t).$$

Consequently one can use F_{2i} as an upper bound of $h(d, t)$ on T_i. F_{2i} can be evaluated by solving the problem

$$\begin{aligned} F_{2i} &= \min_{z \in Z, u} u \qquad (12) \\ \max_{t \in T_i} g_j(d, z, t) &\leq u, \quad j = 1, ..., m. \end{aligned}$$

The ULB method for solving the TSOP is a two-level iterative procedure, which employs a partitioning of T. The upper level serves to partition T using information obtained from the lower level. At iteration k of the upper level the lower level is used to calculate an upper bound $f^{U,(k)}$ and a lower bound $f^{L,(k)}$ of f. Suppose at iteration k of the upper level the set T is partitioned into N_k subregions $T_l, l = 1, ..., N_k$. On each T_l the algorithm will automatically update a set of critical points: $S_{2,l}^{(k)} = \{\theta^i : \theta^i \in T_l, i \in I_l\}$ where I_l is the index set of critical points on the subregion T_l. The upper bound $f^{U,(k)}$ for a new set of subregions is calculated by solving the problem

$$\begin{aligned} f^{U,(k)} &= \min_{d, z^i, z^l} \sum_{i \in I_1} w_i f(d, z^i, t^i) \qquad (13) \\ g_j(d, z^i, t^i) &\leq 0, \quad i \in I_1, \quad j = 1, ..., m, \end{aligned}$$

$$\max_{t \in T_l} g_j(d, z^l, t) \leq 0, \quad l = 1, ..., N_k, \quad j = 1, ..., m, \tag{14}$$

where z^l is a control variables vector corresponding to the subregion T_l and I_1 is the set of approximation points (see (9)). The lower bound $f^{L,(k)}$ is calculated at the lower level as

$$\begin{aligned} f^{L,(k)} &= \min_{d, z^i, z^l} \sum_{i \in I_1} w_i f(d, z^i, t^i) \\ g_j(d, z^i, t^i) &\leq 0, \quad i \in I_1, j = 1, ..., m, \\ g_j(d, z^{lq}, t^{lq}) &\leq 0, \quad \forall t^{lq} \in T_l, l = 1, ..., N_k, \quad j = 1, ..., m, \end{aligned} \tag{15}$$

where z^{lq} is a vector of control variables corresponding to the point t^{lq}.

Now consider the upper level. At each iteration a partitioning of T_l is performed. The partitioning strategy strongly affects the computational complexity of the procedure. The simplest approach is to partition all subregions. However, the dimension of problems (13) and (15) will be very large. To alleviate this problem, we employ the following heuristic: at the k-th iteration, T_l $(l = 1, ..., N_k)$ is partitioned only if for this l the constraints (14) are active for at least one $j, (j = 1, ..., m)$.

Next, consider the lower level. Problem (15) is a standard nonlinear program that can be handled by standard algorithms. Problem (13), however, is not a conventional NLP. Therefore, a two-level iterative procedure is used for its solution. In the first step we solve the problem

$$\begin{aligned} f_2 &= \min_{d, z^i, z^l} \sum_{i \in I_1} w_i f(d, z^i, t^i) \\ g_j(d, z^i, t^i) &\leq 0, \quad i \in I_1, \quad j = 1, \ldots, m, \\ g_j(d, z^l, t^{lq}) &\leq 0, \quad \forall t^{lq} \in S_{2,l}, l = 1, ..., N_k, \quad j = 1, ..., m. \end{aligned} \tag{16}$$

Let $[d^*, z^*]$ be the solution to the problem, and let n be the iteration counter at the lower level. We note that in problem (16), the control vector z^l is associated with all the critical points of T_l whereas in problem (15), there is a control vector z^{lq} for each point t^{lq}.

During the second step, a stopping criterion is checked and some of the sets of critical points are extended. For this, the following mN_k problems are solved

$$\max_{t \in T_l} g_j(d^*, z^{l*}, t), \quad l = 1, ..., N_k; \ j = 1, ..., m. \tag{17}$$

In the case when it is not possible to obtain an explicit expression for state variables as a function of the variables $[d, z, t]$, problem (17) is equivalent to

$$\max_{x,t \in T_l} \bar{g}_j(d^*, x, z^{l*}, t) \tag{18}$$
$$\varphi(d^*, x, z^{l*}, t) = 0.$$

For fixed l and j, let t_j^l be the solution of the problem. If the condition

$$g_j(d^*, z^*, t_j^l) \leq 0, \quad l = 1, ..., N_k; \quad j = 1, ..., m, \tag{19}$$

is met, then the solution of problem (13) was obtained. Otherwise those t_j^l for which conditions (19) are violated are added to the sets of critical points of the corresponding subregions. We will assume that a local optimization method (for example SQP) is used for solving problems (15), (16) and (17). Now we will show that the ULB method gives the global solution of problems (2) and (9) under the assumption of

Condition 3

1 Functions $g_j(d, z, t)$ are jointly quasi-convex in d and z (for problem (2) quasi-convexity in z is sufficient)

2 Functions $g_j(d, z, t)$ are quasi-concave in t.

3 Functions $f(d, z, t)$ is quasi-convex in d and z.

Let us consider problem (2). The ULB method gives the global solution if the upper (see problem (13)) and lower bounds of $h(d, t)$ are global solutions. Rewrite the problem (12) in the form

$$F_{2i}(d) = \min_{z,u} u \tag{20}$$
$$G_j(d, z) \leq u \tag{21}$$
$$G_j(d, z) = \max_{t \in T} g_j(d, z, t) \tag{22}$$

By *Condition 3.2*, a local maximum of problem (22) coincides with the global maximum (Bazaraa and Shetty, 1979). One can show that if g_j is quasi-convex in z (*Condition 3.1*), then G_j is also quasi-convex in z. It follows that the region determined by Eqn. (21) is convex and a local minimum of problem (20) coincides with the global minimum (Bazaraa et. al.,1993). Moreover, the local minimum of problem (8) coincides with its global minimum. Thus, by solving problems (20) and (8), we will obtain the global solutions, i.e., the valid upper and lower bounds for subregion T. Therefore, the branch and bound procedure must give the global solution. Similarly, one can show that the local minimum of problem (9) coincides with the global minimum if *Condition 3* is satisfied.

If Condition 3.2 holds, which is the case in many applications, but *Condition 3.1* is not satisfied then our procedure only guarantees an upper bound for F_1. Indeed, in this case we obtain a local minimum $\tilde{F}_{2i,loc}$ of the problem (20) associated with problem (2) and since $\tilde{F}_{2i,loc} \geq \tilde{F}_{2i} \geq F_{2i}$ our upper bound estimate is less tight.

Let us now consider two scenarios for the TSOP. If, on the one hand, the ULB method yields a solution then this solution corresponds to the global solution of problem (17), since we have assumed that *Condition 3.2* is satisfied. Hence, by (19), we can guarantee the flexibility of the chemical process (CP). However, since the solution is in general a local minimum of problem (9), the design, although flexible may not be the best.

The second scenario corresponds to the case when the ULB method cannot obtain a solution. This means that the following condition is met

$$\max_{t \in T} \min_{z \in Z} \max_{j \in J} g_j(d, z, t) \;\geq\; 0 \quad \forall d$$

This can result from one of the following: (a) the flexibility of the CP cannot be guaranteed or (b) the solution corresponds to a local minimum of $\tilde{F}_2$.

Comparison of several methods with respect to their ability to obtain a global solution and their computational complexity leads us to conclude that they all supplement each other to some extent. Subsequently we recommend the following:

1 if *Condition 1* is met and the dimension of the vector t is small, the HG method should be employed;

2 if the dimension of t is not small and each g_j is monotonic, the SG method is appropriate;

3 if the number of active constraint sets in problem (8) is not large and *Condition 2* is met, it is reasonable to use the ACS method; and

4 if *Conditions 1* and *2* are not met but *Condition 3* is met, one should use the ULB method.

Using the above methods a larger class of problems can be solved. However, there is an inherent drawback of all the methods discussed earlier. For many realistic problems, it is very difficult, if not impossible, to check the convexity (concavity) of f and g_j. Even if it was possible to make such a check, the functions may turn out to be neither convex nor concave. Thus the above methods cannot guarantee the flexibility of most realistic CP's. To address this issue, we will now consider methods with a minimal dependence on knowledge of convexity (concavity) properties of the functions f and g_j.

3. MODIFICATION OF THE ULB METHOD

We have developed two modifications of the ULB method, which will obtain a solution of TSOP with guaranteed flexibility for the case when *Condition 3.2* is not met. The flexibility of the CP can be guaranteed if we can find global solutions to problem (17). In light of this, let us rewrite (17) and (19) in the form

$$\max_{t \in T_l} g_j(d^*, z^{l*}, t) \quad \leq 0 \quad l = 1, ..., N_k; j = 1, ..., m \tag{23}$$

These conditions are equivalent to

$$g_j(d^*, z^{l*}, t), \quad \leq \quad 0 \quad j = 1, ..., m, \quad \forall t \in T_l, l = 1, ..., N_k; \tag{24}$$

The conditions in (24) mean that for each subregion $T_l(l = 1, ..., N)$ we are able to find a vector of control variables z^{l*}, which guarantee satisfaction of all the constraints in (3). It should be noted that the dimensionality of problem (17) is less than the dimensionality of problem (9). However, finding the global maximum at each iteration is expected to be very computationally intensive. Subsequently, we suggest the following modification of the ULB method. First, let us consider the flexibility function F_1. In subregion T_l, we will use as upper bound for $h(d, t)$ the value

$$\begin{aligned} \bar{F}_{2l} &= \min_{u, z \in Z} u \\ \max_{t \in T_l} U(g_j; T_l) &\leq u, \quad j = 1, ..., m, \end{aligned} \tag{25}$$

where $U(g_j; T_l)$ is a concave overestimator of g_j in T_l with respect to t satisfying the conditions

$$U(g_j; T_l) \quad \geq \quad g_j, \forall t \in T_l. \tag{26}$$

If each g_j is bilinear, then McCormick's (1976) technique can be used to construct $U(g_j; T_l)$ whilst for general polynomial functions g_j, Sherali and Alameddine's (1992) linearization-reformulation approach is applicable.

Comparing problems (12) and (25) and taking into account (11), it is easy to obtain the following inequality

$$\bar{F}_{2i} \quad \geq F_{2i} \geq \quad h(d, t)$$

We will assume that $U(g_j;T_i)$ satisfies the condition

$$\lim_{r(T_l)\to 0} \max[U(g_j;T_l) - g_j(d,z,t)] \quad = \quad 0, \tag{27}$$

where $r(T_l)$ is a measure of the size of T_l. This means that, as T_i decreases in size, $U(g_j;T_i)$ gets closer to g_j. Consequently $\bar{F}_{2i}$ becomes a more accurate estimate of $h(d,t)$. In the limit we obtain (27) that

$$\lim_{r(T_i)\to 0} \bar{F}_{2i} \quad = \quad h(d,t_i) \tag{28}$$

and T_i reduces to a point t_i.

It is clear that the modification can be used if *Condition 1* is satisfied. This modification is expected to be superior to the HG method since it does not require full enumeration. It is also expected that it will be superior to the SG method since it can be applied to a broader class of problems.

Let us revisit the ULB method for solving the TSOP. In order to obtain the upper bound $f^{U,(k)}$ of the objective function, we will solve the problem

$$\begin{aligned} \bar{f}_3 \quad &= \quad \min_{d,z^i,z^l} \sum_{i\in I_1} w_i f(d,z^i,t^i) && (29) \\ g_j(d,z^i,t^i) \quad &\leq \quad 0, \qquad i\in I_1, j=1,...,m \\ \max_{t\in T_l} U[g_j(d,z^l,t);T_l] \quad &\leq \quad 0, \qquad l=1,\ldots,N_k, j=1,...,m \end{aligned}$$

and replace (17) by

$$\max_{t\in T_l} U[g_j(d^*,z^{l*},t);T_l] \tag{30}$$

where d^*, z^{l*} is a solution of problem (16). Of course we obtain an upper bound which is worse than the upper bound from (13). It is easy to see that if the following condition

$$\max_{t\in T_l} U[g_j(d^*,z^{l*},t);T_l] \quad \leq \quad 0 \tag{31}$$

is met, then condition (23) is met as well. Therefore, in the ULB method, we only need to replace (17) (i.e. maximization of each g_j) with the maximization of the concave overestimator of each g_j (i.e. by (30)). Now consider the case when it is impossible to obtain an explicit expression for the state variable $x = x(d,z,t)$. For a calculation of an upper bound in this case it is necessary to solve problem (18) instead of problem (17).

To formulate the equivalent of problem (30) we first replace each equality constraint in (18) with two inequality constraints which results in

$$\max_{x,t\in T_l} \bar{g}_j(d^*, x, z^{l*}, t)$$
$$\begin{aligned} \varphi(d^*, x, z^{l*}, t) &\leq 0 \\ -\varphi(d^*, x, z^{l*}, t) &\leq 0. \end{aligned}$$

Now problem (31) for calculating the upper bound will be equivalent to solving the problem

$$\max_{x,t\in T_l} U[\bar{g}_j(d^*, x, z^{l*}, t)]$$
$$\begin{aligned} L[\varphi(d^*, x, z^{l*}, t)] &\leq 0 \\ L[-\varphi(d^*, x, z^{l*}, t)] &\leq 0 \end{aligned}$$

where $L[\varphi(d^*, x, z^{l*}, t)]$ is a convex lower estimator of f with respect to t.

Since the overestimator $U(g_j; T_i)$ tends to g_j as T_i becomes infinitesimal, its use will not result in an increase in the number of iterative levels of the ULB method. Thus the computational complexity is not increased by the modification. We refer to the current modification as ULBG1. Suppose *Condition 3.1* is not met for ULBG1, then it is easy to show that a solution of the TSOP guarantees the flexibility of the CP. In this case, we should note that the solution is in general a local minimum from the point of view of the objective function (4).

Let us consider a second modification ULBG2 to the ULB method. We have noted already that if the ULB method can obtain the solution of the TSOP then in many cases the solution will guarantee flexibility of the CP. However, we need an unambiguous answer with regard to flexibility. Thus, after obtaining the solution, we need to solve problem (17) using a global optimization method, specifically a branch and bound strategy. If condition (23) is met, then the solution guarantees CP flexibility. Otherwise the points t_j^l at which condition (23) is violated are added to the corresponding sets of critical points S_{21}. In the computational experiments we have done so far, this procedure involves only one iteration in most of the cases.

4. COMPUTATIONAL EXPERIMENTS

In this section the proposed algorithms are applied to the optimization of a flowsheet consisting of a reactor and a heat exchanger (Fig. 1 from Halemane and Grossmann, 1983).

The reaction is assumed to be first-order exothermal of the type $\mathbf{A} \Longrightarrow \mathbf{B}$. The flow rate through the heat exchanger loop is adjusted to

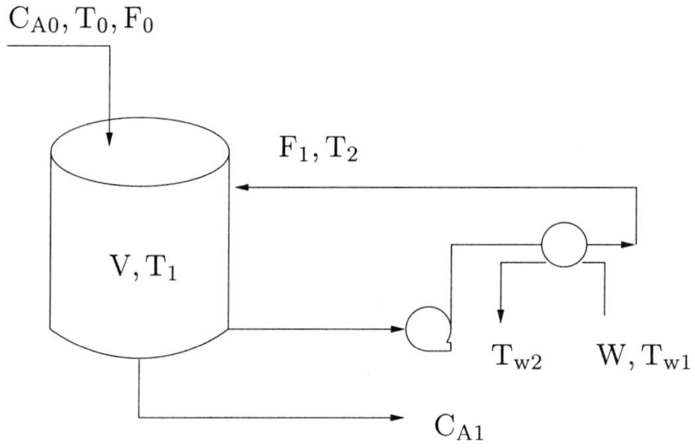

Figure 1 Flowsheet for example.

maintain the reactor temperature below T_{1max} and to get a minimum of 90% conversion. The latter is given by $conv = (c_{A0} - c_{A1})/c_{A0}$. The performance equations of such a system are as follows

Reactor Material and Heat Balance

$$F_0(c_{A0} - c_{A1})/c_{A0} = Vk_R exp(-E/RT_1)C_{A1} \tag{32}$$
$$(-H)F_0(c_{A0} - c_{A1})/c_{A0} = F_0 c_p(T_1 - T_0) + Q_{HE}$$

Heat Exchanger Heat Balance and Design Equations

$$Q_{HE} = F_1 c_p(T_1 - T_2) = c_{pw}(T_{w2} - T_{w1})W \tag{33}$$
$$Q_{HE} = AU\frac{(T_1 - T_{w2}) - (T_2 - T_{w1})}{ln\{(T_1 - T_{w2})/(T_2 - T_{w1})\}}$$

where F_0, T_0, C_{A0} are the feed flow rate ([=] $kgmolh^{-1}$), temperature of the feed ([=] ^oK) and the concentration of the reactant in the feed ([=] $kgmolm^{-1}$), respectively; V, T_1, C_{A1} are the values of the reactor volume ([=] m^3), the reactor temperature ([=] ^oK) and the concentration of the reactant **A** in the product ([=] $kgmolm^{-3}$); H is the heat of the reaction ([=] $kJkg^{-1}mol^{-1}$); F_1 is the flow rate of the recycle ([=] $kgmolh^{-1}$); T_2 is the recycle temperature; $c_p = 167.4kJkgmol^{-1}$ and $c_{pw} = 4.19kJkgmol^{-1}$ are the heat capacities of the recycle mixture and the cooling water, respectively. T_{w1}, T_{w2}, W are the inlet and outlet temperatures and the flow rate (kgh^{-1}) of the cooling water respectively. A ([=] m^2) is the heat transfer area of the heat exchanger and U ([=] $kJm^{-2}h^{-1}k^{-1}$) is the overall heat transfer coefficient.

In this problem the following constraints apply

$$-conv + 0.9 \leq 0 \tag{34}$$
$$conv - 1 \leq 0 \tag{35}$$
$$-(T_2 - T_{w1}) + 11.1 \leq 0 \tag{36}$$
$$T_2 - 389 \leq 0 \tag{37}$$
$$-T_2 + 311 \leq 0 \tag{38}$$
$$T_2 - T_1 \leq 0 \tag{39}$$
$$-T_{w2} + T_{w1} \leq 0 \tag{40}$$
$$311 \leq T_1 \leq 389 \tag{41}$$
$$301 \leq T_{w2} \leq 355. \tag{42}$$

The objective function of the original optimization problem is of the form

$$F = 691.2V^{0.7} + 873.6A^{0.6} + 1.76W + 7.056F_1. \tag{43}$$

The design variables are V and A. The control variables are T_1 and T_{w2}. The vector of state variables is $[C_{A1}, T_2, F_1, W]$. The vector of uncertain parameters is $t = [F_0, T_0, T_{w1}, k_R, U]$. Finally, the uncertainty region T is given by

$$T(\gamma) = [t_i^N(1 - \gamma\delta t_i) \leq t_i \leq t_i^N(1 + \gamma\delta t_i)] \tag{44}$$

where $t_i^N = (45, 333, 300, 9.8, 1635)$ is the nominal value of the uncertain parameters, γ is the parameter to determine the size of the uncertain parameter range, and $\delta t_i = (0.1, 0.02, 0.03, 0.1, 0.1)$ is a deviation fraction. The case $\gamma = 1$ was investigated by Halemane and Grossmann (1983) and Ostrovski and Volin (1994).

After elimination of dependent (i.e. state) variables C_{A1}, T_2, F, w using equations (32) and (33), we obtain

$$conv = \frac{V k_R c_{A0} exp(-E/RT_1)}{F_0 + V k_R c_{A0} exp(-E/RT_1)}$$
$$T_2 = \frac{2(-H)F_0 conv}{AU} - \frac{2F_0 c_p(T_1 - T_0)}{AU} - (T_1 - T_{w2}) + T_{w1}$$

Substituting the expressions into the constraints (34) to (40), we obtain explicit expressions for g_j with respect to the control variables and the uncertain parameters. These expressions contain linear, bilinear and trilinear terms with respect to the uncertain parameters. In Table 1, we give the approximation points from (Halemane and Grossmann,1983). Here, N, L, U designate nominal, lower and upper bounds.

	F_0	T_0	T_{w1}	K_0	U
t^1	N	N	N	N	N
t^2	L	L	L	L	U
t^3	U	U	U	U	L
t^4	U	U	L	U	L
t^5	U	U	U	L	L

Table 1 Approximation points: N - nominal value, L - lower bound, U - upper bound

We solved the problem for different numbers of approximation points and different sizes of the uncertainty region. In Table 2 we present results for (a) the nominal values of the uncertain parameters and (b) different sizes of the uncertainty region. In (9) we used five approximation points from Table 1. In the last column, the number of iterations used in the ULB method is presented.

	γ	f_1	V	A	# Iter
Optimization under Nominal values of Uncertain Parameters	-	9003.62	5.42	5.21	1
Optimization under Uncertainty	1.0	10670.7	6.63	7.77	1
Optimization under Uncertainty	1.25	11187.5	6.97	8.57	2
Optimization under Uncertainty	1.50	11776.5	7.34	9.45	5
Optimization under Uncertainty	1.75	12413.2	7.72	10.42	5

Table 2 Results for different sizes of uncertainty region

In Table 3 we present TSOP results for different numbers (I_5) of approximation points for $\gamma = 1$. We solved two variants corresponding to five approximation points and the first three approximation points from Table 1. If we compare the results of nominal optimization and optimization under uncertainty for $\gamma = 1$ and five approximation points, we conclude that it is necessary to increase the reactor volume by 25% and the heat exchange area by 22%. However, the bilinear and trilinear terms are sources of multiextremality in problem (17). As such we cannot guarantee flexibility of the process (i.e. *Condition 3.2* cannot be met).

In light of this we solved the problem using the ULBG1 and ULBG2 methods. The ULBG1 method requires construction of concave overesti-

	Is	f_1	V	A	# Iter
Optimization under Nominal values of Uncertain Parameters	-	9003.62	5.42	5.21	1
Optimization under Uncertainty	5	10670.7	6.63	7.77	1
Optimization under Uncertainty	3	10033.7	6.05	7.67	1

Table 3 Results for $\gamma = 1$ and different numbers of approximation points

mators $U(g_j)$ of g_j. Since the latter contain bilinear and trilinear terms, we must construct overestimators for these terms. For bilinear terms we used McCormick's (1983) expressions for overestimators and for trilinear terms we used as overestimators expressions from (Maranas and Floudas ,1995) (an extension of the overestimator for a bilinear term). We solved for ($I_5 = 5$) and $\gamma = (1, 1.25, 1.5)$. Results are given in Table 4. In addition, we solved for $\gamma = 1$ and $I_5 = (5, 3)$. Results are given in Table 5.

	γ	f_1	V	A	# Iter
Optimization under Nominal values of Uncertain Parameters	-	9003.62	5.42	5.21	1
Optimization under Uncertainty	1.0	10670.7	6.63	7.77	1
Optimization under Uncertainty	1.25	11187.5	6.97	8.57	1
Optimization under Uncertainty	1.50	11776.5	7.34	9.45	1

Table 4 Results with the ULBG1 method

	Is	f	V	A	# Iter
Optimization under Nominal values of Uncertain Parameters	-	9003.62	5.42	5.21	1
Optimization under Uncertainty	5	10670.7	6.63	7.77	1
Optimization under Uncertainty	3	10033.7	6.05	7.67	1

Table 5 Results with the ULBG2 method

Comparing the results from Table 4 with those of Table 2 and the results from Table 5 with those of Table 3, show that we can ensure flexibility of the process. It is interesting to note that the use of the overestimators does not increase the number of iterations.

In the ULBG2 method, we must take the values of design variables obtained by the ULB method and use them as initial points for global optimization of problems (17). For calculation of an upper bound of the functions g_j we used the overestimators, which were constructed by the technique described above. We solved the TSOP problem for the case when five approximation points are used and $\gamma = (1, 1.25, 1.5)$. Global optimization of all the constraints for all the cases showed that the solutions found by the ULB method are the global maxima of g_j with respect to the uncertain parameters. Consequently, the solutions obtained by the ULB method guarantee flexibility of the process. It is interesting to note that in all the cases the branch and bound global optimization method found global maxima in one iteration.

5. DISCUSSIONS AND CONCLUSIONS

The aim of solving the two step optimization problem consists in determining optimal design margins guaranteeing the preservation of capacity for the operation of chemical processes (flexibility of chemical processes) in spite of model uncertainties at the design stage, and process uncertainties at the operation stage. The design margins obtained can, however, be used in practice only if the solution of the two-stage optimization problem under uncertainty (TSOP) corresponds to the global solution of the flexibility problem. We have therefore carried out a comparative analysis of three deterministic methods of flexibility analysis and optimization of chemical processes under uncertainty.

Our analysis showed that the methods discussed in this paper supplement each other to some degree. Given this range of tools, one can solve a wide class of optimization problems. However there is an inherent drawback of the methods. For many real problems, it is very difficult to check the convexity (concavity) of the constraint and objective functions. Even if it was possible to carry out such a check, the functions may turn out to be neither convex nor concave. As a result, there is a need to develop methods that have minimal requirements on convexity (concavity) of the constraint and objective functions.

We have developed two significant modifications of the ULB method which permit us to obtain the solution of the TSOP guaranteeing flexibility of CP under a single condition, namely the convexity of the constraint functions $g_j(d, z, t)$ in z. Any solution obtained by the algorithm will guarantee flexibility of the CP even though the solution may correspond to a local minimum of the TSOP.

References

[1] M.S. Bazaraa, H.D. Sherali and C.M. Shetty (1993), *Non-linear Programming, Theory and Algorithms*, John Wiley and Sons, New York.

[2] L.T. Biegler, I.E. Grossmann and A.W. Westerberg (1997), *Systematic Methods of Chemical Process Design*, Prentice Hall.

[3] I.E. Grossmann and C.A. Floudas (1987), Active constraint strategy for flexibility analysis in chemical processes, *Comp. & Chem. Eng.*, 11, pp. 675–693.

[4] K.P. Halemane and I.E. Grossmann (1983), Optimal process design under uncertainty, *AIChE Journal*, 29, pp. 425–433.

[5] U. Kabatek and R.E. Swaney (1992), Worst-case identification in structured process systems, *Comp. & Chem. Eng.*, 16, pp. 1063–1071.

[6] G.P. McCormick (1976), Computability of global solutions to factorable nonconvex programs. Part I – convex underestimating problems, *Mathematical Programming*, 10, pp. 147–175.

[7] G.M. Ostrovsky, Yu.M. Volin, I.E. Barit and M.M. Senyavin (1994), Flexibility analysis and optimization of chemical plants under uncertainty, *Comp & Chem. Eng*, 18, pp. 755–767.

[8] G.M. Ostrovsky, Yu.M. Volin, E.I. Barit and M.M. Senyavin (1997), An approach to solving a two step optimization problem, *Comp. & Chem. Eng.*, 21, pp. 317–325.

[9] E.N. Pistikopoulos and I.E. Grossmann (1989), Optimal retrofit design for improving process flexibility in non-linear systems-1, Fixed degree of flexibility, *Comp. & Chem. Eng.*, 12, pp. 1003–1016.

[10] E.N. Pistikopoulos and M.G. Ieraptritou (1995), Novel approach for optimal process design under uncertainty, *Comp & Chem. Eng.*, 19, pp. 1089–1110.

[11] H.D. Sherali and Tunchbilek (1992), A global optimization algorithm for polynomial programming programs using reformulation-linearization technique, *Journal of Global Optimization*, 2, pp. 101–112.

[12] R.E. Swaney and I.E. Grossmann (1985), An index for operational flexibility in chemical process design, *AIChE Journal*, 31, pp. 621–630.

NUMERICAL SOLUTION TO THE ACOUSTIC HOMICIDAL CHAUFFEUR GAME

V.S. Patsko
Institute of Mathematics and Mechanics
S.Kovalevskaya str., 16, Ekaterinburg, 620219, Russia
patsko@imm.uran.ru

V.L. Turova
Center of Advanced European Studies and Research
Friedensplatz 16, 53111 Bonn, Germany
turova@caesar.de

Abstract A well-known differential game in the theory of differential games is the "homicidal chauffeur" problem which was introduced by Isaacs [7]. It is a pursuit-evasion game. In the paper, a variant of this problem proposed by Bernhard [3] is considered. The computation of level sets of the value function in this variant becomes difficult since holes in the "victory domains" of the pursuer can appear. Some results of the computation of level sets of the value function are presented. An explanation of the generation of holes is given, based on the analysis of families of semipermeable curves.

1. INTRODUCTION

A differential game where an inertial object pursues a non-inertial one is considered. The dynamics of the game is similar to those for the classical [7, 4, 11] homicidal chauffeur game of R. Isaacs. The difference is that the evader must apply a reduced speed (in order not to be heard by the pursuer) when the distance between him and the pursuer becomes less than a given value. The idea of such a modification was suggested in [3]. The pursuer minimizes the time of capture and the evader maximizes it. The game is over when the evader gets into a given neighborhood of the state of the pursuer (capture neighborhood).

M.J.D. Powell and S. Scholtes (Eds.), *System Modelling and Optimization: Methods, Theory and Applications.*

In [6], level sets of the value function for particular magnitudes of parameters of the problem were computed using an algorithm based on viability theory. The solution to the problem has a complicated structure: holes in the solvability set (in the victory domain) of the pursuer can arise, the evader being safe from the pursuer within these holes.

The investigation of such complex structures of solutions is of great interest for viability theory and the theory of differential games. In the problem considered, the geometry of level sets of the value function differs from the one that was analyzed for other problems with the homicidal chauffeur dynamics [7, 4, 11, 8, 15].

In this paper, the problem is studied using an algorithm proposed by the authors for computing level sets of the value function. The algorithm is based on the theory of differential games [9, 10]. The dependence of the structure of the solution on the parameters of the problem is investigated. The computations are done in the plane because a change of variables can reduce the dimension of the original problem to two [7]. The algorithm uses specific properties of the plane and is very accurate. It allows one to explore some fine peculiarities of the solution. Additionally, the analysis of families of so-called semipermeable curves is used to explain the occurrence of holes.

2. STATEMENT OF THE PROBLEM

The dynamics of the game in reduced coordinates has the form [7, 6]:

$$\dot{x}_1 = -\frac{w^{(1)}}{R} x_2\, \varphi + v_1, \qquad \dot{x}_2 = \frac{w^{(1)}}{R} x_1\, \varphi + v_2 - w^{(1)}, \tag{1}$$
$$\text{where} \quad |\varphi| \le 1 \quad \text{and} \quad v \in Q(x).$$

Here $(x_1,\ x_2)'$ is the state vector which gives the relative position of the evader E with respect to the pursuer P, and $w^{(1)}$ and R are constants which define the pursuer's velocity and the minimal radius of turn, respectively. The control of player P is φ, and the control of the evader E is $v = (v_1, v_2)'$.

The vector v belongs to the circle $Q(x)$ with center at the origin and radius $w^{(2)}(x) = \min\{(x_1^2 + x_2^2)^{1/2}, s\}w_e/s$, where w_e is the maximal value of the velocity of player E, and s is a fixed positive number. Thus the radius of the constraint $Q(x)$ on the control of player E is constant and equal to w_e outside the circle of radius s with center at the origin, but the radius is proportional to $|x|$ inside this circle.

The terminal set M is the rectangle $\{(x_1, x_2) \in R^2 : -3.5 \le x_1 \le 3.5, -0.2 \le x_2 \le 0\}$. The objective of the control φ is to minimize the time of attaining the terminal set M, but the objective of the control

$v = (v_1, v_2)'$ is to maximize this time. Therefore the payoff of the game is the time of attaining the terminal set.

The statement of the problem was taken from [6]. In the classical statement [7] of the problem, the terminal set (capture neighborhood) is a circle. A circle can be used in the acoustic version too. However, more interesting cases from the mathematical point of view arise when the capture neighborhood is a rectangle with its horizontal side much greater than its vertical side.

The game is treated in frames of formalization from [9, 10]. We are interested in finding level sets $W(T, M)$, $T > 0$. Each of them is the set of all initial states x_0 in the plane such that player P can guarantee the transition of the state vector to the set M within time T.

3. THE ALGORITHM

Here the main idea of the algorithm for computing the level sets $W(T, M)$ of the value function is described.

Let Δ be a time step of the backward procedure. Let the i-th level set of the value function, namely $W(i\Delta, M)$, be available. This is the maximal set from where the pursuer P guarantees the termination of the game within the time $i\Delta$. On the basis of this set, we compute the set $W((i+1)\Delta, M)$, consisting of all states from which player P guarantees the attainment of $W(i\Delta, M)$ within time Δ. As a result of such computations for $i = 0,\ 1,\ 2, \ldots$, we obtain the collection of embedded sets $W(\Delta, M) \subset W(2\Delta, M) \subset \cdots \subset W(i\Delta, M) \subset \cdots \subset W(T, M)$.

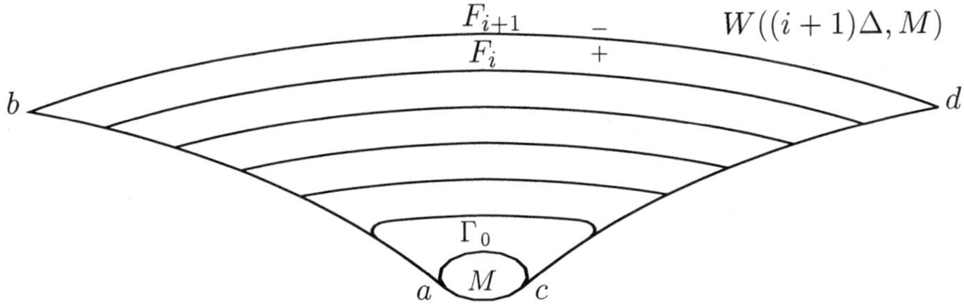

Figure 1 Construction of the sets $W(i\Delta, M)$

This is a dynamic programming method. In the theory of differential games, the fundamental idea of the backward construction of level sets was considered in works of Isaacs, Fleming, Pontryagin, Krasovskii and Pschenichnyi.

The central part of our algorithm is the notion of a front. The front F_{i+1} contains all points on the boundary of the set $W((i+1)\Delta, M)$ with the property that the minimal guaranteed time of attaining the previous set $W(i\Delta, M)$ is precisely Δ. The side of the front in the backward time direction will be called negative, and the opposite side will be called positive, as in Figure 1. The algorithm computes a new front F_{i+1} using the previous front F_i. For the first step of the backward procedure, F_0 coincides with the usable part [7] Γ_0 of the boundary of M. The barrier lines are obtained via connection of the corresponding ends of the fronts. In Figure 1, the lines ab and cd are barriers.

We explain briefly how the fronts are constructed. Using the notation $p(x) = (-x_2, x_1)' \cdot w^{(1)}/R$ and $g = (0, -w^{(1)})'$, we rewrite the equations (1) as $\dot{x} = p(x)\varphi + v + g$. In the computation, each front is stored as an ordered collection of points, so fronts are polygonal lines. An apex of a polygonal line is called a point of local convexity if the angle between the positive sides of the adjoining links is less than π. An apex of a front is a point of local concavity if the above angle is greater than π. A cone K of outer (inner) normal vectors is assigned to each point of local convexity (concavity). At the endpoints of the front, the cone K is defined in a different way: one extreme ray of the cone is the outer normal vector to the front link, and the other extreme ray of the cone is defined by some special relations. For each fixed point $x_* \in F_i$ of local convexity and any vector $\ell \in K(x_*)$, the extremal controls φ°, v° take the values $\varphi^\circ = \text{argmin}\{\ell' p(x_*)\varphi : |\varphi| \leq 1\}$ and $v^\circ = \text{argmax}\{\ell' v : v \in Q(x_*)\}$. Similarly, for the points of local concavity, the extremal controls are $\varphi^\circ = \text{argmax}\{\ell' p(x_*)\varphi : |\varphi| \leq 1\}$ and $v^\circ = \text{argmin}\{\ell' v : v \in Q(x_*)\}$.

The extremal control φ° of player P can switch its value from one extreme value to another, not only at the apexes of the front, but also at inner points of the front links. In the game considered, such a switching occurs at not more than one inner point of each front link, due to the linearity of the dynamics in x and φ. The points where such a switching takes place will be called neutral. The collection of all neutral points is included (with the preservation of ordering) in the collection of apexes defining the front. Each neutral point divides the original link into two parts that are also considered as links of the front.

Other additional division points on the front links may also be introduced, which take into account the dependence of the constraint on the control of player E on x. The cone K for a neutral or additional point contains only the outer normal vector at the point. Further, the extremal controls of the players are given by the above formulae for the local convexity case.

Using the extremal controls, one computes the extremal trajectory $x(\sigma) = x_* - \sigma(p(x_*)\varphi^\circ + v^\circ + g)$, $\sigma \in (0, \Delta]$, in reverse time. If the extremal controls are not unique at x_*, a bundle of extremal trajectories emanating from the point x_* is considered.

As x_* ranges through F_i, the ends of the extremal trajectories at $\sigma = \Delta$ are used to form the next front F_{i+1}. One can divide F_i into regular parts so that the extremal trajectories emanating from the points of one part do not intersect for $\sigma \in (0, \Delta]$. Thus each regular part generates a regular field of extremal trajectories. The ends of these trajectories form an ordered collection of points. Being connected, these points give a polygonal line, which is called the secondary arc. The new front F_{i+1} is obtained by processing the regular secondary arcs, the processing being reduced to the intersection of secondary arcs.

We consider this procedure for the simple case shown in Figure 2. Here the front F_i consists of two regular parts $[z_1 \cdots z_\omega]$ and $[z_\omega \cdots z_r]$. Both parts are composed of local convexity points. The ends of the extremal trajectories computed at $\sigma = \Delta$ give two secondary arcs, namely $[\xi_1\xi_2 \cdots \xi_s]$ and $[\xi_{s+1} \cdots \xi_m]$, as shown in the left half of the figure. The control of player E can be chosen for each of the points ξ_s and ξ_{s+1} so that the trajectories of the system (1) cannot reach the front F_i within time Δ. Therefore, the "swallow tail" $\xi_s\xi_\alpha\xi_{s+1}$ is not included in the front $F_{i+1} = [\xi_1\xi_2 ... \xi_\alpha ... \xi_m]$, which is drawn on the right hand side of Figure 2.

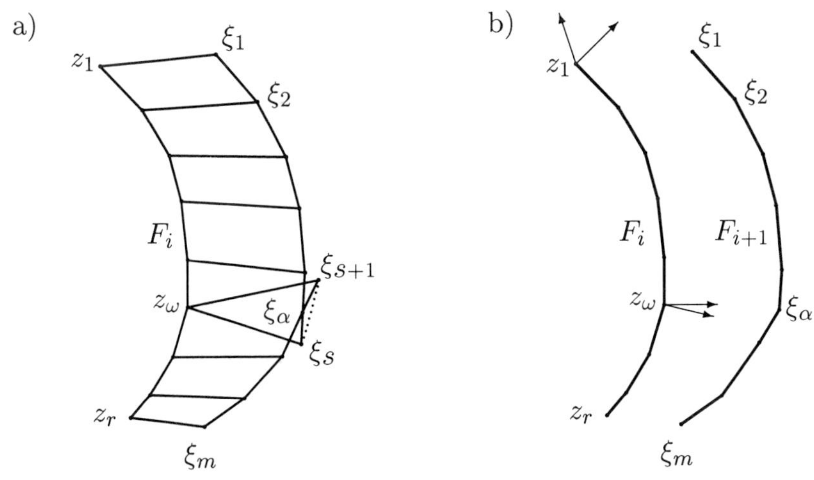

Figure 2 Construction of fronts

Unfortunately, very often, it is not sufficient to intersect neighboring secondary arcs only. In Figure 3, for example, the secondary arcs S_1, S_2 and S_3 are computed sequentially, but the next front is obtained due to the intersection of S_1 and S_3.

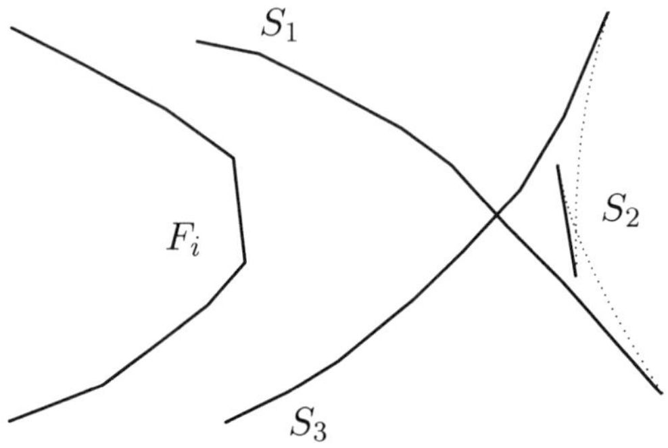

Figure 3 Secondary arcs: complicated case of disposition

Thus the algorithm produces a collection of fronts. In the course of computations, possible self-intersections of fronts and their collisions with the barrier lines are processed. The details of the algorithm can be found in [12, 13, 14].

4. SEMIPERMEABLE CURVES

In this section, the results of some analysis of families of semipermeable curves in differential games with homicidal chauffeur dynamics will be given. Using these results, one can find the solvability sets of the game of kind [7]. Since the set $W(T, M)$ converges to the solvability set of the corresponding game of kind as $T \to \infty$, solutions to this game of kind can be used for verifying the computation of the sets $W(T, M)$. The families of semipermeable curves can also be helpful for checking the computations of level sets of the value function within solvability sets.

The families of semipermeable curves are determined from only the dynamics of the system and the bounds on the controls of the players. We explain now what semipermeable curves mean. Let

$$\begin{aligned} H(\ell, x) &= \min_{\varphi \in [-1,1]} \max_{v \in Q(x)} \ell' f(x, \varphi, v) \\ &= \max_{v \in Q(x)} \min_{\varphi \in [-1,1]} \ell' f(x, \varphi, v), \quad x \in R^2, \ \ell \in R^2, \end{aligned} \tag{2}$$

be the Hamiltonian of the game. Here $f(x, \varphi, v) = p(x)\varphi + v + g$. It is easy to see that the function $\ell \to H(\ell, x)$ is convex in ℓ in the cones $\ell' p(x) \geq 0$ and $\ell' p(x) \leq 0$ for any fixed $x \in R^2$. Fix x and consider ℓ such that $H(\ell, x) = 0$. Letting $\varphi^* = \text{argmin}\{\ell' p(x)\varphi : \varphi \in [-1, 1]\}$ and $v^* = \text{argmax}\{\ell' v : v \in Q(x)\}$, it follows that $\ell' f(x, \varphi^*, v) \leq 0$ holds for any $v \in Q(x)$, and $\ell' f(x, \varphi, v^*) \geq 0$ holds for any $\varphi \in [-1, 1]$. This means that the direction $f(x, \varphi^*, v^*)$, which is orthogonal to ℓ, separates the vectograms $U(v^*) = \{f(x, \varphi, v^*) : \varphi \in [-1, 1]\}$ and $V(\varphi^*) = \{f(x, \varphi^*, v) : v \in Q(x)\}$ of the players P and E as in Figure 4. Such a direction is called semipermeable. A smooth curve is called a semipermeable curve if the tangent vector at any point of this curve is a semipermeable direction.

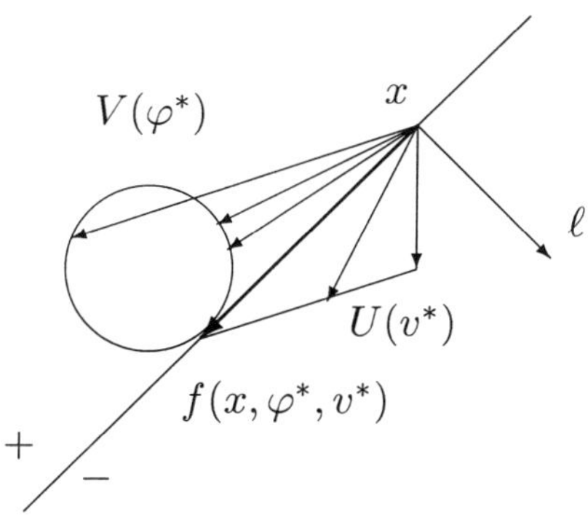

Figure 4 **Semipermeable direction**

We now describe how the families of semipermeable curves can be obtained. The semipermeable directions are derived from the roots of the equation $H(\ell, x) = 0$. We distinguish the roots "−" to "+" and the roots "+" to "−". When classifying these roots, we suppose that $\ell \in \mathcal{E}$, where $\mathcal{E}$ is the boundary of a convex polygon containing the origin. We say that ℓ_* is a root − to + if $H(\ell_*, x) = 0$, and if $H(\ell, x) < 0$ ($H(\ell, x) > 0$) for $\ell < \ell_*$ ($\ell > \ell_*$) that are sufficiently close to ℓ_*, where the notation $\ell < \ell_*$ means that the direction of the vector ℓ can be obtained from the direction of the vector ℓ_* using a counterclockwise rotation through an angle not exceeding π. The roots − to + and the roots + to − are called roots of the first and second type, respectively. Due to the above mentioned property of the piecewise convexity of the function $H(\cdot, x)$, the equation $H(\ell, x) = 0$ can have at most two roots of each type for any given x.

Let us denote the roots by $\ell^{(j),i}(x)$. The left index corresponds to the type of root ($-$ to $+$ or $+$ to $-$). The right index takes the value 1 or 2, and indicates whether the minimum in (2) occurs for $\varphi = 1$ or $\varphi = -1$.

One can find the domains of the functions $\ell^{(j),i}(\cdot)$. They have very simple structures for the classical formulation of the homicidal chauffeur problem. In this case, the constraint Q on the control of player E does not depend on x, so we have $w^{(2)} = w_e$.

Figure 5 shows the domains of $\ell^{(j),i}(\cdot)$ in the classical case $w^{(2)} = w_e \leq w^{(1)}$. Two symmetric cones with a joint apex at the origin are cut by polygonal approximations to circular arcs of radius $w^{(2)}R/w^{(1)}$, the centers of the arcs being at the points $(-R, 0)$ and $(R, 0)$. The regions of values of x where two roots of each type exist are marked A and B in the figure. There is only one root of each type at the points x that are outside A and B.

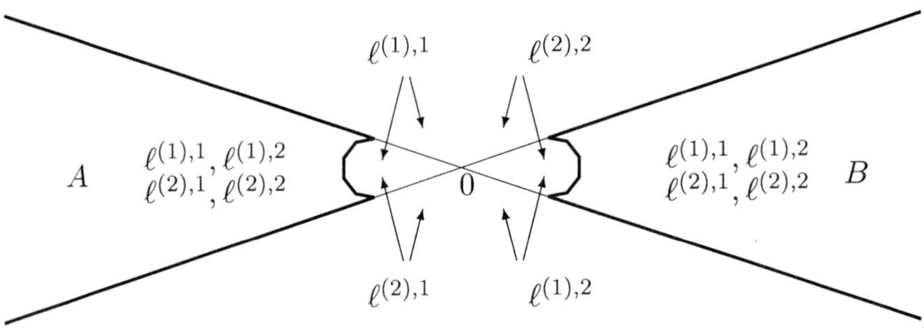

Figure 5 Domains of the functions $\ell^{(j),i}(\cdot)$ when $w^{(2)} = w_e \leq w^{(1)}$

In Figure 6, the domains of $\ell^{(j),i}(\cdot)$ are depicted for $w^{(2)} = w_e > w^{(1)}$. The digits 4, 2 and 0 state the number of roots. In this case, a region C (the intersection of the circles of radius $w^{(2)}R/w^{(1)}$ with centers at $(-R, 0)$ and $(R, 0)$) occurs where roots do not exist. The following property holds true for any point $x \in C$: for any $\varphi \in [-1, 1]$ there exists $v \in Q$ such that $f(x, \varphi, v) = 0$. Therefore, in the region C, player E can counter any control of player P, so the state remains immovable all the time. Further, if a point x with the above property does not belong to the terminal set M, then M cannot be reached from x. Regions of such points are called the superiority sets of player E.

Using the forms of the domains of $\ell^{(j),i}(\cdot)$ in the classical case, one can construct the domains for the case when Q depends on x. We describe schematically how it can be done. First note that $w^{(2)}$ is constant on the circumference of any fixed circle with center $(0, 0)$. Further, $w^{(2)} = w_e$

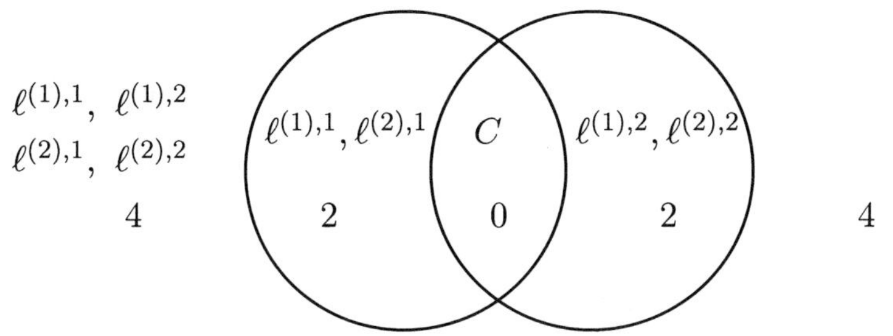

Figure 6 Domains of the functions $\ell^{(j),i}$ when $w^{(2)} = w_e > w^{(1)}$.

holds outside the circle of radius s. Let $\Omega(r)$ be the circumference of the circle of radius r with center at $(0,0)$. Find $w^{(2)}(r) = \min\{r, s\}w_e/s$. If $w^{(2)}(r) \leq w^{(1)}$, then put the points $x \in \Omega(r)$ onto the domains of Figure 5 constructed for $w^{(2)} = w^{(2)}(r)$. Otherwise, if $w^{(2)}(r) > w^{(1)}$, then put these points onto the domains of Figure 6. Thus a division of $\Omega(r)$ into arcs is obtained. The number and the type of roots are the same for all points of each arc. In Figure 7, the division points a, b, c and d, and those symmetric to them in the left half-plane, are shown, $\Omega(r)$ being the dotted line. In Figure 8, the division points e and f, and those symmetric to them, are depicted.

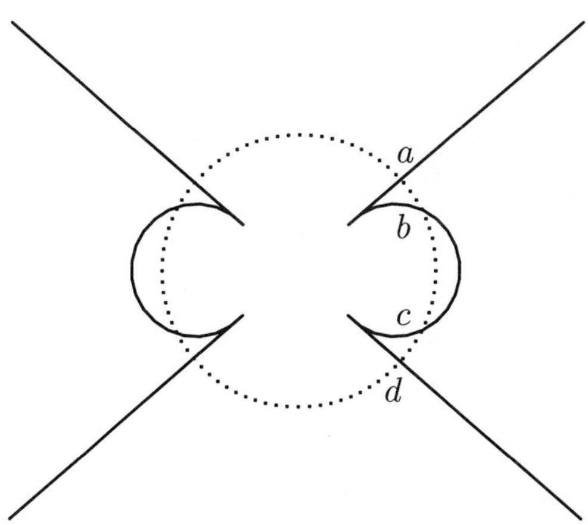

Figure 7 Construction of domains of $\ell^{(j),i}(\cdot)$ when $w^{(2)}(r) \leq w^{(1)}$

This technique is applied for every r in $[0, s]$, and identically named division points are connected. Thus the circle of radius s is divided into parts according to the kinds of roots. Outside this circle, the dividing lines coincide with the lines constructed for the case when Q does not depend on x. We use the lines of Figure 5 or Figure 6, depending on $w_e \leq w^{(1)}$ or $w_e > w^{(1)}$, respectively.

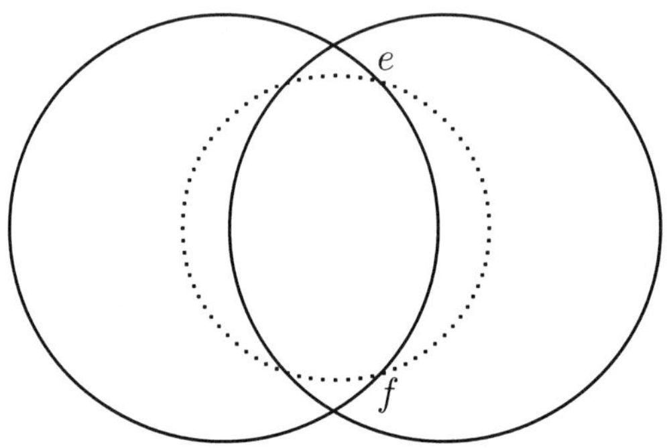

Figure 8 Construction of domains of $\ell^{(j),i}(\cdot)$ when $w^{(2)}(r) > w^{(1)}$

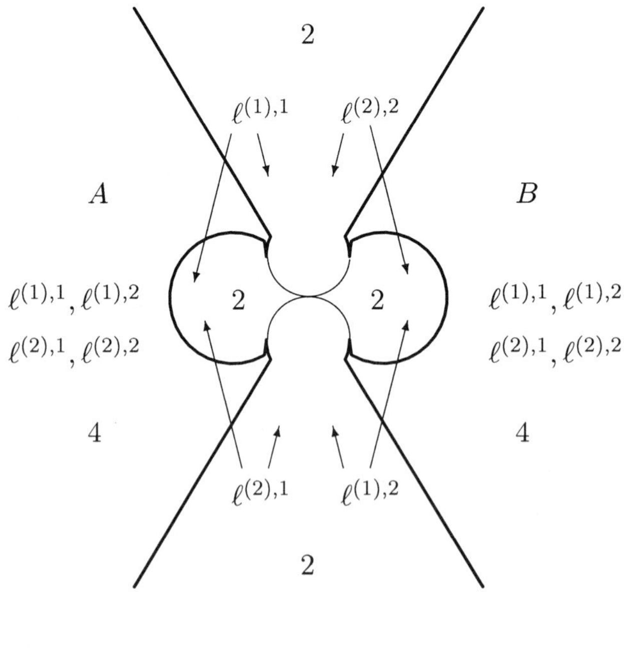

Figure 9 Domains of the functions $\ell^{(j),i}(\cdot)$ when Q depends on x and $w_e = 0.8$

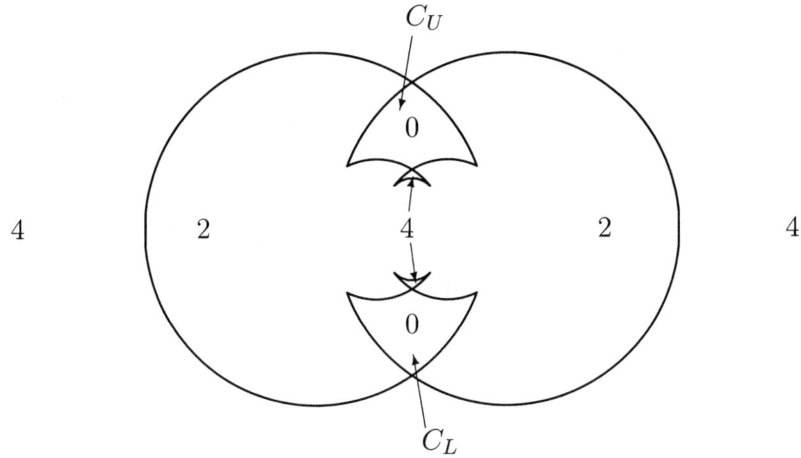

Figure 10 Superiority sets of player E when Q depends on x and $w_e = 1.8$

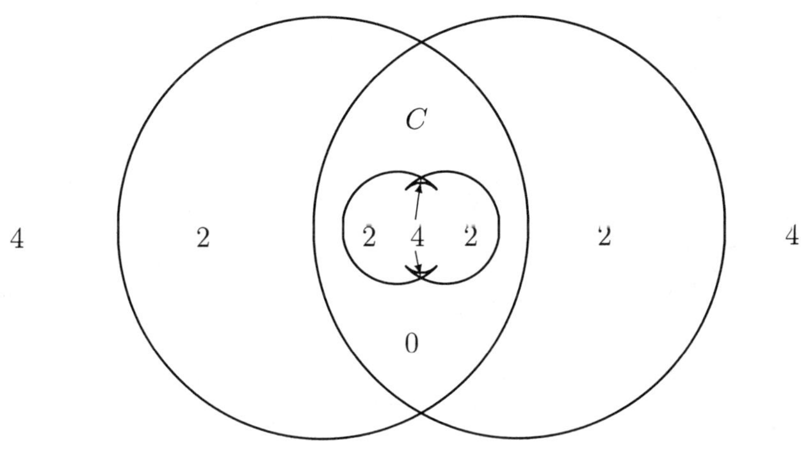

Figure 11 Superiority set of player E when Q depends on x and $w_e = 2$

Figures 9, 10 and 11 were constructed in this way for the parameters $w^{(1)} = 1$, $R = 0.8$, $s = 0.75$ and $w_e = 0.8$, 1.8 and 2. In Figure 9, the domains of the functions $\ell^{(j),i}(\cdot)$ are shown, and also the sets that are analogous to A and B in Figure 5 are marked. In Figure 10, two symmetric superiority sets of player E arise, the upper set being denoted by C_U and the lower set by C_L. If we increase w_e, the sets C_U and C_L

expand and form the double connected region that is denoted by C in Figure 11. The number of roots of the equation $H(\ell, x) = 0$ is also given in Figures 10 and 11.

Figure 12 shows a fragment of the central part of Figure 10. The lines that separate the domains of the functions $\ell^{(j),i}(\cdot)$ are included.

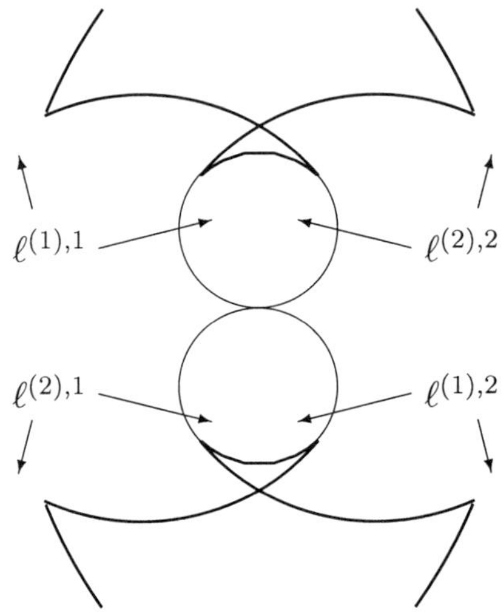

Figure 12 Domains of the functions $\ell^{(j),i}$ in part of Figure 10

The function $\ell^{(j),i}(\cdot)$ is Lipschitz continuous on any closed bounded subset of the interior of its domain. We consider the two-dimensional differential equation

$$dx/dt = \Pi\ell^{(j),i}(x), \tag{3}$$

where Π is the matrix of rotation through the angle $\pi/2$, the rotation being clockwise or counterclockwise if $j = 1$ or $j = 2$, respectively. Since the tangent vector at each point of the trajectory defined by this equation is a semipermeable direction, the trajectories are semipermeable curves. Therefore player P can keep the state vector x on one side of the curve (positive side), and player E can keep x on the other (negative) side. Further, equation (3) specifies a family $\Lambda^{(j),i}$ of semipermeable curves, such that a unique smooth semipermeable curve goes through each point x of the domain of $\ell^{(j),i}(\cdot)$, the root $\ell^{(j),i}(x)$ being the normal vector to the curve at the point x. The notation $p^{(j),i}$ will be used for the curves of the family $\Lambda^{(j),i}$.

The family $\Lambda^{(1),1}$ for the values of the parameters of Figure 10 is depicted in Figure 13. The arrows show the direction of motion in reverse

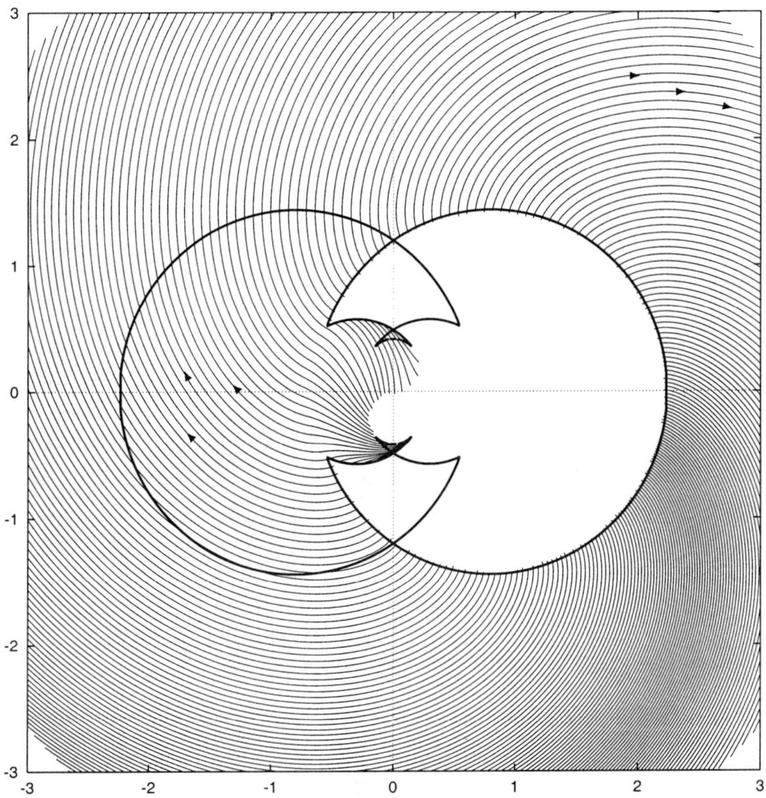

Figure 13 Family of semipermeable curves for the root $\ell^{(1),1}$

time. The families $\Lambda^{(1),2}$, $\Lambda^{(2),1}$ and $\Lambda^{(2),2}$ can be obtained from $\Lambda^{(1),1}$ by reflections in the x_1- and x_2-axes.

5. SUPERIORITY SETS

In this section, the role of superiority sets in the appearance of holes within the solvability sets will be explained. As noted above, there can be one doubly connected superiority set C of player E, or two simply connected sets C_U and C_L, or the superiority set can be empty.

Let D be a connected superiority set of player E, and let the objective of this player be to bring the state of the system to the set D. Denote by D^* the maximal solvability set (victory domain) of player E. It follows from the definition of D^* that E can bring the state of the system to D from any point $x \in D^*$, but player P can prevent the state of the system from approaching the set D for any point $x \notin D^*$. Since D is a superiority set of E, it possesses the property of v-stability [9, 10] (or viability for E [1, 6]), and the set D^* is v-stable too. This means that player E can hold the trajectories of the system in D^* for infinite time. Hence, if $D^* \cap M = \emptyset$, then the time for achieving the terminal set M in the main problem is infinite for any x in D^*.

The boundary of D^* is composed of smooth semipermeable curves of the families $\Lambda^{(j),i}$. The joins of these curves are called "sewing points", and they possess the semipermeability property [5]. In some cases, a part of the boundary of D^* can coincide with a part of the boundary of D.

Due to the simple geometry of the sets D of the problem considered, the sets D^* can be obtained easily using the families of semipermeable curves. For example, Figure 14 shows the configuration of D^* when $D = C_U$ as in Figure 10. The sewing point of the curves $p^{(2),2}$ and $p^{(1),2}$, and the symmetric sewing point of the curves $p^{(1),1}$ and $p^{(2),1}$, lie on the boundary of D.

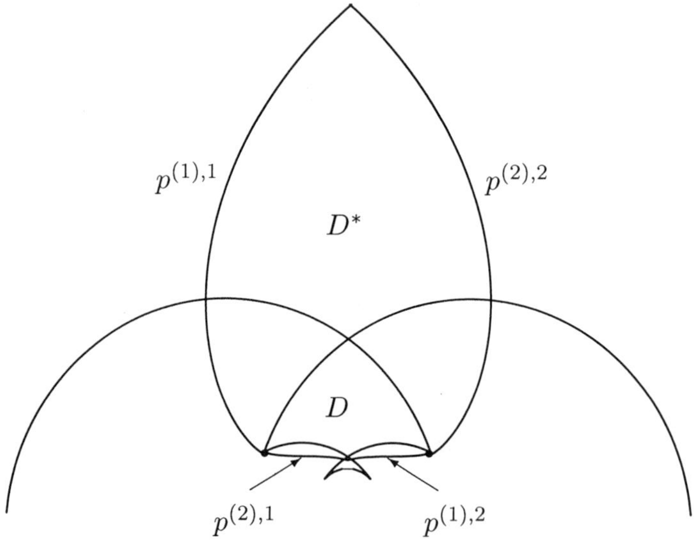

Figure 14 Generation of the hole D^* due to the superiority set D

Since level lines of the value function do not "penetrate" into the sets D^* in the case $D^* \cap M = \emptyset$, one can easily generate examples where holes occur in the solvability sets, using this knowledge of the geometry of the sets D^*.

6. COMPUTATIONAL RESULTS

In this section, the dependence of the solution on the parameter w_e is demonstrated. Other parameters of the problem are fixed and have the values $w^{(1)} = 1$, $R = 0.8$ and $s = 0.75$. The circle $Q(x)$ is approximated by a polygon. Let τ be the reverse time in the backward procedure for the construction of fronts. The optimal time for a given state x is the least time τ such that $x \in W(\tau, M)$.

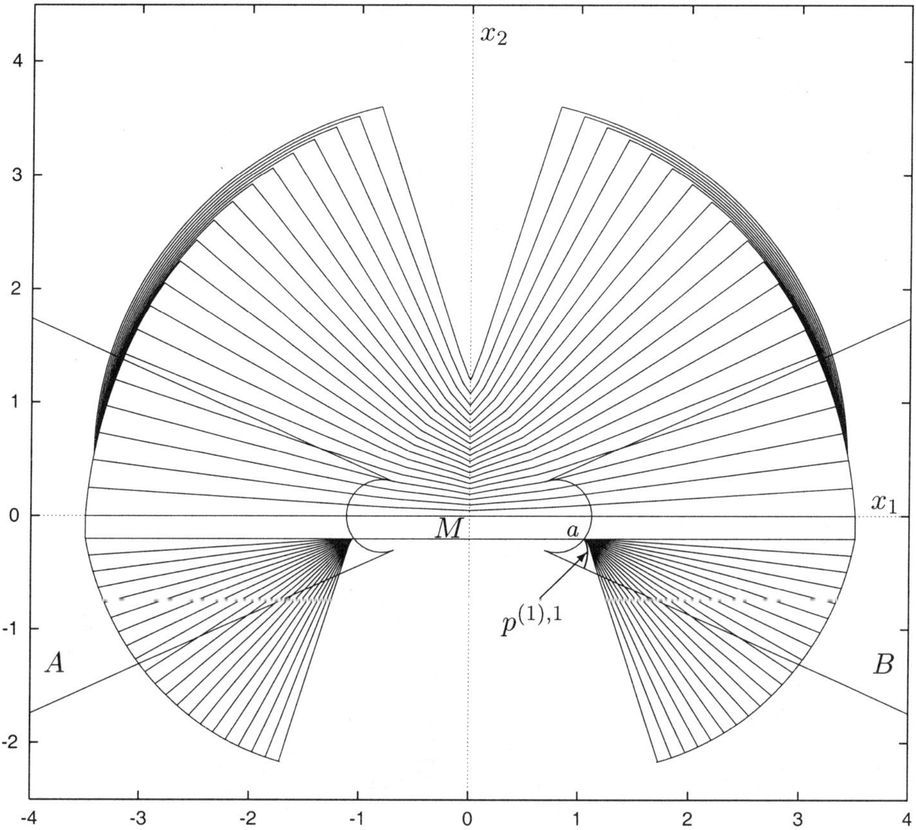

Figure 15 200 upper and lower fronts for $w_e = 0.4$ (every 10th front is plotted)

In Figure 15, the initial computations for $w_e = 0.4$ are shown. The step Δ is 0.005. The usable part of the terminal set M consists of three segments: the upper side of M and two segments on the lower side. The upper fronts that occur until $\tau = 0.29$ are bounded on the left and right by barrier lines. At $\tau = 0.29$, these barrier lines meet the upper boundaries of the sets A and B (see Figure 9), so they terminate. The value function is discontinuous across the barrier lines. For $\tau > 0.29$, the fronts begin to envelop the barrier lines, and left and right corner points arise. The propagation of the front beyond the barrier lines from these corner points is at a very low rate. An enlargement of this development of the fronts on the right hand side is presented in Figure 16.

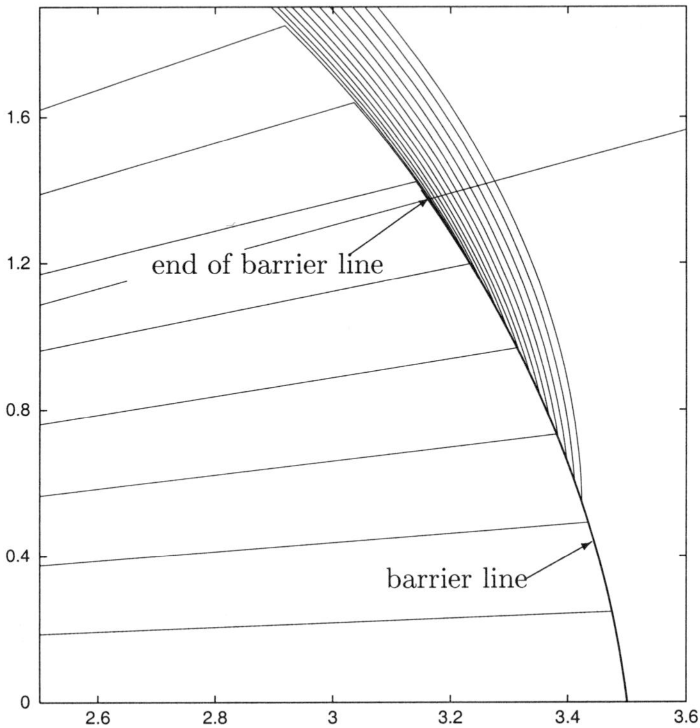

Figure 16 The structure of fronts near the barrier line

The continuation of the computation is shown in Figure 17. The upper and lower fronts are calculated until $\tau = 1.6$ and $\tau = 3.3$, respectively. The left and right lower fronts collide at $\tau = 1.76$. Only one lower front remains after this collision. The greatest value of τ below M occurs on the lower boundary of M at the point $(0, -0.2)$.

An enlargement of the accumulation of the lower fronts is shown in Figure 18. We see that the end of the front moves along the terminal set from the end of the usable part to the point a on the boundary of the set B. The accumulation of fronts begins when they approach the semipermeable curve $p^{(1),1}$ that emanates from the point a, as shown in Figure 15. The value function changes very rapidly in the accumulation region, but it remains continuous.

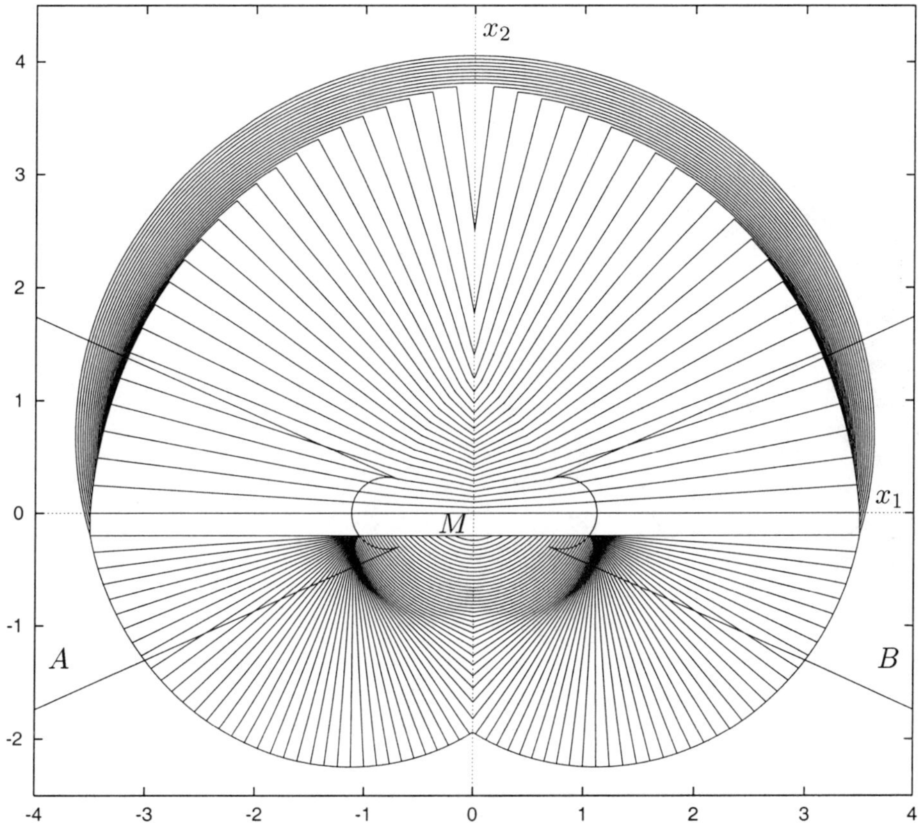

Figure 17 320 upper fronts and 660 lower fronts for $w_e = 0.4$

Figure 19 presents the computational results for $w_e = 0.95$ and $\Delta = 0.005$. As in the previous example, the upper barrier lines end at some moment of reverse time, and the fronts begin to envelop them. The main difference from before is the formation of a loop where the upper fronts from the two sides of the figure meet. In this example, the region within this loop (a "lagoon") is filled out entirely by the further development of the fronts, the filling out being completed at $\tau = 1.68$.

An important feature of the lower part of Figure 19 is that the semipermeable curve $p^{(1),1}$, emanating from the point a, intersects the right barrier which is the semipermeable curve $p^{(2),1}$. This did not happen in the previous example. Thus the right lower fronts are confined to the right side of the curve $p^{(1),1}$. The time of attaining the terminal set becomes infinite as the fronts approach the curve $p^{(1),1}$. A symmetric situation occurs for the left lower fronts. All the fronts are computed until $\tau = 2.4$.

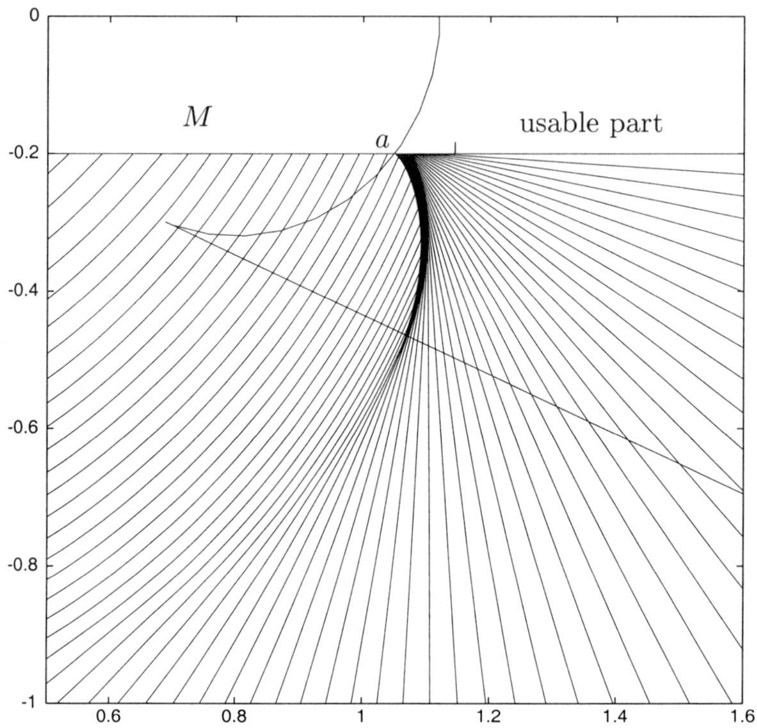

Figure 18 The accumulation of fronts near the point a

The following facts were found experimentally. A lagoon is generated by the upper fronts only if $w_e \geq 0.65$. For $w_e \in [0.65, 1.37)$, a lagoon occurs and is completely filled by the further development of the fronts. For $w_e \in [1.37, 1.61]$, the fronts do not fill the lagoon completely. For $w_e > 1.61$, the lagoon disappears.

Figure 20 presents computational results for $w_e = 1.5$ and $\Delta = 0.005$. The left and right parts of the upper front meet at $\tau = 2.855$. Then the computation within the lagoon begins. The fronts do not penetrate the set D^*, which is a hole inside the solvability set of player P, the value function being infinite for $x \in D^*$. The computation is done until $\tau = 3.73$. The structure of the lower fronts is similar to that in the previous example.

In Figure 21, a three-dimensional graph of the value function of the Figure 20 example is presented. The axes in the horizontal plane are x_1 and x_2, and the vertical axis measures the value function. The picture shows the value function for the region of (x_1, x_2) where the fronts

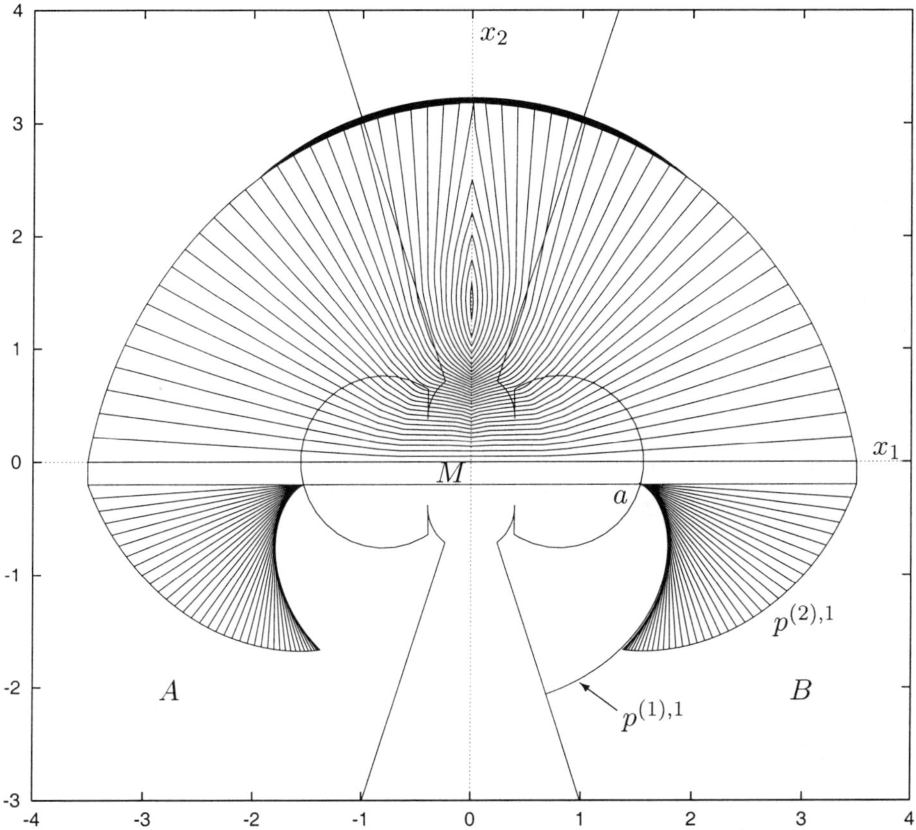

Figure 19 480 upper and lower fronts for $w_e = 0.95$ (every 10th front is plotted)

are computed. The programs for the visualization of such graphs were developed [2] by V.Averbukh and O.Pykhteev, Department of System Support, Institute of Mathematics and Mechanics, Ekaterinburg.

Further increases in the value of w_e extend the set D^*. For example, Figure 22 gives computational results for $w_e = 1.9$ and $\Delta = 0.01$. The upper and lower fronts are computed until $\tau = 8.42$ and $\tau = 1.6$, respectively.

7. CONCLUSION

In this paper, we have studied a variant of the homicidal chauffeur differential game, under the assumption that the constraint on the control of the evader depends on the state. The two-dimensional situation allows a complete description of the families of semipermeable curves

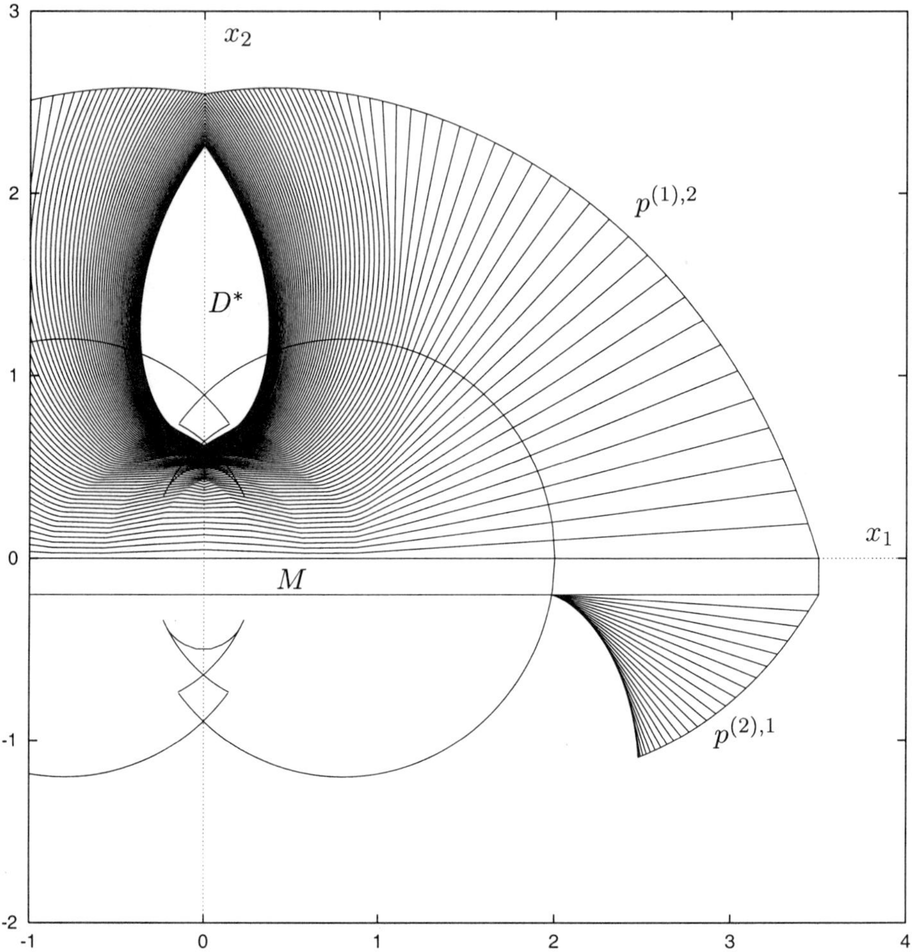

Figure 20 746 upper fronts and 340 lower fronts for $w_e = 1.5$ (every 10th front is plotted)

that occur. The superiority sets of the evader, where semipermeable curves do not exist, are detected. Thus the presence of holes that are strictly inside the "victory domains" of the pursuer is explained. A short description of the backward procedure for the computation of level sets of the value function is also given. This procedure can be employed as a specific algorithm for two-dimensional front propagation.

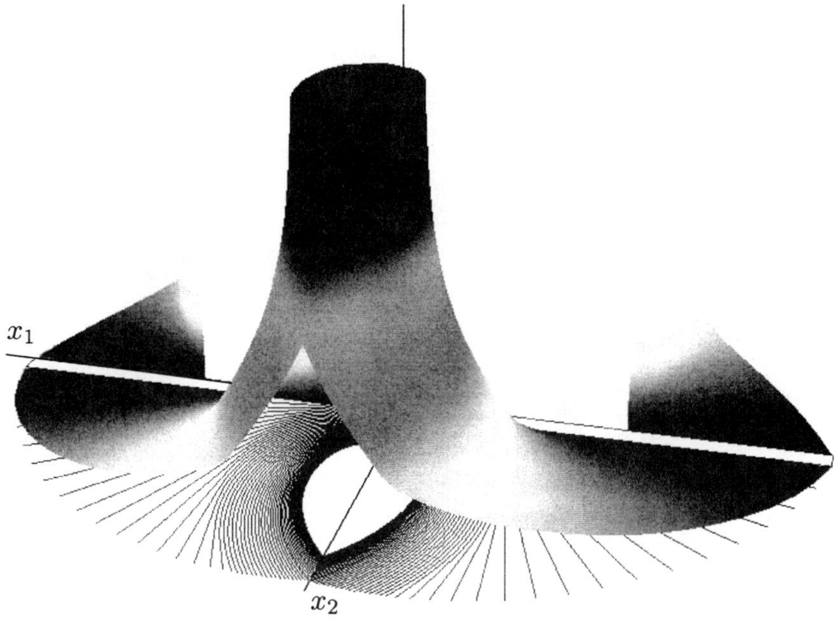

Figure 21 The graph of the value function for $w_e = 1.5$

Acknowledgments

The authors are very grateful to M.J.D. Powell for many valuable remarks on the paper.

References

[1] J.-P. Aubin (1990), A Survey of Viability Theory, SIAM J. Control and Optimization, 28(4), pp. 749–788.

[2] V.L. Averbukh, S.S. Kumkov, V.S. Patsko, O.A. Pykhteev and D.A. Yurtaev (1999), Specialized Vizualization Systems for Differential Games, in *Progress in Simulation, Modeling, Analysis and Synthesis of Modern Electrical and Electronic Devices and Systems*, N.Mastorakis (Ed.), World Scientific and Engineering Society Press, pp. 301–306.

[3] P. Bernhard and B. Larrouturou (1989), Etude de la barriere pour un probleme de fuite optimale dans le plan, Rapport de Recherche,

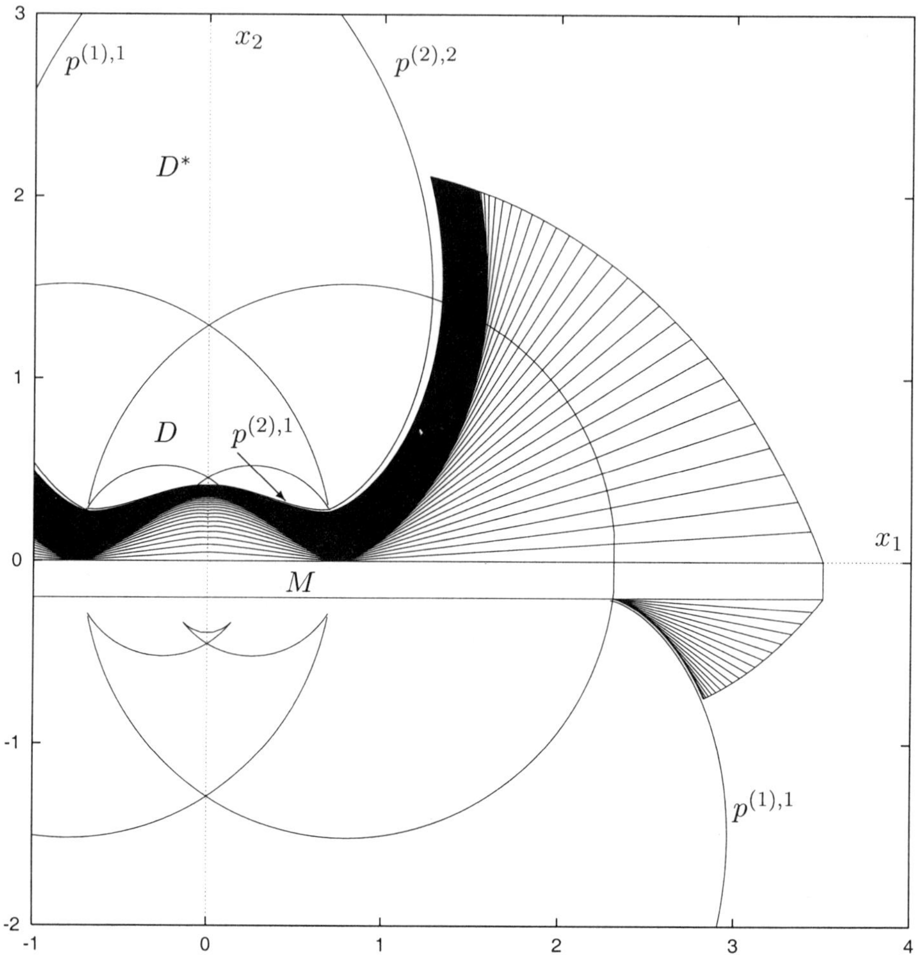

Figure 22 842 upper fronts and 160 lower fronts for $w_e = 1.9$ (every 5th front is plotted)

INRIA, Sophia - Antipolis.

[4] J.V. Breakwell and A.W. Merz (1970), Towards a complete solution of the homicidal chauffeur game, in *Proceedings of the first international conference on the theory and applications of differential games*, Amherst.

[5] P. Cardaliaguet (1997), Nonsmooth Semipermeable Barriers, Isaacs' Equation, and Application to a Differential Game with One Target and Two Players, *Applied Mathematics and Optimization*, 36, pp. 125–146.

[6] P. Cardaliaguet, M. Quincampoix and P. Saint-Pierre (1995), Numerical methods for optimal control and differential games, Ceremade CNRS URA 749, University of Paris - Dauphine.

[7] R. Isaacs (1965), *Differential Games*, John Wiley, New York.

[8] J. Lewin and G.J. Olsder (1979), Conic Surveillance Evasion, *Journal of Optimization Theory and Applications*, 27(1), pp. 107–125.

[9] N.N. Krasovskii and A.I. Subbotin (1974), *Positional Differential Games*, Nauka, Moscow.

[10] N.N. Krasovskii and A.I. Subbotin (1988), *Game-Theoretical Control Problems*, Springer-Verlag, New York.

[11] A.W. Merz (1971), The Homicidal Chauffeur – a Differential Game, PhD Dissertation, Stanford University.

[12] V.S. Patsko and V.L. Turova (1995), Numerical Solution of Two-Dimensional Differential Games, Preprint, Institute of Mathematics and Mechanics, Ekaterinburg.

[13] V.S. Patsko and V.L. Turova (1997), Numerical Solutions to the Minimum-Time Problem for Linear Second-Order Conflict-Controlled Systems, in *Proceedings of the Seventh International Colloquium on Differential Equations, Plovdiv, Bulgaria, 18–23 August 1996*, D.Bainov (Ed.), Utrecht, The Netherlands, pp. 327–338.

[14] V.S. Patsko and V.L. Turova (1998), Numerical study of the Homicidal Chauffeur Game, in *Proceedings of the Eight International Colloquium on Differential Equations, Plovdiv, Bulgaria, 18–23 August 1997*, D.Bainov (Ed.), Utrecht, The Netherlands, pp. 363–371.

[15] V.S. Patsko and V.L. Turova (1998), Homicidal Chauffeur Game. Computation of Level Sets of the Value Function, *Preprints of the 8th Int. Symposium on Differential Games and Applications. Maastricht, 5–7 July, 1998*, pp. 466–473.

APPLICATION OF A MICRO-GENETIC ALGORITHM IN OPTIMAL DESIGN OF A DIFFRACTIVE OPTICAL ELEMENT *

Svetlana Rudnaya
Avant! Corporation,
48671 Bayside Parkway, Fremont, CA 94598, USA

Fadil Santosa
School of Mathematics,
University of Minnesota, Vincent Hall,
206 Church St SE, Minneapolis, MN 55455, USA
santosa@math.umn.edu

Abstract This study is motivated by a need to design a diffractive optical element arising in an application. Under realistic manufacturing constraints, it can be shown that the design problem is an optimization calculation with integer variables. We consider an optimization strategy based on Genetic Algorithms. We show that for a particular variant, called a Micro-Genetic Algorithm, the algorithm converges in a probabilistic sense to the global optimum. We demonstrate the use of the algorithm in the design of a diffractive optical element.

Keywords: Diffractive Optics, Genetic Algorithms, Optimization

1. INTRODUCTION

We study the problem of creating a desired light intensity pattern on a screen. A light source is given. What is required is a thin-film diffractive optical element, a lens, that alters the incoming light to produce the desired pattern. A sketch of the setup is shown in Figure 1. Under the thin-lens approximation, the desired unknown is a thickness distribution over the film. This is a form of inverse problem.

---

*The authors acknowledge support from the 3M Corporation, and from AFOSR through a MURI grant to the University of Delaware.

M.J.D. Powell and S. Scholtes (Eds.), *System Modelling and Optimization: Methods, Theory and Applications.*

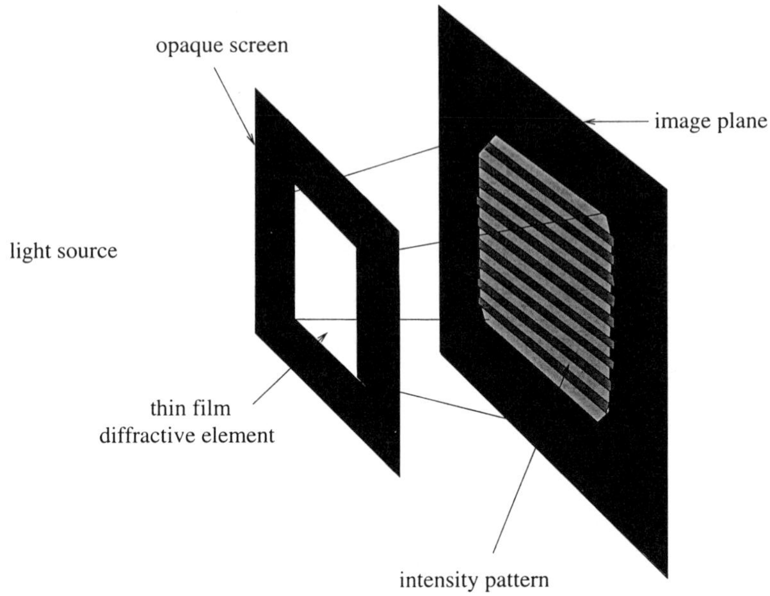

Figure 1 A sketch of the optimal design problem. Light enters the aperture where a thin film element has been placed. The incoming light is altered as it exits the film. As the light strikes the screen, it produces an intensity pattern on the screen. The inverse problem is to design a diffractive optical element that produces a desired intensity pattern.

The so-called 'direct problem' involves finding the intensity pattern given a light source, a diffractive optical element, and the geometry of the problem. We will show that this process is akin to function evaluation, and, under the Kirchhoff approximation, amounts to a quadrature.

The inverse problem then is to find a diffractive optical element which produces an intensity pattern that is as close as possible to the desired pattern. Such a problem is very naturally cast as an optimization calculation. What is somewhat unusual about this work is that we are, in addition, given information about the manufacturing process for the diffractive optical element. The process poses additional constraints, which can be modeled as integer variables with simple bounds. However, the size of the problem, usually involving over 20000 variables in practical applications, makes it extremely difficult to solve using standard optimization methods.

We propose the use of a variant of the Genetic Algorithm [4, 7, 12]. Genetic Algorithms have found success in several areas of optimal design in optics [3, 9], as well as in other areas. We opted to use the Micro-Genetic Algorithm [11]. We study its convergence properties using Markov Chain analysis [15, 16]. The use of the algorithm applied to the optical problem is demonstrated in a numerical example. We end the paper with a discussion of some of the issues that occur in the optimization calculation.

2. OPTIMAL DESIGN OF DIFFRACTIVE OPTICAL ELEMENTS

Consider the 2-dimensional problem that is indicated in Figure 2. A lens has been placed in an opening of width $2w$ in an opaque screen. An incident light propagates from the left and is diffracted by the lens-screen setup. The diffracted light is gathered on the image plane $z = d$.

Under the scalar model of light, the scalar field, which could represent the strength of the electric or magnetic field, satisfies the Helmholtz equation. Specifically, the scalar field $u(x, z)$ has the property

$$\Delta u + k^2 n(x, z)^2 u = 0,$$

where $n(x, z)$, the index of refraction, is normalized to 1 in the air, and is equal to n_0 in the lens. The variable k represents the normalized wavelength in air. To complete the description of the diffraction problem, we would specify boundary conditions on the opaque screen along $\{z = 0, |x| > w\}$, and the Sommerfeld radiation condition for $x^2 + z^2 \to \infty$. By specifying an incident wave, in this case the wave generated by the source, one can determine the scattered wave by solving the partial differential equation with the boundary conditions. Then, after calculating $u(x, z)$, one can easily find the intensity at the image plane $z = d$, because it has the value $|u(x, d)|^2$.

When the features in the diffractive optical element are many times larger than the wavelength of light, we can approximate the field diffracted by the lens using Kirchhoff approximation [2, 5] (see also [14]). Under this approximation, the field at a point (x_0, z_0) is related to its value at the aperture by

$$u(x_0, z_0) = \frac{ik}{2} \int_{-w}^{w} u(x, 0) \, \frac{z_0}{r} \, H_1^{(1)}(kr) \, dx, \tag{1}$$

where $r = \sqrt{(x - x_0)^2 + z_0^2}$, and $H_1^{(1)}(\cdot)$ is the Hankel function of the first kind of order 1. This formula allows us to find the field at (x_0, z_0) if we know the field at the points where the light leaves the lens.

Another approximation, based on the fact that the lens is thin, is applied. Under this approximation, the lens changes only the phase of the incoming light. Specifically, if the incident light at $z = 0$ has

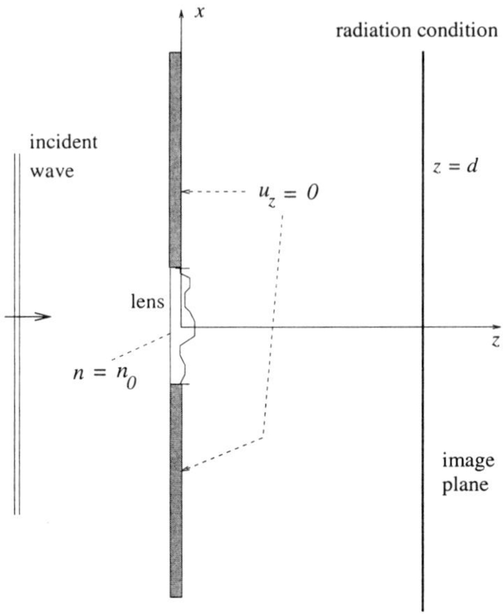

Figure 2 In the scalar model of light, the field $u(x, z)$ satisfies the Helmholtz equation. The lens has an index of refraction different from that of air. There are boundary conditions on the opaque parts of the screen, and where $x^2 + z^2 \to \infty$. The intensity at the image plane is $|u(x, d)|^2$.

amplitude $u(x, 0)$, the lens alters the amplitude to

$$u(x, 0) \exp i\varphi(x).$$

Here φ is the function

$$\varphi(x) = \frac{2\pi(n_0 - 1)}{\lambda} d(x), \qquad -w \leq x \leq w,$$

where $d(x)$ is the thickness of film at x, n_0 is the index of refraction of the lens, and λ is the wavelength of light in air.

The diffractive optical element is produced by a material removal process. The resulting thickness profile $d(x)$ is piecewise constant over regular subintervals of width $\Delta x = 2w/n_x$, where n_x is a prescribed positive integer. Each value of $d(x)$ is an integer multiple of some fixed thickness t. This means that the corresponding phase $\varphi(x)$ is also piecewise constant. Assuming that the material removal rate is constant for each removal process, we choose t, depending on n_0 and λ, so that the phase $\varphi(x)$ is of the form

$$\varphi(x) = f_j \frac{2\pi}{\ell}, \qquad x_j \leq x < x_{j+1}, \quad j = 0, 1, \ldots, n_x - 1, \qquad (2)$$
$$f_j \in A := \{0, 1, 2, \ldots, \ell - 1\},$$

where $x_k = -w + k\Delta x$, $k = 0, 1, \ldots, n_x$. The variable ℓ corresponds to the number of steps in the material removal process. Thus the unknowns of the problem are the components of a vector f of length $I\!R^{n_x}$, representing the thickness profile.

The optimization problem can be posed as follows. Assume the incident amplitude is constant $u(x, 0) = C$. We are given the required intensity $I_{\text{target}}(x)$ on the image plane $z = d$. We wish to solve

$$\min_{f \in A^{n_x}} ||I - I_{\text{target}}||^2, \tag{3}$$

$$I(x) = |u(x, d)|^2, \tag{4}$$

$$u(x, d) = \frac{ik}{2} C \int_{-w}^{w} \exp i\varphi(\xi) \, \frac{d}{r} \, H_1^{(1)}(kr) \, d\xi, \tag{5}$$

where $r = \sqrt{(x - \xi)^2 + d^2}$. The integer vector $f \in I\!R^{n_x}$ is related to $\varphi(x)$ through (2). Conservation of energy allows C to be determined from I_{target}.

The difficulties in this problem stem from the facts that (i) the required components of f are integer variables, (ii) the number of unknowns n_x is large in typical design problems, (iii) the functional in (3) is nonlinear. While there has been much progress in solving nonlinear integer programming problems, we were drawn to use Genetic Algorithms due to their simplicity. We will demonstrate their ability to "solve" this problem in a numerical example, and we will discuss their convergence properties.

3. REVIEW OF GENETIC ALGORITHMS

Genetic Algorithms are "search algorithms based on the mechanism of natural selection and natural genetics" [4]. Principles of evolution and heredity are used for the optimization of an objective function.

Genetic Algorithms are iterative and operate on a set of potential solutions, called population, and utilize concepts such as selection (survival of the fittest), crossover (information exchange), and mutation (adding the genetic diversity) [12]. Genetic Algorithms use only objective function information (no derivatives) and probabilistic transition rules. The origins of Genetic Algorithms go back to the work of Holland [7] and others in the 1970s. In recent years the algorithms have attracted a lot of interest, and have been used successfully for optimization in different areas, in particular, in diffractive optics design problems [3, 9, 10].

Let us consider the most basic version, called the Canonical Genetic Algorithm. Suppose we want to solve

$$\max_{\mathbf{x} \in X} f(\mathbf{x}), \tag{6}$$

where X is some feasible set, and $f(\mathbf{x}) > 0$ for all $\mathbf{x} \in X$. Genetic Algorithms use a *binary representation* of the variable $\mathbf{x}$. If $\mathbf{x}$ is a continuous variable, it is given finite precision, and can thus be expressed in binary form. In our problem, each member of X has integer components, which can be represented easily by a vector of binaries. Therefore we can henceforth assume that $\mathbf{x} = (x^1, x^2, \ldots, x^\nu)$, $x^j \in \{0, 1\}$. Every such string is called a *chromosome* or an *individual.* Binary entries x^j of the string $\mathbf{x}$ are called *genes.* The cost function $f(\mathbf{x})$ is refered to as the *fitness.*

A Genetic Algorithm works on a set of N chromosomes (individuals), $\mathbf{x}_1, \mathbf{x}_2, \ldots, \mathbf{x}_N$, called a *population.* The initial population is formed at random. Each iteration updates a population to the next generation. The basic principle of the update is to increase the average fitness of the population.

At the beginning of each iteration, all individuals of the population are evaluated for fitness, that is, we compute $f(\mathbf{x}_i)$, $i = 1, 2, \ldots, N$. Next the *probability for reproduction* is assigned to each chromosome, according to its relative fitness

$$p_i = \frac{f(\mathbf{x}_i)}{\sum_{j=1}^{N} f(\mathbf{x}_j)}.$$

In this way, the fitter members of the population have a better chance of being chosen, with some fit members having the possibility of being chosen more than once. We pick N chromosomes in this fashion, and they form the *mating pool.*

The mating pool is subdivided at random into pairs. Each pair of parents, with a probability p_{c} say, performs a *crossover* to produce 2 offspring. A crossover is akin to exchanging genetic information. For example, in a 2-point crossover

$$\begin{array}{lcll} \mathbf{x}_1 & = & 1\ 0 & \!\!\!\big|\ 0\ 1\ 1 \\ \mathbf{x}_2 & = & 0\ 1 & \!\!\!\big|\ 1\ 0\ 1 \end{array} \quad \text{becomes} \quad \begin{array}{lcll} \mathbf{x}'_1 & = & 1\ 0 & \!\!\!\big|\ 1\ 0\ 1 \\ \mathbf{x}'_2 & = & 0\ 1 & \!\!\!\big|\ 0\ 1\ 1 \end{array}.$$

The crossover point, which could be chosen at random, is indicated above by a vertical line. If a crossover is performed, 2 *offspring* replace their 2 parents, but otherwise both parents proceed to the next generation unchanged.

Finally, *mutation* is applied to each gene (binary element of the vector $\mathbf{x}$) of each offspring, with a probability p_{m} that is usually very small. The mutated gene is replaced by the opposite value ($0 \to 1$ or $1 \to 0$). Even when the probability of mutation is tiny, its role is important as an additional provider of diversity, so it is a source of valuable information that could be missing in the initial population.

This is the end of a single iteration. A new population is formed and the process is repeated until a convergence condition is satisfied or until the number of iterations reaches a prescribed limit. The process is summarized in the pseudo-code in Figure 3.

```
create initial population of N chromosomes
until convergence, do
        select parent pairs
        crossover
        mutate
        new population
end
```

Figure 3 The Canonical Genetic Algorithm

By convergence, we mean that the highest fitness value of the population has reached an acceptable value. Alternatively, we could stop the algorithm when the maximum of the population is sufficiently close to the average fitness of the population.

There are 3 parameters in the algorithm: (i) size of population N, (ii) crossover probability p_{c}, and (iii) mutation probability p_{m}. These parameters tend to be problem specific, and there are no general rules on how to set them [6]. An evolutionary strategy in which the parameters are allowed to evolve just like the population according to some rule has also been proposed [12].

An important variant to the Canonical Genetic Algorithm is one in which the fittest member of the population in each generation is saved. This guarantees that the maximum fitness value of the population is nondecreasing.

The Micro-Genetic Algorithm, first proposed in [10, 11], works on a small population of $m + 1$ chromosomes (for example, $m = 5$). There is an inner loop and an outer loop. Within the inner loop, *selection* and *crossover* take place, while the fittest member of the population is saved in each generation. After a convergence criterion is satisfied, we go out of the inner loop and regenerate a new population by mutating the fittest member m times. The new population then enters the inner loop. A feature suggested for this scheme is that the mutation probability p_{m} should be made smaller adaptively as we near convergence. The pseudo-code for the algorithm is given in Figure 4.

The crossover in Figure 4 is of a special type called 'uniform crossover'. In this process, we take m members of the population and create m new ones by sequentially picking genes from the pool with uniform probability. That is, in creating a new $\mathbf{x}$ of length ν, we pick the first gene (bit)

```
create initial population of m + 1 chromosomes
for iteration  = 1 to MAX
       until convergence, do
            save best
            generate m more chromosomes for the next
              population by performing uniform crossover
       end
       keep best chromosome
       mutate m times to create the other elements
         of the new population
end
```

Figure 4 The Micro-Genetic Algorithm.

from the first bits of each of the m members, with uniform probability. We pick the second bit from the second bits of the m members similarly, etc.

Two more features of the Micro-Genetic Algorithm are worth noting. Firstly, because we work with a small population, the inner loop converges fairly rapidly. Secondly, the purpose of the outer loop is to introduce diversity via *mutation*, in order to explore other areas of the search space.

One clear drawback of Genetic Algorithms is that they do not have a good stopping rule. Indeed, our numerical experience shows that improvements in the fitness function as we iterate tend to occur abruptly after a few steps that make little or no progress. However, we adopt the point of view that the problem is to get an acceptable design, i.e., a design whose residual in the fitness function is small relative to the norm of the target intensity. Then one can stop the iterations as soon as this threshold is crossed.

4. CONVERGENCE ANALYSIS OF MICRO-GENETIC ALGORITHMS

The application of Markov Chain analysis to Genetic Algorithms was first introduced by Rudolph [15], and developed in his book [16]. A more recent paper by Agapie [1] investigates the minimum criteria for convergence.

The main ideas are: (i) the population at any one point in the iteration can be viewed as a *state*, (ii) the finite binary representation of the variables makes the system a *finite* state machine, (iii) iterations are transitions from one state to another. Basic results from stochastic matrices [8, 13, 17] will be useful.

With this in mind, let $\mathbf{x}$ be a binary vector of length ν, and let $m+1$ be the total number of members in a population. Each state is represented by a concatination of $m+1$ vectors s=$[\mathbf{x}_0, \mathbf{x}_1, \ldots, \mathbf{x}_m]$. We denote the state space by S. The total number of possible states is $|S| := L = 2^{(m+1)\nu}$. Now we index the states from 1 to L as $s_1, s_2, \ldots, s_L$. Of all the possible states, there are $M = 2^\nu$ states in which the members $\mathbf{x}_i$, $i = 0, 1, \ldots, m$, are *identical.* We define

$$U = \{s_i : \mathbf{x}_0 = \mathbf{x}_1 = \cdots = \mathbf{x}_m\} \; = \; \{s_i : i = i_1, i_2, \ldots i_M\}, \qquad (7)$$

say. Let $P\{\bullet\}$ denote the probability of the assertion $\bullet$. Consider the inner loop in the Micro-Genetic Algorithm, sketched in Figure 4, and let $s^{(k)}$ be the population after the k-th iteration.

Theorem 1 *The inner loop of the Micro-Genetic Algorithm in Figure 4 converges to a uniform population in the sense that*

$$\lim_{k\to\infty} P\{s^{(k)} \in U\} = 1.$$

Proof Let $\mathbf{P}_{\text{inner}} = \{p_{ij}\}$ be the transition matrix of the inner loop for each iteration (generation); i.e.,

$$p_{ij} = P\{s_i \to s_j\},$$

by the process of saving the best member and performing uniform crossover. Note that $\mathbf{P}_{\text{inner}}$ will be stochastic (rows summing to 1) and have nonnegative entries. If we order the states so that the first M states are in U, then the transition matrix $\mathbf{P}_{\text{inner}}$ has the form

$$\mathbf{P}_{\text{inner}} = \begin{bmatrix} I & 0 \\ A & B \end{bmatrix},$$

where I is the $M \times M$ identity matrix. This is because, if $s^{(k)} \in U$, then $s^{(k)}$ is not altered by the uniform crossover of the inner loop.

Next we study the properties of A and B. Note that the block A is of size $(L-M) \times M$, and corresponds to the transition from $s^{(k)} \notin U$ to $s^{(k+1)} \in U$. This means that it has only a single nonzero entry per row, corresponding to the element of U that is composed of $m+1$ copies of the 'best' chromosome that was saved at the beginning of the inner iteration. The probability of reproducing m copies of the first member of $s^{(k)}$ is at least $(1/m)^{m\nu}$, which provides the crude lower bound

$$P\{s^{(k+1)} \in U\} \geq a := \left(\frac{1}{m}\right)^{m\nu}.$$

Therefore, if a_{ij} is the nonzero entry in row i of A, then $a_{ij} \geq a > 0$.

The submatrix B is of size $(L-M) \times (L-M)$. Since $\mathbf{P}_{\text{inner}}$ is stochastic

$$\sum_j b_{ij} \leq 1 - a < 1.$$

By direct calculation, we find that

$$\mathbf{P}_{\text{inner}}^k = \begin{bmatrix} I & 0 \\ A_k & B^k \end{bmatrix},$$

where $A_k = \left(\sum_{i=0}^{k-1} B^i\right) A = (I-B)^{-1}(I-B^k)A$. Now, if a matrix B has rows whose sums are less than one, then $B^k \to 0$ as $k \to \infty$. It follows that

$$\lim_{k\to\infty} \mathbf{P}_{\text{inner}}^k = \mathbf{P}_{\text{inner}}^\infty = \begin{bmatrix} I & 0 \\ A_\infty & 0 \end{bmatrix},$$

with $A_\infty = (I-B)^{-1}A$.

Therefore, starting with any distribution $p^{(0)}$, the final distribution

$$p^{(\infty)} = p^{(0)}\mathbf{P}_{\text{inner}}^\infty$$

has the property $p_j^{(\infty)} = 0$ for $j = M+1, \cdots, L$, thus proving the assertion of the theorem. ■

A convergence rate for the inner loop can be estimated. From the probability transition $p^{(k)} = p^{(0)}\mathbf{P}_{\text{inner}}^k$, we deduce

$$P\{s^{(k)} \in U\} = \sum_{\{i:s_i \in U\}} p_i^{(k)} = 1 - P\{s^{(k)} \notin U\} = 1 - \sum_{\{i:s_i \notin U\}} p_i^{(k)}.$$

Further, the nature of the transition matrix implies

$$p_i^{(k)} = \sum_{\{j:s_j \notin U\}} p_j^{(0)} b_{ji}^{(k)}, \quad \{i : s_i \notin U\},$$

where $B^k = [b_{ij}^{(k)}]$. By using $\sum_j b_{ij} \leq 1 - a$ for all i, and by summing both sides of the equation, we get

$$\begin{aligned} \sum_{\{i:s_i \notin U\}} p_i^{(k)} &= \sum_{\{i:s_i \notin U\}} \sum_{\{j:s_j \notin U\}} p_j^{(0)} b_{ji}^{(k)} \\ &= \sum_{\{j:s_j \notin U\}} p_j^{(0)} \sum_{\{i:s_i \notin U\}} b_{ji}^{(k)} \leq (1-a)^k \sum_{\{j:s_j \notin U\}} p_j^{(0)} \leq (1-a)^k. \end{aligned}$$

Therefore we have established

$$P\{s^{(k)} \in U\} \geq 1 - (1-a)^k,$$

where $a = (1/m)^{m\nu}$.

Each single step of the outer loop of the Micro-Genetic Algorithm can be expressed using the transition matrix

$$\mathbf{P}_{\text{outer}} = \mathbf{M}\mathbf{P}_{\text{inner}}^{\infty}, \tag{8}$$

where $\mathbf{M}$ represents the mutation process, and where the entire inner loop has been encapsulated in $\mathbf{P}_{\text{inner}}^{\infty}$. Now we can state the convergence result.

Theorem 2 *Assume that there is a unique global maximizer of the optimization problem (6). Let the probability of mutation in the outer loop be fixed with* $p_m \in (0,1)$. *Then the iterations of the Micro-Genetic Algorithm of Figure 4 converge to a population consisting of* $m+1$ *copies of the global maximizer of the cost function.*

Proof Recall that an unknown binary vector of length ν can take 2^ν values. When these 2^ν different vectors are arranged in descending order of fitness, we denote them by $\mathbf{x}_1, \mathbf{x}_2, \cdots, \mathbf{x}_{2^\nu}$. Moreover, we order the populations of size $(m+1)$ by *class*. The first class contains populations whose first member is $\mathbf{x}_1$, the second class contains populations whose first member is $\mathbf{x}_2$, and so on, where the members of each population need not be in order of fitness. We have a total of 2^ν classes. In each class, we let the first population be the one that has $m+1$ copies of its leading member. This process is nothing more than renumbering the states in a particular order.

Next, recall that the inner loop takes any population into one that has identical members. Consider the transition

$$p^{(k+1)} = p^{(k)}\mathbf{P}_{\text{inner}}^{\infty}.$$

The new vector $p^{(k+1)}$ has only 2^ν nonzeros, corresponding to the populations whose members are identical in the 2^ν classes. Note also that the populations can only go up in class, or remain in the same class, because we always save the fittest member of the population. Therefore we deduce that the matrix $\mathbf{P}_{\text{inner}}^{\infty}$ must be lower triangular, and of the form

$$\mathbf{P}_{\text{inner}}^{\infty} = \begin{bmatrix} v^{(1,1)} & & & \\ v^{(2,1)} & v^{(2,2)} & & \\ \vdots & \vdots & \ddots & \\ v^{(2^\nu,1)} & v^{(2^\nu,2)} & \cdots & v^{(2^\nu,2^\nu)} \end{bmatrix}. \tag{9}$$

The nonzero elements of the matrix are confined to the 2^ν columns indicated by thin rectangles and labelled $v^{(i,j)}$. From our indexing scheme, we know that

$$v^{(1,1)} = \begin{bmatrix} 1 \\ 1 \\ \vdots \\ 1 \end{bmatrix},$$

because $\mathbf{P}_{\text{inner}}^{\infty}$ is stochastic and each row of the block corresponding to the first class has a single nonzero entry. Therefore the entry must be 1. Let us consider $v^{(i,1)}$. Each element of this vector corresponds to the probability of a transition to the first class from class i. We argue that at least one element of $v^{(i,1)}$ must be equal to 1. These elements correspond to the existence of populations in class i with probability 1 of exiting the inner loop in class 1. They have $\mathbf{x}_1$ as a member, and therefore would immediately be promoted to class 1 by the inner loop, and would stay in that class thereafter.

Now we look at the transition matrix $\mathbf{M}$ corresponding to mutation. The population, upon exit from the inner loop, has $m+1$ identical members. A total of m new members are created by sequentially 'flipping' bits of the chromosome with probability p_{m}. The probability that every bit is flipped is $p_{\text{m}}^{m\nu}$, whereas the probability that every bit remains the same is $(1-p_{\text{m}})^{m\nu}$. Therefore the nonzero part of the transition matrix $\mathbf{M}$ satisfies

$$\min_{ij}[\mathbf{M}]_{ij} \geq a := \min\{p_{\text{m}}^{m\nu}, (1-p_{\text{m}})^{m\nu}\}.$$

We keep the original member before mutation and label it as the first member, thus defining the class of the population, so the mutation process cannot change the class. Therefore the transition matrix corresponding to mutation has the block diagonal structure

$$\mathbf{M} = \begin{bmatrix} \mathbf{M}_1 & & & \\ & \mathbf{M}_2 & & \\ & & \ddots & \\ & & & \mathbf{M}_{2^\nu} \end{bmatrix},$$

where each block is a $2^{m\nu}$ by $2^{m\nu}$ stochastic matrix. Using (9) and the above, the product (8) takes the form

$$\mathbf{P}_{\text{outer}} = \begin{bmatrix} w^{(1,1)} & & & \\ w^{(2,1)} & w^{(2,2)} & & \\ \vdots & \vdots & \ddots & \\ w^{(2^\nu,1)} & w^{(2^\nu,2)} & \dots & w^{(2^\nu,2^\nu)} \end{bmatrix},$$

where $w^{(i,j)} = \mathbf{M}_i\, v^{(i,j)}$. The fact that $\mathbf{M}_1$ is stochastic implies

$$w^{(1,1)} = \begin{bmatrix} 1 \\ 1 \\ \vdots \\ 1 \end{bmatrix} = v^{(1,1)}.$$

Observe also that

$$w_j^{(i,1)} = \left[\mathbf{M}_i\, v^{(i,1)}\right]_j = \sum_l [\mathbf{M}_i]_{jl}\, v_l^{(i,1)} \geq a \sum_l v_l^{(i,1)} \geq a, \qquad (10)$$

the last inequality being valid because we have deduced that at least one element of $v^{(i,1)}$ is unity. Further, showing only the first row of $\mathbf{P}_{\text{outer}}$ explicitly, we can write

$$\mathbf{P}_{\text{outer}} = \left[\begin{array}{c|c} 1 & 0 \cdots 0 \\ \hline \vdots & \\ \vdots & Q \end{array}\right],$$

which implies

$$\mathbf{P}_{\text{outer}}^k = \left[\begin{array}{c|c} 1 & 0 \cdots 0 \\ \hline \vdots & \\ \vdots & Q^k \end{array}\right].$$

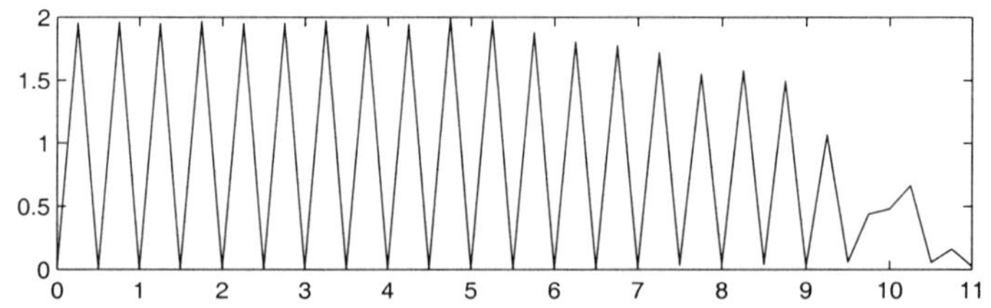

Figure 5 The target intensity pattern.

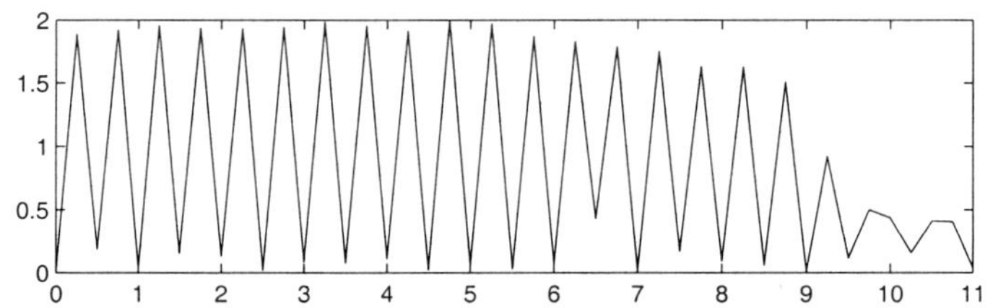

Figure 6 The intensity pattern of the optimized diffractive optical element.

The matrix Q has the property $\sum_j Q_{ij} \leq 1 - a < 1$ by (10), so $Q^k \to 0$ as $k \to \infty$. It follows from the stochastic property of $\mathbf{P}_{\text{outer}}$ that

$$\lim_{k\to\infty} \mathbf{P}_{\text{outer}}^k = \mathbf{P}_{\text{outer}}^\infty = \left[\begin{array}{c|cccc} 1 & 0 & 0 & \cdots & 0 \\ 1 & 0 & 0 & \cdots & 0 \\ \vdots & \vdots & \vdots & \ddots & \vdots \\ 1 & 0 & \cdots & \cdots & 0 \end{array}\right].$$

Therefore, the probability that the population reaches a state in which all the members are $\mathbf{x}_1$ is 1 in the limit. ■

5. NUMERICAL EXPERIMENT

We applied a version of the Micro-Genetic Algorithm to the problem (3). In the example, we chose a symmetric profile with 1000 unknowns. The half-aperture is $x \in [0, 10]$, so the width of each 'bump' is 0.001. The diffractive optical element has 4 levels, so $A = \{0, 1, 2, 3\}$. The desired target image was derived from a rough sampling of the intensity at the image plane $d = 1$ over the interval $x \in [0, 11]$. It was obtained by solving the forward problem with a known diffractive optical element at wavenumber $k = 25$, and is shown in Figure 5. The image found by optimization is given below it in Figure 6.

We used a version of the optimization algorithm that changes the mutation probability p_{m} adaptively. The algorithm ran for a total 500

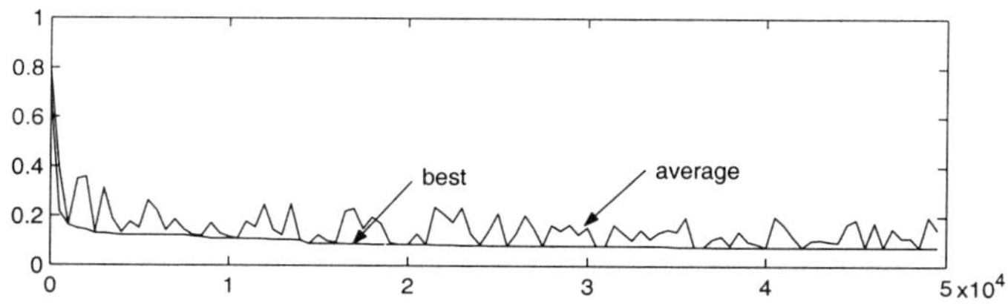

Figure 7 The progress of the optimization process. Since we fix the number of inner loop iterations to 100, the exiting population is not uniform. This graph shows the average fitness and the fitness of the best member upon exit from the inner loop.

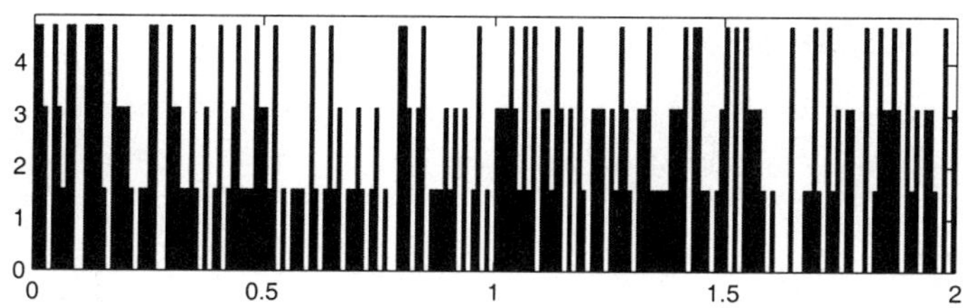

Figure 8 Part of the optimized diffractive optical element. The phase is shown as a function of position.

iterations of the outer loop, and each iteration ran for 100 generations. The total number of function evaluations with the population size $(m+1) = 5$ is 2×10^5. Note that the search space has over 10^{600} elements. The reduction in the residual at the end of each iteration in the outer loop is shown in Figure 7. Part of the calculated profile found is shown in Figure 8.

We were able to run larger examples, with 20000 unknowns, with some success. A disturbing feature of the Micro-Genetic Algorithm, however, is that many iterations may fail to reduce the cost function, which makes it difficult to decide when to stop.

We devised a simple way to obtain crude bounds on the value of the cost function at the optimum. To obtain a lower bound, we relax the search space to continuous variables. This is an unconstrained optimization problem and can be solved by standard methods. In order to get an answer $\varphi(x)$ that is between 0 and 2π, we simply add or subtract integer multiples of 2π if necessary. To obtain a crude upper bound, we take the values of $\varphi(x)$ calculated in the manner described, and round them to the nearest integer multiples of the phase increment. While these bounds are crude, they do provide some guidance for stopping the iterations.

6. DISCUSSION

We have described a problem of designing diffractive optical elements using an optimization strategy. Because of the integer constraints imposed by the manufacturing process, and the size of the problem, we chose a rather unconventional method for solving the optimization calculation. The method is a Micro-Genetic Algorithm. We have analyzed its convergence properties, finding that, while there is convergence, there is no easy way to estimate the convergence rate. Numerical experiments with the method yield satisfactory results. A good match of the target intensity profile is usually achieved by this method.

Acknowledgments

The authors are grateful to Drs. David Misemer and Alessandra Chiareli of 3M for valuable discussions and support of this work.

References

[1] A. Agapie (1998), Genetic Algorithms: Minimal conditions for convergence, *Lecture Notes in Computer Science*, 1363, pp. 183–193.

[2] M. Born and E. Wolf (1964), *Principles of Optics*, Pergamon Press.

[3] D. Brown and A. Kathman (1995), Multi-element diffractive optical designs using evolutionary programming, *SPIE*, 2404, pp. 17–27.

[4] D.E. Goldberg (1989), *Genetic Algorithms in Search, Optimization, and Machine Learning*, Addison-Wesley Publishing Company.

[5] J.W. Goodman (1968), *Introduction to Fourier Optics*, McGraw-Hill, New York.

[6] J.J. Grefenstette (1986), Optimization of control parameters for Genetic Algorithms, *IEEE Transactions on Systems, Man, and Cybernetics*, 16(1), pp. 122–128.

[7] J.H. Holland (1975), *Adaptation in Natural and Artificial Systems*, The University of Michigan Press, Ann Arbor.

[8] M. Iosifescu (1980), *Finite Markov Processes and Their Applications*, Wiley, Chichester.

[9] E.G. Johnson, A.D. Kathman, D.H. Hochmuth, A.L. Cook, D.R. Brown and W.F. Delaney (1993), Advantages of Genetic Algorithm optimization methods in diffractive optic design, *SPIE*, CR49, pp. 54–74.

[10] E.G. Johnson and M.A.G. Abushagur (1995), Microgenetic-algorithm optimization methods applied to dielectric gratings, *JOSA A*, 12(5), pp. 1152–1160.

[11] K. Krishnakumar (1989), Micro-Genetic algorithms for stationary and non-stationary function optimization, *SPIE*, 1196, pp. 289–296.

[12] Z. Michalewicz (1992), *Genetic Algorithms + Data Structures = Evolution Programs*, Springer-Verlag, Berlin.

[13] J.S. Rosenthal (1995), Convergence rates for Markov chains, *SIAM Review*, 37(3), pp. 387–405.

[14] S. Rudnaya, F. Santosa and A.O. Chiareli (1998), Optimal design of a diffractive optical element, in *Proceedings of the 4th International Conference on Mathematical and Computational Aspects of Wave Propagation*, SIAM, Philadelphia.

[15] G. Rudolph (1994), Convergence analysis of Canonical Genetic Algorithms, *IEEE Transactions on Neural Networks*, 5(1), pp. 96–101.

[16] G. Rudolph (1997), *Convergence Properties of Evolutionary Algorithms*, Verlag Dr. Kovac, Hamburg.

[17] E. Seneta (1981), *Non-negative Matrices and Markov Chains*, Springer-Verlag, New York.

SEMIDEFINITE AND LAGRANGIAN RELAXATIONS FOR HARD COMBINATORIAL PROBLEMS*

Henry Wolkowicz
Department of Combinatorics and Optimization
University of Waterloo
Waterloo, Ontario, Canada
hwolkowi@orion.math.uwaterloo.ca

Abstract Semidefinite Programming is currently a very exciting and active area of research. Semidefinite relaxations generally provide very tight bounds for many classes of numerically hard problems. In addition, these relaxations can be solved efficiently by interior-point methods.

In this paper we study these semidefinite relaxations using the equivalent Lagrangian relaxations. In particular, the theme of the paper is to show that the Lagrangian relaxation is, in some respects, best. In all instances we consider, we show that whenever we have a tractable bound (relaxation), then the same bound can be obtained from a Lagrangian relaxation.

Keywords: Semidefinite Programming, Lagrangian Duality, Relaxations, Quadratic Constrained Quadratic Programs, Hard Combinatorial Problems.

1. INTRODUCTION

Semidefinite Programming (denoted SDP and sometimes called linear matrix inequalities, LMIs) is a generalization of linear programming (denoted LP), where the nonnegativity constraints on vector variables are replaced by positive semidefinite constraints on symmetric matrix variables. These problems have an old history dating back more than 100 years to Lyapunov's theory for stability of differential equations, e.g. [82, 71]. They were studied and applied in engineering applications as early as the 1960s, e.g. [114, 116, 115] and continued into the 1980's (see e.g. the historical outline in [20] and the work on matrix completion

*Research supported by The Natural Sciences and Engineering Research Council of Canada.

M.J.D. Powell and S. Scholtes (Eds.), *System Modelling and Optimization: Methods, Theory and Applications.*

problems in [27, 37, 60]). In addition, SDP is a special case of *optimization over cone constraints* (generalized linear programming), which dates back more than 30 years to e.g. Bellman and Fan [11], and was an ongoing active area of research, e.g. [12, 41, 23, 24, 55, 122, 79, 19].

The last ten years have seen an enormous interest in the SDP area, due to many new and important applications in, e.g. combinatorial optimization, engineering (systems and control), statistics, etc. This interest increased greatly due to the fact that SDP problems can be solved efficiently (are tractable) by the new interior-point methods, e.g. [77]. One of the interesting side effects of the activity is that it has brought various different areas of research into contact. For example, the people working on numerical issues for large scale problems are now using sophisticated techniques in convex analysis, and using adjoint operators rather than matrix representations is becoming common. We are all benefiting from the elegance and applicability of this area.

Combinatorial and discrete optimization problems often involve binary ($0,1$ or ± 1) decision variables. These can be modelled using quadratic constraints $x^2 - x = 0$, or $x^2 = 1$, respectively. Thus many hard combinatorial problems can be modelled using quadratically constrained quadratic programs, denoted $\mathrm{Q^2P}$. However, these latter problems can be just as hard (intractable) to solve. Therefore, relaxations are used to find approximate bounds and solutions. One can use linear approximations and obtain models that can be solved efficiently (tractable models). Moreover, using the positive semidefinite matrix construction $X = xx^T$, we can *lift* the problem into matrix space and obtain a *Semidefinite Programming Relaxation* by ignoring the rank one restriction on X, see e.g. [65, 38, 66, 102, 8, 7, 78]. This lifting process provides surprisingly stronger bounds, both empirically and theoretically, than have previously been found, e.g. [35, 47, 3]. Thus SDP provides a means of finding an approximate solution to quadratic models for hard problems.

In this paper we study SDP relaxations as well as exploring the relationship between the SDP and Lagrangian relaxations of various classes of $\mathrm{Q^2P}$s. (This continues on the work in e.g. [103, 91].) We then consider the strength of these relaxations, in the sense of strong duality. We see that in the simplest case of one constraint (the so-called trust region subproblem, TRS), strong duality holds. However, even two convex constraints (the CDT problem) can result in a duality gap. Therefore, it is surprising that there is a class of matrix problems with orthogonal constraints for which there is a zero duality gap. This motivates adding certain nonlinear redundant constraints in order to derive a strengthened SDP relaxation for the max-cut problem.

Throughout this paper we emphasize the theme (or conjecture) that Lagrangian relaxation is somehow best. Though this question is very vague, so perhaps an answer may not be available, it does give the flavour of the approach used in the paper.

The paper is organized as follows. We begin in Section 2.1 with a well known problem in this area, the Max-Cut problem. We present several different relaxations. Following our theme, all these bounds, including the SDP bound, end up being equivalent to the Lagrangian relaxation. We then discuss the TRS and the CDT problem and the difference in strong duality for them. This is followed by finding the SDP relaxation for general Q^2P in Section 2.2. It includes descriptions of the relationships between the SDP relaxation and the Lagrangian relaxation via convex quadratic valid inequalities, following [33, 52]. Occurrences of strong duality for nonconvex quadratic programs are studied in Section 2.3. In every instance where one has a tractable bound, we find a Q^2P such that the bound is attained by the Lagrangian relaxation. This follows the work in [6, 5].

We conclude with a strengthened SDP bound based on a second lifting procedure. This illustrates a *recipe* for constructing semidefinite relaxations using the Lagrangian dual of the Lagrangian dual of the original Q^2P. In addition, we see the advantages of this approach as redundant constraints added at the start provide strengthened bounds, but do not result in redundancy in the final SDP relaxation.

1.1. LAGRANGE MULTIPLIERS FOR Q^2P

The Q^2P in x is

$$(Q^2P_x) \qquad \begin{array}{rll} \mu^* := & \min\limits_{x \in \Re^n} & q_0(x) := x^TQ_0x + 2g_0^Tx + \alpha_0 \\ & \text{s.t.} & q_k(x) := x^TQ_kx + 2g_k^Tx + \alpha_k \leq 0, \\ & & \qquad k \in \mathcal{I} := \{1, \ldots, m\}, \end{array}$$

where $Q_k \in \mathcal{S}^n$, the space of symmetric matrices. The Lagrangian of Q^2P_x is

$$L(x, \lambda) := q_0(x) + \sum_{k \in \mathcal{I}} \lambda_k q_k(x),$$

where $\lambda = (\lambda_k) \geq 0$ are nonnegative Lagrange multipliers.

Lagrange multipliers can be used in two ways. First, if a constraint qualification holds for Q^2P_x at the optimum x (e.g. the Mangasarian–Fromovitz constraint qualification), then the Karush–Kuhn–Tucker necessary conditions for optimality are satisfied, i.e.

$$\nabla L(x, \lambda) = 0 \quad \text{and} \quad \lambda_k q_k(x) = 0, \ \forall k \in \mathcal{I},$$

for some $0 \leq \lambda \in \Re^m$. Therefore the search for the optimum x can be restricted to the points satisfying stationarity of the Lagrangian and complementary slackness. Moreover, if the Lagrangian is also convex, then these (and primal feasibility) are sufficient conditions for optimality.

Lagrange multipliers can also be used to derive the Lagrangian dual (or relaxation) of $\mathrm{Q}^2\mathrm{P}_x$

$$(\mathrm{DQ}^2\mathrm{P}_x) \qquad \mu^* \geq \nu^* := \max_{\lambda \geq 0} \min_x q_0(x) + \sum_{k \in \mathcal{I}} \lambda_k q_k(x),$$

i.e. each inner unconstrained minimization problem provides a lower bound for $\mathrm{Q}^2\mathrm{P}_x$; and we then choose the best of these lower bounds. A zero duality gap holds if $\mu^* = \nu^*$, but this condition can fail in the nonconvex case. Strong duality holds if $\mu^* = \nu^*$ and also ν^* is attained.

Remark 1.1 *Unfortunately, the term* strong duality *is ambiguous in the literature as it is sometimes used to define a zero duality gap with* both *primal and dual attainment.*

Remark 1.2 *Let $q := (q_k)$. Then the sum in the Lagrangian can be rewritten as $\langle \lambda, q(x) \rangle = \lambda^T q(x)$. This appears to be too trivial to mention. However, it does emphasize how Lagrange multipliers arise when one is dealing with matrix valued constraints. In particular, if one has a constraint $Q(x) = 0$, where the image of Q is a symmetric matrix, then the Lagrange multiplier will be a symmetric matrix, say $\Lambda = \Lambda^T$, and the term in the Lagrangian will be the inner product $\langle Q(x), \Lambda \rangle$. With a nonnegativity constraint, there will be a sign restriction on the Lagrange multiplier.*

1.2. SEMIDEFINITE PROGRAMMING PRELIMINARIES

SDP is an extension of LP in that matrix variables replace vector variables and nonnegativity elementwise is replaced by positive semidefiniteness. In addition, it is a special case of the cone programming problem

$$\begin{array}{ll} \min & f(x) \\ \text{s.t.} & g(x) \succeq_K 0, \end{array} \tag{1}$$

where K is a convex cone ($K + K \subset K, \quad \alpha K \subset K, \ \forall \alpha \geq 0$) and $g(x) \succeq_K 0$ denotes the cone partial order $g(x) \in K$. If $K = \Re^n_+$, then we get the usual elementwise ordering $g(x) \geq 0$. This notation allows comparisons with many results that hold for the elementwise ordering, e.g. Jensen's inequality for the definition of convexity. In the case that $K = \mathcal{P}$, the cone of positive semidefinite matrices, then we have the *Löwner partial order*, e.g. [69]. In fact, this is a very general mathematical program and includes standard inequality, equality, and semidefinite constraints, when $K = \Re^n_+ \otimes \{0\} \otimes \mathcal{P}$, where $\{0\}$ is the set containing the 0 vector. We now look at some of the relations between LP and SDP.

Geometry. Much of the elegant geometry of polyhedral sets developed for LP can be extended to SDP. This was studied as early as 1948 by Bohnenblust [15] and later by Tausky [108] and also by Barker and Carlson [10]. More recently, motivated by the high interest in SDP, many new results on differentiability and multiplicity of eigenvalues have appeared, see e.g. Lewis [63, 64] and Pataki [83, 84], respectively. In addition, results on characterizing different types of homogeneous and self-scaled cones appear in [42, 111, 44].

Let $\mathcal{P}^n$ (or $\mathcal{P}$ when the meaning is clear) denote the cone of positive semidefinite matrices in $\mathcal{S}^n$, the space of $n \times n$ symmetric matrices. The following remarks have analogues in LP. The SDP cone is *self-polar*, i.e.

$$\mathcal{P} = \mathcal{P}^+ := \{Y : X \bullet Y \geq 0, \ \forall X \in \mathcal{P}\},$$

where $X \bullet Y = \operatorname{Trace} XY$ is the trace inner product. Also, $\mathcal{P}$ is *homogeneous*, i.e., for any $X, Y \in \operatorname{int}(\mathcal{P})$, there exists an invertible linear operator $\mathcal{A}$ from $\mathcal{S}^n$ to $\mathcal{S}^n$ that leaves $\mathcal{P}$ invariant and that has the property $\mathcal{A}(X) = Y$.

A face of a cone K is defined as

$$\mathcal{F} = \{x \in K : y, z \in K, \, x = \alpha y + (1-\alpha)z, \, 0 < \alpha < 1 \Rightarrow y, z \in \mathcal{F}\},$$

i.e. $x \in \mathcal{F}$ can be an interior point of the line segment that joins $y, z \in K$ only if $y, z \in \mathcal{F}$. The faces of $\Re^n_+$ can be characterized using any vector in its relative interior, i.e. for y in the relative interior, we define the vector z by $z_i = 1$ if $y_i = 0$, and $z_i = 0$ if $y_i > 0$. Thus z is in the relative interior of the complementary face. Then the corresponding face that contains y in its relative interior is

$$\left\{x \in \Re^n_+ : \langle x, z \rangle = 0\right\} = \left\{x \in \Re^n_+ : y_i = 0 \Rightarrow x_i = 0\right\}.$$

The *faces* $\mathcal{F}$ of $\mathcal{P}$ can be similarly characterized using any $Y \in$ relint $\mathcal{F}$, i.e.

$$\mathcal{F} = \{X : \mathcal{N}(X) \supset \mathcal{N}(Y)\},$$

where $\mathcal{N}(\cdot)$ denotes nullspace. The faces of $\mathcal{P}$ are *exposed*, i.e. $\mathcal{F} = \mathcal{P} \cap \varphi^\perp$, where $\varphi \in \mathcal{P} \cap \mathcal{F}^\perp$ is a *conjugate face*. (Here the subscript $\perp$ denotes the orthogonal complement.) Moreover, they are *projectionally exposed*, i.e. there exists a projection matrix P such that $\mathcal{F} = P\mathcal{P}P$; note that $P \cdot P$ is a projection on $\mathcal{S}^n$. In fact, we can choose P (and so

$P \cdot P$) to be an orthogonal projection. See e.g. [87, 40, 39] for results on faces and [18, 9, 93, 107] for results on projectionally exposed cones.

Duality and Optimality Conditions. Extensions of the optimality conditions and duality theory from LP to SDP appeared in [11]. However, unlike the LP case, strong duality theorems required a Slater-type constraint qualification (denoted CQ). This CQ is strict feasibility, i.e. there exists a feasible point in the interior of $\mathcal{P}$. In [95] it is shown that difficulties in the duality theory can arise due to a property of the faces of $\mathcal{P}$, namely that

$$\mathcal{P} + \operatorname{span}(\mathcal{F}) \quad \text{is never closed.} \tag{2}$$

On the other hand, these difficulties can be corrected by taking advantage of the fact that

$$\mathcal{P} + \mathcal{F}^{\perp} \quad \text{is always closed.} \tag{3}$$

We amplify these statements in Remark 1.3 below.

Now, consider the typical primal SDP

$$\text{(P)} \qquad \begin{array}{rl} \mu^* := \min & C \bullet X \\ \text{s.t.} & \mathcal{A}(X) = b \ \text{ and } \ X \succeq 0, \end{array}$$

where $C \in \mathcal{S}^n$, $b \in \mathbb{R}^m$, and $\mathcal{A} : \mathcal{S}^n :\rightarrow \mathbb{R}^m$, is a linear operator. Thus the components of the constraint equations are

$$(\mathcal{A}(X))_i = \operatorname{Trace} A_i X = b_i, \quad i = 1, 2, \ldots, m,$$

for some given symmetric matrices A_i.

We can define the dual and prove weak duality using the notion of a hidden constraint. Specifically we use the form

$$\begin{array}{rcl} \mu^* & = & \min\limits_{X \succeq 0} \max\limits_{y} C \bullet X + y^T(b - \mathcal{A}(X)) \\ & \geq & \max\limits_{y} \min\limits_{X \succeq 0} y^T b + (C - \mathcal{A}^*(y)) \bullet X \ = \ \nu^*, \end{array} \tag{4}$$

where $\mathcal{A}^*$ is the adjoint operator of $\mathcal{A}$, so it satisfies

$$\mathcal{A}(X)^T y = X \bullet \mathcal{A}^*(y), \quad \forall X, \forall y,$$

which implies $\mathcal{A}^*(y) = \sum_i y_i A_i$, the symmetric matrices A_i being defined above. The dual program is

$$\text{(D)} \qquad \begin{array}{rl} \nu^* := \max & b^T y \\ \text{s.t.} & \mathcal{A}^*(y) \preceq C. \end{array}$$

The first equation in (4) follows from the fact that the inner maximization is finite (the hidden constraint) if and only if $\mathcal{A}(X) = b$. The inequality follows from taking the minimization first. The last equation follows from the hidden constraint, because y has to be such that the inner minimization is finite valued.

The duality theory gives rise to the elegant characterization of optimality

$$\begin{array}{rcll} \mathcal{A}^*(y) + Z - C & = & 0 & \text{dual feasibility} \\ b - \mathcal{A}(X) & = & 0 & \text{primal feasibility} \\ ZX & = & 0 & \text{complementary slackness} \\ Z, X & \succeq & 0 & \text{positive semidefiniteness.} \end{array} \tag{5}$$

The variables $X, (y, Z)$ are called a primal-dual optimal pair, Z being the (dual) slack variable. If only complementary slackness fails, then the variables are called a primal-dual feasible pair. Note that complementary slackness can be written in the equivalent form $\operatorname{Trace} ZX = 0$, which follows from an orthogonal diagonalization of Z and X. This is also equivalent to having a zero duality gap between primal and dual values for a feasible pair since

$$\operatorname{Trace} ZX = \operatorname{Trace} (C - \mathcal{A}^*(y))X = \operatorname{Trace} CX - y^T b.$$

Remark 1.3 *We mentioned above that the optimality conditions require a CQ, i.e. strict feasibility, and that condition (2) can cause difficulties. Consider the SDP P and its dual D. Dual feasibility can be written in the form*

$$\mathcal{A}^*(y) + Z = C, \quad Z \succeq 0.$$

If we choose the linear operator $\mathcal{A}^*$ so that its range satisfies $\mathcal{R}(\mathcal{A}^*) = span(\mathcal{F})$, for some face $\mathcal{F}$ of $\mathcal{P}$, then we can choose any C in the set $\mathcal{R}(\mathcal{A}^*) + \mathcal{P}$, but not its closure, because choices in the closure can force failure of dual feasibility and a duality gap. For example, let $A := \begin{pmatrix} 0 & 0 \\ 0 & 1 \end{pmatrix}$, and so $\mathcal{A}^*(y) = yA$. Then $C = \begin{pmatrix} \varepsilon & 1 \\ 1 & 0 \end{pmatrix}$ is admissible if and only if ε is positive. Therefore we address $C = \begin{pmatrix} 0 & 1 \\ 1 & 0 \end{pmatrix}$ and $b = 0$. In this case, the primal problem constraint $\mathcal{A}(X) = b$ implies $X_{22} = 0$, and then the other constraint $X \succeq 0$ implies $X_{12} = X_{21} = 0$. Thus $\mu^* = C \bullet X = 0$ is optimal in P. We have noted, however, that the dual is infeasible for the given choice of C.

The condition (3) can correct this problem by allowing a larger set of dual variables. Modified optimality conditions without any CQ can be obtained. This was done in [17] and also in [97, 95]. In our primal problem

P, we let $\mathcal{F}$ be the minimal face, i.e. the smallest face containing the feasible set. Then the strengthened dual is $\mu^* = \max b^T y$ s.t. $\mathcal{A}^* y \preceq_{\mathcal{F}^+} C$. For our example, the minimal face is $\mathcal{F} = span \begin{pmatrix} 1 & 0 \\ 0 & 0 \end{pmatrix}$, which implies that $C - (-1)A$ is in $\mathcal{F}^+$. Hence the new dual optimal value is 0, and we have a zero duality gap.

If we perturb the complementary slackness conditions,

$$ZX = \mu I, \quad \mu > 0,$$

where μ is a parameter, then (5) becomes the (modified) optimality conditions of a log-barrier problem. These are the equations that are used in interior-point methods. However, unlike linear programming, we have an interesting subtle complication that is also discussed in [113]. One cannot apply Newton's method directly since ZX is not necessarily symmetric, and so we end up with an overdetermined system of equations. There are various ways of modifying this system in order to get good search directions, see e.g. [110, 72, 57]. Many of these directions work very well in practice, which is clear from empirical evidence and the derivation of several public domain codes, e.g. [47, 1, 105, 106, 16]. The SDP problems are tractable because they are convex programs and fall into the class of problems that can be approximately solved to a desired accuracy in polynomial time. This follows from the seminal work of Nesterov and Nemirovski [76, 77]. The algorithms that currently work well are the primal-dual interior-point algorithms. This area of research is ongoing, however, and there are many classes of problems with special structure where dual algorithms based on a bundle trust approach perform better; this is especially true if it is too expensive to evaluate the primal matrix X explicitly, see e.g. [13, 46, 61].

First and second order optimality conditions for SDP are given in [100, 101] and in the survey article [113], for example. Nondegeneracy and strict complementarity are discussed in [2, 85]. Both nondegeneracy and strict complementarity can fail, though they are generic conditions. In addition the theorem of Goldman and Tucker [36], about the existence of an optimal primal-dual pair that satisfies strict complementary slackness, does not apply to SDP. Note that strict complementarity for SDP is the condition $Z + X \succ 0$.

2. RELAXATIONS OF Q²P

We now look at a particular instance of Q^2P, namely the quadratic model for the *Max-Cut problem*. We start with several different tractable relaxations for this problem that have appeared in the literature. We show that, surprisingly, they are all equal to the Lagrangian (and SDP) relaxation.

We then consider trust region type problems and discuss when strong duality holds. We include problems where orthogonal constraints arise, e.g. orthogonal relaxations of the quadratic assignment and graph partitioning problems. Thus this part of the chapter emphasizes the theme about the strength of Lagrangian relaxation.

2.1. RELAXATIONS FOR THE MAX-CUT PROBLEM

One of the problems for which the SDP relaxation has been particularly successful, both empirically and theoretically, is the *Max-Cut Problem*, e.g. [45, 35, 34]. Let $G = (V, E)$ be an undirected graph with vertex set $V = \{v_i\}_{i=1}^n$ and weights w_{ij} on the edges $(v_i, v_j) \in E$. We seek the index set $\mathcal{I} \subset \{1, 2, \dots n\}$, that maximizes the sum of the weights of the edges with one end point with index in $\mathcal{I}$ and the other in the complement. This is equivalent to

$$\text{(MC)} \quad \max \ \tfrac{1}{2}\textstyle\sum_{i<j} w_{ij}(1 - x_i x_j), \quad x \in \mathcal{F},$$

where $\mathcal{F} := \{\pm 1\}^n$, and $x_i = 1$ if $i \in \mathcal{I}$, and $x_i = -1$ otherwise. The objective function is a (homogeneous) quadratic form, $x^T Q x$.

Several *Different* Relaxations. We rewrite MC as the more general problem

$$\text{(MCQ)} \quad \mu^* := \max_{x \in \mathcal{F}} q_0(x), \quad \text{where } q_0(x) := x^T Q x - 2c^T x. \tag{6}$$

The MCQ (and MC) problem is intractable, but there are many different ways to relax the problem and find approximate solutions and/or bounds. The simplest way is to relax the constraints to the interval conditions $x \in [-1, 1]^n$. This *bound constrained quadratic problem* is NP-hard if Q is not negative semidefinite [81]; while, if Q is negative semidefinite and $c = 0$, then the solution is the trivial 0 solution.

Other relaxations are also geometric in nature and involve perturbations of the objective function. For example, one method relaxes the constraints to the unit ball of radius $\sqrt{n}$, while another relaxes the constraints to their convex hull, i.e. to the unit cube.

The relaxations yield bounds which are derived by making changes to the objective function q_0 that are zero when $x_i^2 = 1$, i.e. on the feasible set $\mathcal{F}$. In particular, for every $u \in \Re^n$ we have

$$\begin{aligned} q_u(x) &:= x^T(Q + \mathrm{Diag}\,(u))x - 2c^T x - u^T e \\ &= q_0(x), \ \ \forall x \in \mathcal{F}, \end{aligned} \tag{7}$$

where e is the vector of ones (of the appropriate dimension), and Diag (u) denotes the diagonal matrix formed from the vector u. For each u we get an upper bound by ignoring the constraints and by allowing the diagonal perturbations, i.e. we have

$$\mu^* \leq f_0(u) := \max_x q_u(x). \tag{8}$$

We then find the bound

$$\mu^* \leq B_0 := \min_u f_0(u). \tag{9}$$

Let

$$S := \left\{u : u^T e = 0,\, Q + \text{Diag}\,(u) \preceq 0\right\}.$$

Note that, if the set S is not empty, then we can minimize over the unconstrained parameter u, or we can add the restriction $u^T e = 0$, i.e.

$$B_0 = \min_{u^T e=0} f_0(u), \text{ if } S \neq \emptyset.$$

This can be seen from the equivalence of the optimality conditions for min-max problems, e.g. [26, Pg 188,Theorem 2.1], or from a perturbation analysis of min-max problems, e.g. [29]. For example, if the solution of the inner maximization problem is attained at a unique x, then f_0 is differentiable and $\nabla f_0(u) = x \circ x - e$, where $x \circ x$ is the *Hadamard* (elementwise) product. In general, the function $f_0(u)$ is directionally differentiable at each point u (with $u^T e = 0$ in the restricted case), in any direction h. Assuming $\|h\| = 1$, the directional derivative has the value

$$\partial f_0(u;h) = \max_{x\in\mathcal{X}(u)} \left\langle \frac{\partial q_u(x)}{\partial u}, h\right\rangle = \max_{x\in\mathcal{X}(u)} \langle x \circ x - e, h\rangle,$$

where $\mathcal{X}(u)$ denotes the set of values of x that solve the inner problem (8) for given u. One can now compare stationarity for f_0 in the unconstrained case with the use of a Lagrange multiplier for the constraint $u^T e = 0$. This involves the subgradient of the convex function f_0 and its domain, i.e. the set where f_0 is finite valued. Details can be found in [92]. The comment about the possible use of $u^T e = 0$ is also true for the bounds that follow.

But the function f_0 can take on the value $+\infty$. We can avoid these infinite values by restricting the parameter u, using a hidden semidefinite constraint. Specifically, this constraint depends on the remark that a quadratic function is bounded above if and only if its Hessian is negative semidefinite and its stationarity equation is consistent. Thus we obtain the following bound, which is tractable since we minimize a convex function over a convex set:

$$\mu^* \le B_0 = \min_{Q+\mathrm{Diag}\,(u)\preceq 0} f_0(u). \tag{10}$$

Next we relax the conditions on x to the sphere of radius $\sqrt{n}$, which provides

$$\mu^* \le f_1(u) := \max_{||x||^2=n} q_u(x). \tag{11}$$

Hence our next bound is

$$\mu^* \le B_1 := \min_u f_1(u). \tag{12}$$

The inner maximization problem is called a trust region subproblem and is tractable, as shown in Section 2.3 below. Thus we have our second tractable bound.

We can replace the spherical constraint with a box constraint, which gives

$$\mu^* \le f_2(u) := \max_{|x_i|\le 1} q_u(x). \tag{13}$$

After adding the semidefinite constraint on u to make the calculation of f_2 tractable, we obtain our next bounds

$$\mu^* \le \min_u f_2(u) \tag{14}$$

and

$$\mu^* \le B_2 := \min_{Q+\mathrm{Diag}\,(u)\preceq 0} f_2(u). \tag{15}$$

Given Q and c of the function (7), we define the $(n+1)\times(n+1)$ matrix Q^c by adding a leading row and column, so Q^c has the elements

$$\begin{array}{l} q^c_{00} = 0, \quad q^c_{0i} = q^c_{i0} = -c_i \text{ for } i > 0, \text{ and} \\ q^c_{ij} = q_{ij} \text{ for } i, j > 0, \end{array}$$

i.e.

$$Q^c := \begin{bmatrix} 0 & -c^T \\ -c & Q \end{bmatrix}. \tag{16}$$

Further, in order to have functions $q^c_u(y)$ and $f_i(u)$ that are analogous to the previous cases, we introduce

$$q^c_u(y) := y^T(Q^c + \mathrm{Diag}\,(u))\,y - u^T e, \quad y \in \Re^{n+1}, \tag{17}$$

where u and e are also in $\Re^{n+1}$. Note that q_u^c reduces to q_u if the first component of y is $y_0 = 1$. The equivalent relaxed problem is

$$\mu^* \leq f_1^c(u) := \max_{||y||^2=n+1} q_u^c(y) = (n+1)\,\lambda_{\max}(Q^c + \mathrm{Diag}\,(u)) - u^T e, \quad (18)$$

where $\lambda_{\max}(A)$ denotes the maximum eigenvalue of A, say. Thus another bound on μ^* is

$$\mu^* \leq B_1^c := \min_u f_1^c(u). \quad (19)$$

Similarly, we get equivalent bounds B_0^c and homogenized bounds for the other models.

The above argument shows that we can homogenize the problem by moving into a higher dimension. Therefore we can assume the special case $c = 0$. We now look at the SDP bound. The relaxation comes from the fact that the trace has the commutative property

$$x^T Q x = \mathrm{Trace}\, x^T Q x = \mathrm{Trace}\, Q x x^T,$$

and, for $x \in \mathcal{F}$, the matrix Y with elements $y_{ij} = x_i x_j$ is symmetric, rank one and positive semidefinite, with ones on the diagonal. Therefore we can *lift* the problem into the higher dimensional space of symmetric matrices and relax the rank one constraint. Thus we obtain a relaxation that gives the bound

$$\text{(MCSDP)} \qquad \begin{array}{rll} B_3 := & \displaystyle\max_{Y \in \mathcal{S}^n} & \mathrm{Trace}\, QY \\ & \text{s.t.} & \mathrm{diag}\,(Y) = e \ \text{ and } \ Y \succeq 0, \end{array} \quad (20)$$

where $\mathrm{diag}\,(Y)$ is the vector formed from the diagonal of Y. This SDP is a convex programming problem and is tractable.

We have presented several different tractable bounds that have simple geometric interpretations. It is not at all clear which bounds are better or how to compare them. We now do something that may seem meaningless; we replace the ± 1 constraints with $x_i^2 = 1$, $i = 0, \ldots, n$, which does not change the feasible set of the original problem. In [92, 91] it is shown, however, that all the above relaxations and bounds for MC come from the Lagrangian dual of the following equivalent problem to MCQ:

$$(\mathrm{P}_E) \qquad \begin{array}{ll} \max & q_0(x) = x^T Q x - 2c^T x \\ \text{s.t.} & x_i^2 = 1, \quad i = 1, \ldots, n. \end{array} \quad (21)$$

Thus we enforce our theme about the strength of Lagrangian relaxation. The strong duality result for the trust region subproblem is the key to the proofs. Note that the Lagrangian dual of P_E yields precisely our first bound B_0 on μ^*, given in (9).

Theorem 2.1 *All the bounds for MCQ discussed above are equal to the optimal value of the Lagrangian dual of the equivalent program P_E.* ■

2.2. GENERAL Q²P

We now move on to applying the Lagrangian relaxation to *general* quadratic constrained quadratic problems, denoted Q²P. The general Q²P problem is also studied in [33, 52, 96, 58, 56, 68, 59], for example.

Quadratic bounds using a Lagrangian relaxation have received much attention and been applied in the literature, for example in [53] and, more recently, in [54]. The latter calls the Lagrangian relaxation the "best convex bound". Discussions on Lagrangian relaxation for nonconvex programs also appear in [31]. More references are given throughout this paper.

Remark 2.1 *Any equality constraints are written as two inequality constraints; any linear equality constraints, $Ax = b$, are transformed to the quadratic constraint $\|Ax - b\|^2 = 0$. The reason for these transformations for linear equality constraints is discussed in [91]. It is that the Lagrangian dual essentially ignores linear constraints, as can be seen from: $-\infty = \max_\lambda \min_x -x^2 + \lambda x$, which is the dual of the problem* $\min\{-x^2 : x = 0\}$.

We now recall from Section 1.1 the Q²P in x:

$$(\mathrm{Q^2P}_x) \qquad \begin{array}{rll} \mu^* := & \min\limits_{x \in \Re^n} & q_0(x) := x^T Q_0 x + 2g_0^T x + \alpha_0 \\ & \text{s.t.} & q_k(x) := x^T Q_k x + 2g_k^T x + \alpha_k \leq 0, \\ & & k \in \mathcal{I} := \{1, \ldots, m\}, \end{array} \tag{22}$$

where the matrices Q_k are symmetric. The *feasible set* is

$$\mathcal{F} := \{x \in \Re^n : q_k(x) \leq 0, \forall k \in \mathcal{I}\}.$$

Note that, though the feasible set $\mathcal{F}$ may be empty, the feasible set of the relaxation may not be. The objective function and the constraints need not be convex. Therefore the feasible set can have "nasty" features that cause the problem to be very hard to solve in general, see e.g. [81].

Let

$$P_k := \begin{bmatrix} \alpha_k & g_k^T \\ g_k & Q_k \end{bmatrix}, \tag{23}$$

and, by abuse of notation, define

$$q_k(y) := y^T P_k y, \quad k = 0, 1, \ldots, m.$$

Then an equivalent formulation of Q^2P_x is the homogenized problem

$$(Q^2P_y) \qquad \begin{array}{rll} \mu^* = & \min & q_0(y) \\ & \text{s.t.} & q_k(y) \leq 0, \quad k \in \mathcal{I} \\ & & y_0^2 = 1, \quad y = \begin{pmatrix} y_0 \\ x \end{pmatrix} \in \Re^{n+1}. \end{array}$$

We see that the optimal values of Q^2P_x and Q^2P_y are equal. Further, if $y_0 = -1$ is optimal, then we can replace y by $-y$, because the objective function and all but the last constraint are homogeneous.

We will refer to both equivalent formulations of Q^2P in the sequel. The relevant one will be clear from the context.

Remark 2.2 *Note that we could replace the constraint $y_0^2 = 1$ by $y_0 = 1$ as in [33]. In the $y_0 = 1$ case, the feasible sets of the two formulations coincide exactly, while in the former case they can differ by a sign. Specifically, $x \in \mathcal{F}$ implies that both $\begin{pmatrix} -1 \\ -x \end{pmatrix}$ and $\begin{pmatrix} 1 \\ x \end{pmatrix}$ are in $\mathcal{F}_y$, where $\mathcal{F}_y$ is the feasible region of the homogenized problem Q^2P_y.*

The Lagrangian Relaxation of a General Q^2P. The Lagrangian relaxation of the homogenized problem Q^2P_y provides a simple technique for obtaining the SDP relaxation. In addition, an application of the strong duality result for the trust region subproblem shows that the SDP and Lagrangian relaxation are equal. The problem Q^2P_y has the Lagrangian

$$L(y, \sigma, \lambda) := y^T P_0 y - \sigma(y_0^2 - 1) + \sum_{k \in \mathcal{I}} \lambda_k y^T P_k y,$$

and the Lagrangian relaxation of Q^2P_y is

$$(DQ^2P_y) \qquad \nu^* := \max_{\lambda \geq 0, \sigma} \min_y y^T P_0 y - \sigma(y_0^2 - 1) + \sum_{k \in \mathcal{I}} \lambda_k y^T P_k y.$$

Note that

$$\begin{aligned} \nu^* &= \max_{\lambda \geq 0} \max_{\sigma} \min_y y^T P_0 y - \sigma(y_0^2 - 1) + \sum_{k \in \mathcal{I}} \lambda_k y^T P_k y \\ &= \max_{\lambda \geq 0} \min_{y : y_0^2 = 1} y^T P_0 y + \sum_{k \in \mathcal{I}} \lambda_k y^T P_k y, \end{aligned}$$

from strong duality of the trust region subproblem [104]. Therefore, we get equivalence of the dual values for the problems DQ^2P_y and

$$(DQ^2P_x) \qquad \nu^* = \max_{\lambda \geq 0} \min_x q_0(x) + \sum_{k \in \mathcal{I}} \lambda_k q_k(x),$$

which is similar to the approaches in [116, 103]. It follows that *weak duality*

$$\nu^* \leq \mu^* = \min_y \max_{\lambda \geq 0, \sigma} y^T P_0 y - \sigma(y_0^2 - 1) + \sum_{k \in \mathcal{I}} \lambda_k y^T P_k y$$

holds. Therefore, if the optimal σ^* and λ^* can be found, we have derived a single quadratic function whose minimal value approximates the original minimal value μ^* of $\mathrm{Q^2P}_y$, i.e.

$$\mu^* \geq \nu^* = \min_y y^T P_0 y - \sigma^*(y_0^2 - 1) + \sum_{k \in \mathcal{I}} \lambda_k^* y^T P_k y. \tag{24}$$

Moreover, in the dual program $\mathrm{DQ^2P}_y$, the Lagrangian is a quadratic function of y. Therefore the outer maximization problem has not only nonnegativity constraints but also the hidden semidefinite constraint

$$P_0 - \sigma E_{00} + \sum_{k \in \mathcal{I}} \lambda_k P_k \succeq 0, \quad \lambda \geq 0, \tag{25}$$

where E_{00} is the zero matrix except for 1 in the top left corner. The solution of the minimization subproblem is attained at $y = 0$. Therefore the Lagrangian dual is equivalent to the SDP problem

$$(\text{DSDP}) \qquad \begin{array}{rll} \nu^* := & \max & \sigma \\ & \text{s.t.} & \sigma E_{00} - \sum_{k \in \mathcal{I}} \lambda_k P_k \preceq P_0 \\ & & \lambda \geq 0. \end{array}$$

Valid Inequalities. Using the above approach, we see that more constraints $q_k(y)$ give a stronger dual. Therefore the addition of redundant constraints to get new valid inequalities can strengthen the relaxation. We will see how this occurs when we look at orthogonally constrained problems below. Another approach is also specified in detail in [33] and [52, 51].

For problems that also have linear inequality constraints, one can use the notion of copositivity to strengthen the SDP relaxation. However, this does not result in a tractable relaxation in general [94].

Specific instances of these relaxations (graph partitioning and quadratic assignment problems) appear in [121, 112]. We will present a *recipe* for generating a relaxation in Section 3.

2.3. STRONG DUALITY

In the case of strong duality (zero duality gap and dual attainment), our bounds are exact. As expected, this holds (generically) in the general convex case. Surprisingly, there are several cases of nonconvex quadratic programs where it holds as well. In this Section 2.3 we amplify on our theme that confirms the strength of Lagrangian relaxation, namely that a tractable bound implies a Lagrangian relaxation is at work.

We recall the general quadratically constrained quadratic program (22), where for simplicity we have replaced each equality constraint by two inequality constraints. We will use equality constraints when absolutely required. We let $\mathcal{F}$ denote the feasible set.

We define the Lagrangian

$$L(x,\lambda) := q_0(x) + \sum_{k=1}^{m} \lambda_k q_k(x),$$

and the dual functional

$$\varphi(\lambda) := \min_x L(x,\lambda).$$

The Lagrangian is linear in λ, and so the dual functional is a concave function of λ. Thus the calculation of the maximum of this concave function is a tractable problem if $\varphi(\lambda)$ can be evaluated efficiently. For each $\lambda \geq 0$, we have the lower bound

$$\begin{array}{rcl} \mu^* & = & \min\limits_{x\in\mathcal{F}} q_0(x) \ \geq \ \min\limits_{x\in\mathcal{F}} L(x,\lambda) \\ & \geq & \min\limits_{x} L(x,\lambda) \ = \ \varphi(\lambda). \end{array}$$

Thus we deduce the dual problem

$$\mu^* \geq \nu^* = \max_{\lambda\geq 0} \varphi(\lambda),$$

which provides a lower bound for the primal problem. If, in addition, we find a feasible $\bar{x} \in \mathcal{F}$ with attainment in the Lagrangian $\bar{x} \in \operatorname{argmin}_x L(x,\bar{\lambda})$ and with complementary slackness $\sum_k \bar{\lambda}_k q_k(\bar{x}) = 0$, then

$$\begin{array}{rcl} \mu^* & \geq & \nu^* \ \geq \ L(\bar{x},\bar{\lambda}) \\ & = & q_0(\bar{x}) \ \geq \ \mu^*. \end{array}$$

Therefore $\bar{x}$ is optimal and the duality gap is zero when these sufficiency conditions (feasibility, attainment, complementary slackness) hold. Note that, since we are dealing with an unconstrained minimum of a quadratic Lagrangian, we obtain the interesting statement: the given conditions hold only if the Lagrangian is stationary and its Hessian is positive semidefinite. Further, when these two conditions are incompatible, we lose strong duality, and can even expect a duality gap.

We now present several Q^2P problems where the Lagrangian relaxation is important and well known. In all these cases, the Lagrangian dual provides an important theoretical tool for algorithmic development, even if the duality gap may be nonzero. We continue to emphasize our theme that the Lagrangian relaxation is best.

Convex Quadratic Programs. We start with the easy case; consider the convex quadratic program

$$\text{(CQP)} \qquad \mu^* := \begin{array}{ll} \min & q_0(x) \\ \text{s.t.} & q_k(x) \leq 0, \quad k = 1, \ldots, m, \end{array}$$

where all the functions $q_i(x)$ are convex and quadratic. The following remarks show that Lagrangian duality can always solve this problem.

The dual is

$$\text{(DCQP)} \qquad \nu^* := \max_{\lambda \geq 0} \min_x \; q_0(x) + \sum_{k=1}^{m} \lambda_k q_k(x).$$

If ν^* is attained at $\lambda = \lambda^*$ and $x = x^*$, then *sufficient* conditions for x^* to be optimal for CQP are primal feasibility and complementary slackness

$$\sum_{k=1}^{m} \lambda_k^* q_k(x^*) = 0.$$

It is also well known that the Karush–Kuhn–Tucker (KKT) conditions are sufficient for global optimality, and, under an appropriate constraint qualification, they are also necessary. Therefore strong duality holds if a constraint qualification is satisfied, and then there is no duality gap and the dual is attained.

Further, *if the primal value of CQP is bounded then it is attained and there is no duality gap*, see [109, 89, 90, 88]. This assertion can be regarded as an extension of the Frank–Wolfe Theorem [68]. Surprisingly, however, the dual may not be attained. For example, the convex program

$$0 = \min\{x : x^2 \leq 0\} \qquad (26)$$

has the (unattained) dual

$$0 = \max_{\lambda \geq 0} \min_x \; x + \lambda x^2 = \max_{\lambda > 0} \min_x \; x + \lambda x^2.$$

Algorithmic approaches based on Lagrangian duality appear in [48, 67, 77], for instance.

Rayleigh Quotients. Suppose that $A = A^T \in \mathcal{S}^n$. It is well known that the smallest eigenvalue λ_1 of A is the Rayleigh quotient

$$\lambda_1 = \min\{x^T A x : x^T x = 1\}. \tag{27}$$

Since A is not necessarily positive semidefinite, this form may require the minimization of a nonconvex function on a nonconvex set. The Rayleigh quotient, however, forms the basis of many very efficient algorithms for finding the smallest eigenvalue. It is easy to deduce that there is no duality gap for this nonconvex problem, by using the equation

$$\lambda_1 = \max_{\lambda} \min_{x} x^T A x - \lambda(x^T x - 1) = \max_{A-\lambda I \succeq 0} \lambda. \tag{28}$$

Specifically, the inner minimization problem in (28) is unconstrained. Therefore the outer maximization problem has the hidden semidefinite constraint (an ongoing theme in this paper)

$$A - \lambda I \succeq 0,$$

which requires λ to be at most the smallest eigenvalue of A. With λ set to the smallest eigenvalue, the inner minimization yields either $x = 0$ or the eigenvector corresponding to λ_1, the corresponding value of the inner objective function being λ_1 in both cases. Thus, we have an example of a *nonconvex problem for which strong duality holds.* Note that the problem (27) has a special norm constraint, and a homogeneous quadratic objective. Thus, this nonconvex problem can be solved efficiently. Furthermore, strong duality holds for the Lagrangian dual, which supports our theme.

Trust Region Subproblems. We are going to see that strong duality holds for a larger class of apparently nonconvex problems. The trust region subproblem, TRS, is the minimization of a quadratic function subject to a norm constraint, where no convexity or homogeneity of the objective function is assumed. We make a further extension, i.e. we do not assume convexity of the constraint, so both the objective and constraint functions are allowed to be indefinite quadratics. Some applications of indefinite quadratic forms are given in [22]. This problem is important in nonlinear programming, e.g. [75, 74]. Specifically, TRS has the form

$$\text{(TRS)} \qquad \begin{array}{rll} \mu^* := & \min & q_0(x) = x^T Q_0 x - 2c_0^T x \\ & \text{s.t.} & x^T x - \delta^2 \leq 0 \ \ (\text{or} = 0), \end{array}$$

and the generalized trust region subproblem [104, 73] is

$$\text{(GTRS)} \qquad \begin{array}{rll} \mu^* := & \min & q_0(x) = x^T Q_0 x - 2c_0^T x \\ & \text{s.t.} & q_1(x) \leq 0 \ \ (\text{or} = 0), \end{array}$$

where q_1 is another quadratic function. Further, one can have two sided constraints $\alpha \leq q_1(x) \leq \beta$, which occur in some trust region algorithms as well.

For TRS, assuming that the constraint is written "$\leq$," the Lagrangian dual is:

$$\text{(DTRS)} \qquad \nu^* := \max_{\lambda \geq 0} \min_x \; q_0(x) + \lambda(x^T x - \delta^2).$$

An equivalent problem (see [104]) is the (concave) nonlinear semidefinite program

$$\text{(DTRS)} \qquad \begin{array}{rll} \nu^* := & \sup & c_0^T(Q_0 + \lambda I)^\dagger c_0 - \lambda\delta^2 \\ & \text{s.t.} & Q_0 + \lambda I \succ 0, \quad \lambda \geq 0, \end{array}$$

where the superscript $\dagger$ denotes the Moore–Penrose generalized inverse. It is shown in [104] that there is a zero duality gap for TRS, $\mu^* = \nu^*$. (The primal is attained though the dual may not be, as in example (26) below.) Thus, as in the eigenvalue calculation, we have an example of a nonconvex program where strong duality holds. Therefore this problem can be solved efficiently, polynomial time results being presented in [117].

We include a short proof of strong duality for the inequality constrained case, based on the outline in [62], which applies a convex case result after a perturbation. Note that the key to the proof is being able to pass between the inequality and equality constraints.

Proof.
Without loss of generality, we assume that TRS is nonconvex, because otherwise we apply the convex results discussed above. Therefore μ^* is attained on the boundary of the feasible set and the smallest eigenvalue of Q_0, denoted γ, is negative. Thus TRS has the required property

$$\begin{array}{rll} \mu^* & = & \min\limits_{x^T x \leq \delta^2} \; x^T(Q_0 - \gamma I)x - 2c_0^T x + \gamma x^T x \\ & = & \min\limits_{x^T x = \delta^2} \; x^T(Q_0 - \gamma I)x - 2c_0^T x + \gamma x^T x \quad (Q_0 \text{ is indefinite}) \\ & = & \min\limits_{x^T x = \delta^2} \; x^T(Q_0 - \gamma I)x - 2c_0^T x + \gamma \delta^2 \\ & = & \min\limits_{x^T x \leq \delta^2} \; x^T(Q_0 - \gamma I)x - 2c_0^T x + \gamma \delta^2 \;\; (Q_0 - \gamma I \text{ is singular}) \\ & = & \max\limits_{\lambda \geq 0} \min\limits_x \; x^T(Q_0 - \gamma I)x - 2c_0^T x + \lambda(x^T x - \delta^2) + \gamma\delta^2 \\ & & \qquad\qquad\qquad\qquad\qquad\qquad\qquad\qquad\qquad \text{(convex case)} \\ & = & \max\limits_{\lambda \geq 0} \min\limits_x \; x^T Q_0 x - 2c_0^T x + (\lambda - \gamma)(x^T x - \delta^2) \\ & \leq & \max\limits_{\lambda \geq \gamma} \min\limits_x \; x^T Q_0 x - 2c_0^T x + (\lambda - \gamma)(x^T x - \delta^2) \quad (\gamma < 0) \\ & = & \nu^* \leq \mu^*. \end{array} \qquad (29)$$

■

As mentioned above, extensions of this result to a two-sided, general, possibly nonconvex, constraint are discussed in [104, 73]. An algorithm based on Lagrangian duality appears in [98] and (implicitly) in [74, 99]. Such algorithms are highly efficient for the TRS problem, since they solve it almost as quickly as an eigenvalue problem.

This efficiency when the objective and constraint may be nonconvex is surprising. In fact, Martinez [70] shows that the TRS can have at most one local and nonglobal optimum, and that the Lagrangian at this point has one negative eigenvalue. Therefore, it is even more surprising that the Lagrangian dual (relaxation) allows one to find the global minimum without ever getting trapped near the local minimum.

In fact, for GTRS we still have a zero duality gap, though strong duality may fail, as shown in (26). The results in [104] provide strong duality for GTRS with a two sided constraint $\alpha \leq q_1(x) \leq \beta$, the constraint qualification being $\alpha < \beta$. In [73], necessary and sufficient optimality conditions are presented for GTRS with the constraint qualification $\min q_0(x) < \max q_0(x)$. A combination of these results with the extension of the Frank–Wolfe theorem [68] gives the following properties.

Theorem 2.2 *Consider GTRS: a zero duality gap always holds and, if the optimal value is finite, then it is attained.* ■

Two Trust Region Subproblem. The two trust region subproblem, TTRS, is the minimization of a (possibly nonconvex) quadratic function subject to a norm and a least squares constraint, i.e. two convex quadratic constraints. This problem arises in algorithms for general nonlinear programs that use a sequential quadratic programming approach, and is often called the CDT problem, because it was introduced by Celis, Dennis and Tapia [21].

In contrast to the above single TRS, the TTRS can have a nonzero duality gap [86, 118, 119, 120], which is closely related to quadratic theorems of the alternative [25]. In addition, if the constraints are not convex, then the primal may not be attained (see [68]).

As mentioned above, Martinez [70] shows that the TRS can have at most one local optimum that is nonglobal, the Lagrangian there having one negative eigenvalue. Therefore, if we have such a case and add another ball constraint that contains the local, nonglobal, optimum in its interior, so that this point becomes the global optimum, then we obtain a TTRS that does not have a zero duality gap due to the negative eigenvalue. It is uncertain what additional constraints are successful at closing this duality gap. In fact, it is still an open problem whether TTRS is an NP-hard or a polynomial time problem.

General Q^2P. The general, possibly nonconvex, Q^2P has many applications in modeling and approximation theory, see e.g. the applications to SQP methods in [58]. Examples of approximations to Q^2P also appear in [32].

The Lagrangian relaxation of a Q^2P is equivalent to the SDP relaxation, and is sometimes called the Shor relaxation [103]. The Lagrangian relaxation can be written as an SDP if one takes into the account the hidden semidefinite constraint, the inner quadratic objective function being bounded below only if the Hessian is positive semidefinite. The SDP relaxation is then the Lagrangian dual of this semidefinite program. It can also be obtained directly by *lifting* the problem into matrix space using the identity $x^TQx = \operatorname{Trace} x^TQx = \operatorname{Trace} Qxx^T$, and relaxing xx^T to a semidefinite matrix X.

The geometry of the original feasible set of Q^2P can be related to the feasible set of the SDP relaxation. The connection is through *valid quadratic inequalities*, i.e. nonnegative (convex) combinations of the quadratic constraints; see [33, 52] and our Section 2.2.

Orthogonally Constrained Programs with Zero Duality Gaps. Let $\mathcal{M}_{m,n}$ denote the space of $m \times n$ real matrices. We now follow the approach in [6, 5, 4] and consider the *orthonormal type constraint*

$$X^TX = I, \qquad X \in \mathcal{M}_{m,n},$$

sometimes known as the Stiefel manifold [28], and the trust region type constraint

$$X^TX \preceq I, \qquad X \in \mathcal{M}_{m,n}.$$

Applications and algorithms for optimization on orthonormal sets of matrices are discussed in [28]. In this section we will show that, for $m = n$, strong duality holds for a certain (nonconvex) quadratic program defined over orthonormal matrices. Because of the similarity of the orthonormality constraint to the norm constraint $x^Tx = 1$, the given results can be viewed as a matrix generalization of the strong duality property (27) for the Rayleigh Quotient problem.

Let A and B be $n \times n$ symmetric matrices, and consider the orthonormally constrained homogeneous Q^2P

$$(\mathrm{QQP_O}) \qquad \mu^O := \begin{array}{ll} \min & \operatorname{Trace} AXBX^T \\ \text{s.t.} & XX^T = I. \end{array} \tag{30}$$

This problem can be solved exactly using Lagrange multipliers [43], or using the classical Hoffman–Wielandt inequality [14], the solution being as follows.

Proposition 2.1 *Let the orthogonal diagonalizations of A and B be $A = V\Sigma V^T$ and $B = U\Lambda U^T$, respectively, where the eigenvalues of Σ and Λ are in nonincreasing and nondecreasing order, respectively. Then the optimal value of* $\mathrm{QQP_O}$ *is $\mu^O = \mathrm{Trace}\,\Sigma\Lambda$, and the optimal matrix X is the product VU^T, which is obtained from the orthogonal matrices of the diagonalizations.* ■

The Lagrangian dual of $\mathrm{QQP_O}$ is

$$\max_{S=S^T} \min_X \ \mathrm{Trace}\, AXBX^T - \mathrm{Trace}\, S(XX^T - I). \tag{31}$$

There can be a nonzero duality gap for this Lagrangian dual, however, as shown by example in [121]. The inner minimization of problem (31) is an unconstrained quadratic minimization in the elements of X, with hidden constraint on the Hessian

$$B \otimes A - I \otimes S \succeq 0.$$

On the other hand, the first order stationarity conditions at an optimum X, for the original problem $\mathrm{QQP_O}$, are equivalent to $AXB = SX$ or, by orthogonality, $AXBX^T = S$, which yields not only mutual diagonalizability of A and XBX^T but also the characterization of the optimum. One can easily construct examples with a duality gap, caused by a conflict between the semidefinite condition and stationarity. In order to close the duality gap, we need more constraints on X.

Note that in $\mathrm{QQP_O}$ the constraints $XX^T = I$ and $X^TX = I$ are equivalent. Adding the redundant constraints $X^TX = I$, we arrive at

$$(\mathrm{QQP_{OO}}) \qquad \mu^O := \begin{array}{ll} \min & \mathrm{Trace}\, AXBX^T \\ \text{s.t.} & XX^T = I,\ X^TX = I. \end{array} \tag{32}$$

Using symmetric matrices S and T to relax the constraints $XX^T = I$ and $X^TX = I$, respectively, we obtain the dual problem

$$(\mathrm{DQQP_{OO}}) \qquad \mu^O \geq \mu^D := \begin{array}{ll} \max & \mathrm{Trace}\, S + \mathrm{Trace}\, T \\ \text{s.t.} & (I \otimes S) + (T \otimes I) \preceq (B \otimes A) \\ & S = S^T,\ T = T^T, \end{array}$$

because, by homogeneity, the inner minimization over X is achieved at $X = 0$.

Theorem 2.3 *Strong duality holds for* QQP$_{OO}$ *and* DQQP$_{OO}$, *i.e.* $\mu^D = \mu^O$, *and both primal and dual are attained.* ■

A further relaxation of the above orthogonal relaxation is the trust region relaxation, studied in [49],

$$\text{(QAPT)} \qquad \mu^*_{QAPT} := \begin{array}{ll} \min & \operatorname{Trace} AXBX^T \\ \text{s.t.} & XX^T \preceq I. \end{array} \tag{33}$$

The constraints are convex with respect to the Löwner partial order, i.e. the partial order induced by the cone of positive semidefinite matrices. This problem is visually similar to the TRS problem discussed above, and we hope that methods for its solution will be useful. Therefore we would like to find a characterization of optimality. The set $\{X : XX^T \preceq I\}$ is studied separately in [80, 30], and is useful in eigenvalue variational principles.

It is shown in [5] that the following generalization of the Hoffman–Wielandt inequality holds,

Theorem 2.4 *Let* $V^TAV = \Sigma$ *and* $U^TBU = \Lambda$ *be the orthogonal diagonalizations of* A *and* B, *respectively, and let their eigenvalues be in nonincreasing order, say* $\sigma_1 \geq \sigma_2 \geq \cdots \geq \sigma_n$ *and* $\lambda_1 \geq \lambda_2 \geq \cdots \geq \lambda_n$. *Then, for any* $XX^T \preceq I$, *we have the inequality*

$$\sum_{i=1}^{n} \min\{\lambda_i \sigma_{n-i+1}, 0\} \leq \operatorname{Trace} AXBX^T \leq \sum_{i=1}^{n} \max\{\lambda_i \sigma_i, 0\}.$$

The upper bound is attained for $X = V \operatorname{Diag}(\varepsilon) U^T$, *where* $\varepsilon_i = 1$ *if* $\sigma_i\lambda_i \geq 0$, *and* $\varepsilon_i = 0$ *otherwise. The lower bound is attained for* $X = V \operatorname{Diag}(\varepsilon) JU^T$, *where* $\varepsilon_i = 1$ *if* $\sigma_i \lambda_{n+1-i} \leq 0$ *and* $\varepsilon_i = 0$ *otherwise. Further,* J *is the permutation matrix* $(e_n, e_{n-1}, \cdots, e_1)$, *where* e_i *is the* i*-th coordinate vector.* ■

For a scalar ξ, let $[\xi]^- := \min\{0, \xi\}$. The lower bound in the above Theorem 2.4 establishes $\mu^*_{QAPT} = \sum_{i=1}^n [\lambda_i \sigma_{n-i+1}]^-$. Since the theorem also provides the feasible point of attainment, i.e. an upper bound for the relaxation problem, the theorem can be proved by showing that the value μ^*_{QAPT} is also attained by a Lagrangian dual program. However, there can be a duality gap if we use the Lagrangian dual of the trust region type relaxation with the constraint (33). Note that XX^T and X^TX have the same eigenvalues, which implies $XX^T \preceq I$ if and only if $X^TX \preceq I$.

Explicitly, using both sets of constraints as in [6], we write QAPT in the form

$$\text{(QAPTR)} \qquad \mu^*_{QAPT} := \begin{array}{ll} \min & \operatorname{Trace} AXBX^T \\ \text{s.t.} & XX^T \preceq I, \quad X^TX \preceq I. \end{array}$$

Next we apply Lagrangian relaxation to QAPTR, using the symmetric matrices $S \succeq 0$ and $T \succeq 0$, say, to relax the constraints $XX^T \preceq I$ and $X^TX \preceq I$, respectively. This gives the dual problem

$$\text{(DQAPTR)} \quad \mu^*_{QAPT} \geq \mu^D_{QAPT} := \begin{array}{ll} \max & -\operatorname{Trace} S - \operatorname{Trace} T \\ \text{s.t.} & -(I \otimes S) - (T \otimes I) \preceq (B \otimes A) \\ & S \succeq 0,\ T \succeq 0. \end{array}$$

The following properties are proved in [5].

Theorem 2.5 *Strong duality holds for QAPTR and DQAPTR, i.e.* $\mu^D_{QAPT} = \mu^*_{QAPT}$, *and both primal and dual are attained.* ■

3. A STRENGTHENED BOUND FOR MC

From the results in Section 2.1, it appears that we might have the strongest possible tractable bound for the max-cut problem MC (or its equivalent quadratic model MCQ). Bounds can be strengthened, however, by adding redundant constraints. We now present a strategy (recipe) for constructing relaxations.

Algorithm 3.1 (Recipe for SDP relaxations)

- *add redundant constraints*
- *homogenize*
- *take Lagrangian dual*
- *use hidden semidefinite constraint to obtain SDP equivalent*
- *take Lagrangian dual of SDP*
- *check Slater's CQ — project if it fails*
- *delete redundant constraints from final SDP*

We will use the above recipe to strengthen our bound for MC. Therefore we introduce the following notation for linear operators and adjoints. For $S \in \mathcal{S}^n$, the vector $s = \operatorname{svec}(S) \in \Re^{t(n)}$ is formed (columnwise) from

S, ignoring the strictly lower triangular part of S, so $t(n) = n(n+1)/2$. The inverse operator, which constructs $S \in \mathcal{S}^n$ from $s \in \Re^{t(n)}$ is $S = \operatorname{sMat}(s)$. The adjoint of svec is the operator hMat, where $\operatorname{hMat}(v)$ is like $\operatorname{sMat}(v)$, $v \in \Re^{t(n)}$, except that the off-diagonal elements are halved, in order to provide the adjoint equation

$$\operatorname{svec}(S)^T v = \operatorname{Trace} S \operatorname{hMat}(v), \qquad \forall S \in \mathcal{S}^n, \;\; v \in \Re^{t(n)}.$$

The adjoint of sMat is the operator $\operatorname{dsvec}(S)$, which works like svec, except that the off diagonal elements are multiplied by 2, in order to satisfy

$$\operatorname{dsvec}(S)^T v = \operatorname{Trace} S \operatorname{sMat}(v), \qquad \forall S \in \mathcal{S}^n, \;\; v \in \Re^{t(n)}.$$

For notational convenience, we define the vectors

$$\operatorname{sdiag}(s) := \operatorname{diag}(\operatorname{sMat}(s)) \in \Re^n, \quad \forall s \in \Re^{t(n)},$$

and

$$\operatorname{vsMat}(s) = \operatorname{vec}(\operatorname{sMat}(s)) \in \Re^{n^2}, \quad \forall s \in \Re^{t(n)}.$$

The adjoint of vsMat is then given by

$$\operatorname{vsMat}^*(v) = \operatorname{dsvec}\left(\left(\operatorname{Mat}(v) + \operatorname{Mat}(v)^T\right)/2\right) \in \Re^{t(n)}, \quad \forall v \in \Re^{n^2}.$$

The following bound is motivated by the strong duality results of Section 2.3, and is studied in depth in [6]. We recall that the SDP bound (20) for MCQ arises from a lifting procedure that employs

$$0 \preceq X = xx^T \quad \text{and} \quad x^T Q x = \operatorname{Trace} QX.$$

Discarding the rank one condition on X provides the tractable SDP bound, but it is not clear what constraints one can add to the P_E problem (21), in order to strengthen the Lagrangian relaxation. Linear combinations of the constraints will not help since they are already included in the Lagrangian. Nor is it clear what linear constraints to add to the SDP relaxation (20). One can try including the so-called triangle inequalities, which are standard in branch and bound methods for MC. This choice is sometimes very successful in practice [47], but one cannot guarantee an improvement [50]. Instead, we see that the matrices $X = xx^T$ have the property

$$X^2 = xx^T xx^T = nX$$

in the case $x_i = \pm 1$, $i = 1, \ldots, n$. Therefore the quadratic matrix model

$$\begin{array}{rll} \mu^* := & \max & \operatorname{Trace} QX \\ & \text{s.t.} & \operatorname{diag}(X) = e \\ & & X^2 - nX = 0, \end{array} \tag{34}$$

where X is a symmetric matrix, is equivalent to MCQ. A common diagonalization of X and X^2 shows that the only eigenvalues of X are 0 and n, while (34) shows $\operatorname{Trace} X = n$. Therefore the rank of X is one and $X \succeq 0$.

To illustrate the recipe (Algorithm 3.1), we add an additional redundant constraint, $X \circ X = E$, where E is the matrix of ones, obtaining the program

$$\text{(MC2)} \qquad \begin{array}{rll} \mu^* = & \max & \operatorname{Trace} QX \\ & \text{s.t.} & \operatorname{diag}(X) = e \\ & & X \circ X = E \\ & & X^2 - nX = 0, \end{array} \tag{35}$$

which is equivalent to (34) and so to MC. To apply Lagrangian relaxation efficiently, without losing information from the linear constraint, we replace $\operatorname{diag}(X) = e$ by the norm constraint $\|\operatorname{diag}(X) - e\|^2 = 0$. Then we introduce homogeneity by adding the variable y_0 and the constraint $1 - y_0^2 = 0$. (Because of the homogenization, if x and $y_0 = -1$ are optimal, then we can just multiply both x and y by -1 without changing the optimal value or feasibility.) Thus, because $X = \operatorname{sMat}(x)$ is a symmetric matrix, we get the equivalent program

$$\begin{array}{ll} \max\limits_{x,y_0} & y_0 \operatorname{Trace}(Q \operatorname{sMat}(x)) \\ \text{s.t.} & \operatorname{sdiag}(x)^T \operatorname{sdiag}(x) - 2y_0\, e^T \operatorname{sdiag}(x) + n = 0 \\ & \operatorname{sMat}(x) \circ \operatorname{sMat}(x) = E \\ & \operatorname{sMat}(x)^2 - ny_0 \operatorname{sMat}(x) = 0 \\ & 1 - y_0^2 = 0. \end{array} \tag{36}$$

In fact, letting $z = \begin{pmatrix} y_0 \\ x \end{pmatrix}$, we see that we have another max-cut problem with the same objective function as MCQ, additional linear constraints $\operatorname{sdiag} x = e$ (these nodes must be grouped together), and the extra nonlinear constraints $\operatorname{sMat}(x)^2 - ny_0 \operatorname{sMat}(x) = 0$.

As discussed in Remark 1.2, the Lagrange multipliers for symmetric matrix valued constraint functions are symmetric matrices. Therefore problem (36) has the Lagrangian dual

$$\begin{aligned} \mu^* \leq \nu_2^* := \min_{w,T,S} \max_{x, y_0^2 = 1} \quad & y_0 \operatorname{Trace}(Q \operatorname{sMat}(x)) \\ & + w(\operatorname{sdiag}(x)^T \operatorname{sdiag}(x) - 2y_0\, e^T \operatorname{sdiag}(x) + n) \\ & + \operatorname{Trace} T(E - \operatorname{sMat}(x) \circ \operatorname{sMat}(x)) \\ & + \operatorname{Trace} S((\operatorname{sMat}(x))^2 - ny_0 \operatorname{sMat}(x)), \end{aligned}$$

where w, T and S are the Lagrange multipliers, the matrices T and S being symmetric. We can move the constraint on y_0 into the Lagrangian without increasing the duality gap, which provides the form

$$\begin{aligned}\nu_2^* = \min_{w,T,S,t} \max_{x,y_0} \quad & y_0 \,\mathrm{Trace}\,(Q\,\mathrm{sMat}\,(x)) \\ & + w\,(\mathrm{sdiag}\,(x)^T \mathrm{sdiag}\,(x) - 2y_0\, e^T \mathrm{sdiag}\,(x) + n) \\ & + \mathrm{Trace}\, T(E - \mathrm{sMat}\,(x) \circ \mathrm{sMat}\,(x)) \\ & + \mathrm{Trace}\, S((\mathrm{sMat}\,(x))^2 - n y_0\, \mathrm{sMat}\,(x)) + t(1 - y_0^2).\end{aligned}$$

The inner maximization of the above relaxation is an unconstrained pure quadratic maximization, so the optimal value is infinity unless the Hessian is negative semidefinite **(hidden constraint)**, in which case $x = 0$ and $y_0 = 0$ are optimal. Therefore we need to evaluate the Hessian of the Lagrangian to find the first SDP in the recipe.

Using $\mathrm{Trace}\,(Q\,\mathrm{sMat}\,(x)) = x^T \mathrm{dsvec}\,(Q)$, and adding a 2 for convenience, we find that the constant part (no Lagrange multipliers) of the Hessian is

$$2H_c := 2\begin{pmatrix} 0 & \frac{1}{2}\mathrm{dsvec}\,(Q)^T \\ \frac{1}{2}\mathrm{dsvec}\,(Q) & 0 \end{pmatrix}.$$

We now specify and manipulate some of the other linear operators that appear in the Lagrangian, which facilitates the derivation of the Hessian and the adjoints.

The matrix $\mathrm{sdiag}^* \mathrm{sdiag} \in \mathcal{S}^{t(n)}$, where sdiag^* is the adjoint of sdiag, is diagonal and has the elements

$$\begin{aligned} e_i^T (\mathrm{sdiag}^* \mathrm{sdiag}\,)\, e_j &= \mathrm{sdiag}\,(e_i)^T \mathrm{sdiag}\,(e_j) \\ &= \begin{cases} 1 & \text{if } i = j = t(k) \\ 0 & \text{otherwise.} \end{cases} \end{aligned} \tag{37}$$

Moreover, if $T = \sum_{ij} t_{ij} E_{ij}$, where the matrices $E_{ij} \in \mathcal{S}^n$ are the elementary matrices $e_i e_j^T + e_j e_i^T$, then linearity gives

$$\mathrm{dsvec}\,\left(T \circ \mathrm{sMat}\,\right) = \sum_{ij} t_{ij}\, \mathrm{dsvec}\,\left(E_{ij} \circ \mathrm{sMat}\,\right). \tag{38}$$

Also we have

$$\mathrm{dsvec}\,\mathrm{Diag}\,\mathrm{diag}\,\mathrm{sMat} = \mathrm{sdiag}^* \mathrm{sdiag} = \mathrm{Diag}\,\mathrm{svec}\,(I_n), \tag{39}$$

where the first two terms can be equated to a matrix because they are linear operators from $\Re^{t(n)}$ to $\Re^{t(n)}$, and where I_n is the $n \times n$ unit matrix. Using these relations, we rewrite the quadratic forms in the Lagrangian as

$$\begin{aligned} \mathrm{sdiag}\,(x)^T \mathrm{sdiag}\,(x) &= x^T (\mathrm{dsvec}\,\mathrm{Diag}\,\mathrm{diag}\,\mathrm{sMat}\,)\, x, \\ y_0\, e^T \mathrm{sdiag}\,(x) &= y_0\, (\mathrm{dsvec}\, I_n)^T x, \end{aligned} \tag{40}$$

$$\begin{aligned}\operatorname{Trace} T(\operatorname{sMat}(x) \circ \operatorname{sMat}(x)) &= x^T \{\operatorname{dsvec}(T \circ \operatorname{sMat}(x))\} \\ &= x^T (\operatorname{dsvec}(T \circ \operatorname{sMat}))\, x,\end{aligned} \tag{41}$$

$$\begin{aligned}\operatorname{Trace} S(\operatorname{sMat}(x))^2 &= \operatorname{Trace} \operatorname{sMat}(x)\, S \operatorname{sMat}(x) \\ &= x^T \operatorname{dsvec}(S \operatorname{sMat}(x)) \\ &= x^T (\operatorname{dsvec} S \operatorname{sMat})\, x.\end{aligned} \tag{42}$$

Because $S \operatorname{sMat}(x)$ may not be symmetric in (42), we let $\operatorname{dsvec}(B)$ denote $\operatorname{dsvec}(\frac{1}{2}B + \frac{1}{2}B^T)$ if B is any unsymmetric $n \times n$ matrix. It is easy to find the Hessian of a quadratic form written as $x^T \cdot x$. Therefore, we can now write down the *negative* of the nonconstant part of the Hessian of the inner maximization of ν_2^*. Splitting it into four linear operators with the factor 2, we find the formula

$$\begin{aligned}2\mathcal{H}(w,T,S,t) &:= 2\mathcal{H}_1(w) + 2\mathcal{H}_2(T) + 2\mathcal{H}_3(S) + 2\mathcal{H}_4(t) \\ &:= 2w \begin{pmatrix} 0 & (\operatorname{dsvec} I_n)^T \\ (\operatorname{dsvec} I_n) & -\operatorname{sdiag}^* \operatorname{sdiag} \end{pmatrix} \\ &\quad +2 \begin{pmatrix} 0 & 0 \\ 0 & \operatorname{dsvec}(T \circ \operatorname{sMat}) \end{pmatrix} \\ &\quad +2 \begin{pmatrix} 0 & \frac{n}{2}\operatorname{dsvec}(S)^T \\ \frac{n}{2}\operatorname{dsvec}(S) & -\operatorname{dsvec} S \operatorname{sMat} \end{pmatrix} + 2t \begin{pmatrix} 1 & 0 \\ 0 & 0 \end{pmatrix}.\end{aligned}$$

After cancelling the 2, it follows that our calculation is the semidefinite program

$$(\text{MCDSDP2}) \qquad \begin{array}{rcll} \nu_2^* = & \min\limits_{w,T,S,t} & nw + \operatorname{Trace} ET + \operatorname{Trace} 0S + t \\ & \text{s.t.} & \mathcal{H}(w,T,S,t) \succeq H_c. \end{array}$$

By taking T to be sufficiently positive definite and t to be sufficiently large, we can guarantee Slater's constraint qualification. Therefore there is no duality gap between this SDP and its dual, which is the following strengthened SDP relaxation of MC

$$(\text{MCPSDP2}) \qquad \begin{array}{rcll} \nu_2^* = & \max\limits_{Y \succeq 0} & \operatorname{Trace} H_c Y \\ & \text{s.t.} & \mathcal{H}_1^*(Y) = n, & \mathcal{H}_2^*(Y) = E, \\ & & \mathcal{H}_3^*(Y) = 0, & \mathcal{H}_4^*(Y) = 1, \end{array}$$

where the superscripts * denote adjoint operators as usual.

To find the explicit form of the SDP relaxation, we now derive the adjoint operators explicitly. We partition Y as

$$Y = \begin{pmatrix} Y_{00} & x^T \\ x & \bar{Y} \end{pmatrix}. \tag{43}$$

From (41) and

$$e_i^T \left(\operatorname{dsvec}\left(T \circ \operatorname{sMat}\right)\right) e_j = \left(\operatorname{dsvec}(T) \circ e_j\right)^T e_i, \quad 1 \le i, j \le t(n),$$

we deduce

$$\mathcal{H}_2^*(Y) = \operatorname{sMat}\operatorname{diag}(\bar{Y}).$$

Therefore, $\mathcal{H}_2^*(Y) = E$ and $\mathcal{H}_4^*(Y) = 1$ are equivalent to

$$\operatorname{diag}(Y) = e. \tag{44}$$

Also, $\mathcal{H}_1^*(Y)$ is twice the sum of the elements in the first row of Y whose positions correspond to the diagonal elements of sMat $(\cdot)$ minus the sum of the elements in these positions in the diagonal of $\bar{Y}$, so the notation (43) provides

$$\mathcal{H}_1^*(Y) = 2\left(\operatorname{svec} I_n\right)^T x - \operatorname{Trace}\operatorname{Diag}\left(\operatorname{svec} I_n\right)\bar{Y}.$$

It follows from equation (44) that the constraint $\mathcal{H}_1^*(Y) = n$ requires

$$x_{t(i)} = Y_{0,t(i)} = 1, \quad \forall\, i = 1, \ldots, n. \tag{45}$$

Finally, we find $\mathcal{H}_3^*(Y)$. Recall that the definition of $\mathcal{H}_3$ is made up of four blocks, with the bottom right block defined by minus the quadratic form (42). Thus this block can be seen in two ways: either as a quadratic form or as its $t(n) \times t(n)$ symmetric matrix representation $-\operatorname{dsvec} S \operatorname{sMat}$. We regard dsvec $\cdot$ sMat as the linear operator that maps any $n \times n$ symmetric matrix S to the $t(n) \times t(n)$ symmetric matrix $\operatorname{dsvec} S \operatorname{sMat}$. Further, we write $\operatorname{vsMat}^* \operatorname{vec}$ instead of dsvec ($= \operatorname{sMat}^*$), in order to avoid applying dsvec to unsymmetric matrices, as mentioned after (42). Therefore the definition of $\mathcal{H}_3(S)$ gives

$$\begin{aligned}
\langle \mathcal{H}_3(S), Y\rangle &= n \operatorname{dsvec}(S)^T x - \left\langle \operatorname{vsMat}^* \operatorname{vec} S \operatorname{sMat}, \bar{Y}\right\rangle \\
&= n \operatorname{Trace} S \operatorname{sMat}(x) - \left\langle S, \operatorname{sMat} \bar{Y} \operatorname{vsMat}^* \operatorname{vec}\right\rangle \\
&= \left\langle S, n \operatorname{sMat}(x) - \operatorname{sMat} \bar{Y} \operatorname{vsMat}^* \operatorname{vec}\right\rangle,
\end{aligned}$$

which provides the adjoint

$$\mathcal{H}_3^*(Y) = n \operatorname{sMat}(x) - \operatorname{sMat} \bar{Y} \operatorname{vsMat}^* \operatorname{vec}. \tag{46}$$

We now have another linear operator that maps symmetric matrices into symmetric matrices. Specifically, $\operatorname{sMat} \cdot \operatorname{vsMat}^* \operatorname{vec}$ maps any $t(n) \times t(n)$ symmetric matrix $\bar{Y}$ to $\operatorname{sMat} \bar{Y} \operatorname{vsMat}^* \operatorname{vec}$, which is construed as an $n \times n$ symmetric matrix by invoking the quadratic form

$$\begin{aligned}
v^T \operatorname{sMat} \bar{Y} \operatorname{vsMat}^* \operatorname{vec} v &= \left\langle vv^T, \operatorname{sMat} \bar{Y} \operatorname{vsMat}^* \operatorname{vec}\right\rangle \\
&= \operatorname{Trace}\left\{\operatorname{sMat} \bar{Y} \operatorname{dsvec}\left(vv^T\right)\right\}.
\end{aligned}$$

Indeed, the k, l element of this quadratic form, which becomes the k, l element of $\mathrm{sMat}\,\bar{Y}\mathrm{vsMat}^*\,\mathrm{vec}$, has the value

$$e_k^T \mathrm{sMat}\,\bar{Y}\mathrm{vsMat}^*\mathrm{vec}\, e_l = \tfrac{1}{2}\,\mathrm{Trace}\,\left\{\mathrm{sMat}\,\bar{Y}\mathrm{dsvec}\,(e_k e_l^T + e_l e_k^T)\right\}.$$

The explicit form of $\mathcal{H}_3^*(Y)$ is given in the sums of MCPSDP3 below, and also we have removed many of the redundant constraints. We see that the space of the variables of the new problem has increased to $\mathcal{S}^{t(n)+1}$, from $\mathcal{S}^n$ in (20), and that there are now $2t(n) - 1$ constraints. They come from (44), (45), and the upper triangular part of (46). The latter constraints can be related to the original condition $\mathrm{sMat}\,(x) = \frac{1}{n}\mathrm{sMat}\,(x)\,\mathrm{sMat}\,(x)$. As in the original SDP relaxation MCSDP (20), we still require the matrix of variables to have ones on the diagonal, but we now have additional constraints on the first row of Y. Specifically, the new problem has the form

$$(\mathrm{MCPSDP3}) \qquad \begin{array}{ll} \max\limits_{Y\in\mathcal{S}^{t(n)+1}} & \mathrm{Trace}\, H_c Y \\ \text{s.t.} & \mathrm{diag}\,(Y) = e \\ & Y_{0,t(i)} = 1, \quad \forall i = 1, \ldots, n \\ & Y_{0,t(j-1)+i} = \frac{1}{n}\Big\{\sum_{k=1}^{i} Y_{t(i-1)+k,t(j-1)+k} \\ & \qquad + \sum_{k=i+1}^{j} Y_{t(k-1)+i,t(j-1)+k} \\ & \qquad + \sum_{k=j+1}^{n} Y_{t(k-1)+i,t(k-1)+j}\Big\}, \\ & \qquad\qquad \forall 1 \le i < j \le n \\ & Y \succeq 0. \end{array}$$

Remark 3.1 *Since the first row of Y has some off-diagonal elements equal to 1, and since all the diagonal elements are 1, the Slater constraint qualification fails for this problem, i.e. we cannot have a positive definite feasible matrix. Therefore we can project this problem onto a face of the semidefinite cone, and then project into a smaller dimensional space to get a new further simplified problem. The final program has constraints that are linearly independent and that satisfy Slater's condition. Details are given in [3].*

The constraint $\mathcal{H}_3^*(Y) = 0$ is the key to proving the following useful result. It is remarkable because the nonlinear constraint that $\mathrm{sMat}\,(x)$ be positive semidefinite (and in fact feasible for MCSDP) was discarded in MC2. A proof is included to illustrate the usefulness of using adjoints.

Lemma 3.1 *Suppose that Y is feasible in MCPSDP3. Then its first row satisfies*

$$\mathrm{sMat}\,\left(Y_{0,1:t(n)}\right) \succeq 0,$$

and so is feasible in the original SDP relaxation MCSDP (20).

Proof. Let Y be feasible for MCPSDP3, and write

$$Y = \begin{pmatrix} 1 & x^T \\ x & \bar{Y} \end{pmatrix}.$$

The fact that $\bar{Y}$ is a principal submatrix of Y implies $\bar{Y} \succeq 0$. We have previously established that

$$\mathcal{H}_3^*(Y) = n \,\mathrm{sMat}\,(x) - \mathrm{sMat}\,\bar{Y} \mathrm{vsMat}^*\,\mathrm{vec}\,,$$

where vsMat*vec is essentially sMat* except that it acts on possibly non-symmetric matrices. Therefore the constraint $\mathcal{H}_3^*(Y) = 0$ is equivalent to

$$\mathrm{sMat}\,(x) = \frac{1}{n}\,\mathrm{sMat}\,\bar{Y} \mathrm{vsMat}^*\,\mathrm{vec}\,.$$

Thus $\mathrm{sMat}\,(x)$ is a congruence of the positive semidefinite matrix $\bar{Y}$. The result follows. ■

The term $X^2 - nX$, from the added nonlinear constraint $X^2 - nX = 0$ in the original max-cut problem (34) or (35), has the following interesting and useful properties in the SDP relaxation.

Lemma 3.2 *Suppose that both X and $\bar{X}$ are feasible for MCSDP. Then*

$$\mathrm{Trace}\,(X^2 - nX)\,(\bar{X}^2 - n\bar{X}) \geq 0. \tag{47}$$

Suppose, in addition, that

$$(X^2 - nX) \neq 0 \quad \textit{and} \quad (\bar{X}^2 - n\bar{X}) \neq 0$$

hold, and that both X and $\bar{X}$ are in $\mathcal{F}$, a face of $\mathcal{P}$, with $\bar{X} \in \mathrm{relint}\,\mathcal{F}$. Then

$$\mathrm{Trace}\,(X^2 - nX)\,(\bar{X}^2 - n\bar{X}) > 0. \tag{48}$$

Proof. See [3]. ■

The above two lemmas can be used to show the strength of this new bound. Indeed, unless there is no gap between MCSDP and MC, the relaxation MCPSDP3 **always** provides a strict improvement over MCSDP. The proof is given in [3].

Theorem 3.1 *The optimal values satisfy*

$$\nu_2^* \leq \nu^* \quad \textit{and} \quad \nu_2^* = \nu^* \Rightarrow \nu_2^* = \mu^*. \tag{49}$$

■

Thus we see that this new bound is always strictly better than the previous one. Numerical results on small problems are presented in [3]. They consistently provide a significant improvement over the already strong original SDP bound.

4. CONCLUSION

We have looked at several problems where strong relaxations exist. In each case we have shown that our theme holds; one cannot do better than the Lagrangian relaxation. In particular, this has led us to a recipe for finding strong relaxations for hard, discrete optimization problems and a strengthened relaxation for the max-cut problem.

Acknowledgments

The author would like to thank Mike Powell for many helpful comments that significantly improved the readability of this paper.

References

[1] F. Alizadeh, J-P.A. Haeberly and M.L. Overton (1994), A new primal-dual interior-point method for semidefinite programming, in J.G. Lewis, ed., *Proceedings of the Fifth SIAM Conference on Applied Linear Algebra*, SIAM, pp. 113–117.

[2] F. Alizadeh, J-P.A. Haeberly and M.L. Overton (1997), Complementarity and nondegeneracy in semidefinite programming, *Math. Programming*, 77, pp. 111–128.

[3] M. Anjos and H. Wolkowicz (1999), A strengthened SDP relaxation via a second lifting for the Max-Cut problem, Technical Report Research Report, CORR 99-55, University of Waterloo, Waterloo, Canada.

[4] K.M. Anstreicher (1999), Eigenvalue bounds versus semidefinite relaxations for the quadratic assignment problem, Technical report, University of Iowa, Iowa City, IA.

[5] K.M. Anstreicher, X. Chen, H. Wolkowicz and Y. Yuan (1999), Strong duality for a trust-region type relaxation of QAP, *Linear Algebra Appl.*, 301, pp. 121–136.

[6] K.M. Anstreicher and H. Wolkowicz (1999), On Lagrangian relaxation of quadratic matrix constraints, to appear in *SIAM J. Matrix Anal. Appl.*

[7] E. Balas (1997), A modified lift-and-project procedure, *Math. Programming*, 79(1-3, Ser. B), pp. 19–31. Lectures on Mathematical Programming ismp97 (Lausanne, 1997).

[8] E. Balas and E. Zemel (1984), Lifting and complementing yields all the facets of positive zero-one programming polytopes, in *Mathematical programming (Rio de Janeiro, 1981)*, North-Holland, Amsterdam, pp. 13–24.

[9] G. P. Barker, M. Laidacker and G. Poole (1987), Projectionally exposed cones, *SIAM J. Algebraic Discrete Methods*, 8(1), pp. 100–105.

[10] G.P. Barker and D. Carlson (1975), Cones of diagonally dominant matrices, *Pacific J. of Math.*, 57, pp. 15–32.

[11] R. Bellman and K. Fan (1963), On systems of linear inequalities in Hermitian matrix variables, in *Proceedings of Symposia in Pure Mathematics, Vol 7, AMS.*

[12] A. Ben-Israel, A. Charnes and K. Kortanek (1969), Duality and asymptotic solvability over cones, *Bull. Amer. Math. Soc.*, 75(2), pp. 318–324.

[13] S. J. Benson, Y. Ye, and X. Zhang (1997), Solving large-scale sparse semidefinite programs for combinatorial optimization, Technical report, The University of Iowa.

[14] R. Bhatia (1987), *Perturbation Bounds for Matrix Eigenvalues: Pitman Research Notes in Mathematics Series 162*, Longman, New York.

[15] F. Bohnenblust (1948), Joint positiveness of matrices, Manuscript.

[16] B. Borchers (1999), SDPLIB 1.2, a library of semidefinite programming test problems, Technical report, New Mexico Tech, Soccorrow, NM, to appear in *Optim. Methods Softw.*

[17] J.M. Borwein and H. Wolkowicz (1980/81), Characterization of optimality for the abstract convex program with finite-dimensional range, *J. Austral. Math. Soc. Ser. A*, 30(4), pp. 390–411.

[18] J.M. Borwein and H. Wolkowicz (1981), Regularizing the abstract convex program, *J. Math. Anal. Appl.*, 83(2), pp. 495–530.

[19] J.M. Borwein and H. Wolkowicz (1982), Characterizations of optimality without constraint qualification for the abstract convex program, *Math. Programming Stud.*, 19, pp. 77–100. Optimality and stability in mathematical programming.

[20] S. Boyd, V. Balakrishnan, E. Feron, and L. El Ghaoui (1993), Control system analysis and synthesis via linear matrix inequalities, *Proc. ACC*, pp. 2147–2154.

[21] M.R. Celis, J.E. Dennis Jr. and R.A. Tapia (1984), A trust region strategy for nonlinear equality constrained optimization, in *Proceedings of the SIAM Conference on Numerical Optimization, Boulder, CO.* Also available as Technical Report TR84-1, Department of Mathematical Sciences, Rice University, Houston, TX.

[22] S. Chandrasekaran, M. Gu and A. H. Sayed (1999), The sparse basis problem and multilinear algebra, *SIAM J. Matrix Anal. Appl.*, 20(2), pp. 354–362.

[23] B. Craven and J.J. Koliha (1977), Generalizations of Farkas' thoerem, *SIAM J. Matrix Anal. Appl.*, 8, pp. 983–997.

[24] B. Craven and B. Mond (1981), Linear programming with matrix variables, *Linear Algebra Appl.*, 38, pp. 73–80.

[25] J.-P. Crouzeix, J.-E. Martinez-Legaz and A. Seeger (1995), An alternative theorem for quadratic forms and extensions, *Linear Algebra Appl.*, 215, pp. 121–134.

[26] Dem'yanov and Malozemov (1974), *Introduction to Minimax*, Keter Publishing House, Jerusalem.

[27] H. Dym and I. Gohberg (1981), Extensions of band matrices with band inverses, *Linear Algebra Appl.*, 36, pp. 1–24.

[28] A. Edelman, T. Arias and S.T. Smith (1999), The geometry of algorithms with orthogonality constraints, *SIAM J. Matrix Anal. Appl.*, 20(2), pp. 303–353 (electronic).

[29] A.V. Fiacco (1983), *Introduction to Sensitivity and Stability Analysis in Nonlinear Programming*, volume 165 of *Mathematics in Science and Engineering*, Academic Press.

[30] P.A. Fillmore and J.P. Williams (1971), Some convexity theorems for matrices, *Glasgow Mathematical Journal*, 10, pp. 110–117.

[31] F. Forgó (1988). *Nonconvex Programming*, Akadémiai Kiadó, Budapest.

[32] M. Fu, Z. Luo and Y. Ye (1998), Approximation algorithms for quadratic programming, *J. Comb. Optim.*, 2(1), pp. 29–50.

[33] T. Fujie and M. Kojima (1997), Semidefinite programming relaxation for nonconvex quadratic programs. *J. Global Optim.*, 10(4), pp. 367–380.

[34] M.X. Goemans (1997), Semidefinite programming in combinatorial optimization, *Math. Programming*, 79, pp. 143–162.

[35] M.X. Goemans and D.P. Williamson (1995), Improved approximation algorithms for maximum cut and satisfiability problems

using semidefinite programming, *J. Assoc. Comput. Mach.*, 42(6), pp. 1115–1145.

[36] A. J. Goldman and A. W. Tucker (1956), Theory of linear programming, in *Linear inequalities and related systems*, Annals of Mathematics Studies, no. 38, Princeton University Press, pp. 53–97..

[37] B. Grone, C.R. Johnson, E. Marques de Sa and H. Wolkowicz (1984), Positive definite completions of partial Hermitian matrices, *Linear Algebra Appl.*, 58, pp. 109–124.

[38] M. Grötschel, L. Lovász and A. Schrijver (1988), *Geometric Algorithms and Combinatorial Optimization*, Springer Verlag, Berlin.

[39] Branko Grünbaum (1965), On the facial structure of convex polytopes, *Bull. Amer. Math. Soc.*, 71, pp. 559–560.

[40] Branko Grünbaum (1967), *Convex polytopes*, Pure and Applied Mathematics, Vol. 16, Interscience Publishers John Wiley & Sons, Inc., New York. With the cooperation of Victor Klee, M. A. Perles and G. C. Shephard.

[41] M. Guignard (1969), Generalized Kuhn–Tucker conditions for mathematical programming problems in a Banach space, *SIAM J. of Control*, 7, pp. 232–241.

[42] O. Güler and L. Tuncel (1998), Characterization of the barrier parameter of homogeneous convex cones, *Math. Programming*, 81, pp. 55–76.

[43] S.W. Hadley, F. Rendl and H. Wolkowicz (1992), A new lower bound via projection for the quadratic assignment problem, *Math. Oper. Res.*, 17(3), pp. 727–739.

[44] R. Hauser (1999), Self-scaled barrier functions: Decomposition and classification, Technical Report DAMTP 1999/NA13, Department of Applied Mathematics and Theoretical Physics, University of Cambridge.

[45] C. Helmberg (1994), *An interior point method for semidefinite programming and max-cut bounds*, PhD thesis, Graz University of Technology, Austria.

[46] C. Helmberg and F. Rendl (1997), A spectral bundle method for semidefinite programming, Technical Report ZIB Preprint SC-97-37, Konrad-Zuse-Zentrum Berlin, Berlin, Germany, to appear in *SIAM J. Optim.*

[47] C. Helmberg, F. Rendl, R. J. Vanderbei and H. Wolkowicz (1996), An interior-point method for semidefinite programming, *SIAM J. Optim.*, 6(2), pp. 342–361.

[48] F. Jarre (1990), On the convergence of the method of analytic centers when applied to convex quadratic programs, *Math. Programming*, 49(3), pp. 341–358.

[49] S.E. Karisch, F. Rendl and H. Wolkowicz (1994), Trust regions and relaxations for the quadratic assignment problem, in *Quadratic assignment and related problems (New Brunswick, NJ, 1993)*, Amer. Math. Soc., Providence, RI, pp. 199–219.

[50] H. Karloff (1999), How good is the Goemans-Williamson MAX CUT algorithm?, *SIAM J. Comput.*, 29(1), pp. 336–350.

[51] M. Kojima and L. Tuncel (1998), Discretization and localization in successive convex relaxation methods for nonconvex quadratic optimization problems, Technical Report CORR98-34, University of Waterloo, Waterloo, Canada.

[52] M. Kojima and L. Tuncel (1999), Cones of matrices and successive convex relaxations of nonconvex sets, to appear in *SIAM J. Optim.*.

[53] F. Körner (1988), A tight bound for the Boolean quadratic optimization problem and its use in a branch and bound algorithm, *Optimization*, 19(5), pp. 711–721.

[54] F. Körner (1992), Remarks on a difficult test problem for quadratic Boolean programming, *Optimization*, 26, pp. 355–357.

[55] K. O. Kortanek and J. P. Evans (1968), Asymptotic Lagrange regularity for pseudoconcave programming with weak constraint qualification, *Operations Res.*, 16, pp. 849–857.

[56] S. Kruk (1996), Semidefinite programming applied to nonlinear programming, Master's thesis, University of Waterloo, Waterloo, Canada.

[57] S. Kruk, M. Muramatsu, F. Rendl, R.J. Vanderbei and H. Wolkowicz (1998), The Gauss–Newton direction in linear and semidefinite programming, Technical Report CORR 98-16, University of Waterloo, Waterloo, Canada.

[58] S. Kruk and H. Wolkowicz (1998), SQ^2P, sequential quadratic constrained quadratic programming, in Ya-xiang Yuan, ed., *Advances in Nonlinear Programming*, volume 14 of *Applied Optimization*, Kluwer, Dordrecht, pp. 177–204. Proceedings of Nonlinear Programming Conference in Beijing in honour of Professor M.J.D. Powell.

[59] S. Kruk and H. Wolkowicz (2000), Sequential, quadratic constrained, quadratic programming for general nonlinear programming, in H. Wolkowicz, R. Saigal and L. Vandenberghe, eds,

Handbook of Semidefinite Programming: Theory, Algorithms and Applications, Kluwer Academic Publishers, Boston, MA.

[60] M. Laurent (1998), A tour d'horizon on positive semidefinite and Euclidean distance matrix completion problems, in *Topics in Semidefinite and Interior-Point Methods*, volume 18 of *The Fields Institute for Research in Mathematical Sciences, Communications Series*, Amer. Math. Soc., Providence, Rhode Island.

[61] C. Lemaréchal and F. Oustry (1999), Nonsmooth algorithms to solve semidefinite programs, in L. EL Ghaoui and S-I. Niculescu, eds, *Recent Advances on LMI methods in Control*, Advances in Design and Control series, SIAM, to appear.

[62] A.S. Lewis (1994), *Take-home final exam, Course CO663 in Convex Analysis*, University of Waterloo, Waterloo, Canada.

[63] A.S. Lewis (1996), Derivatives of spectral functions, *Math. Oper. Res.*, 21(3), pp. 576–588.

[64] A.S. Lewis and M.L. Overton (1996), Eigenvalue optimization, *Acta Numerica*, 5, pp. 149–190.

[65] L. Lovász (1979), On the Shannon capacity of a graph, *IEEE Transactions on Information Theory*, 25, pp. 1–7.

[66] L. Lovász and A. Schrijver (1991), Cones of matrices and set-functions and 0-1 optimization, *SIAM J. Optim.*, 1(2), pp. 166–190.

[67] Z-Q. Luo and J. Sun (1995), An analytic center based column generation algorithm for convex quadratic feasibility problems, Technical report, McMaster University, Hamilton, Ontario.

[68] Z-Q. Luo and S. Zhang (1997), On the extension of Frank–Wolfe theorem, Technical report, Erasmus University, Rotterdam, The Netherlands.

[69] A.W. Marshall and I. Olkin (1979), *Inequalities: Theory of Majorization and its Applications*, Academic Press, New York.

[70] J.M. Martinez (1994), Local minimizers of quadratic functions on Euclidean balls and spheres, *SIAM J. Optim.*, 4(1), pp. 159–176.

[71] V.M. Matrosov and A.I. Malikov (1993), The development of the ideas of A. M. Lyapunov over one hundred years: 1892–1992, *Izv. Vyssh. Uchebn. Zaved. Mat.*, 4, pp. 3–47.

[72] R.D.C. Monteiro and Y. Zhang (1998), A unified analysis for a class of long-step primal-dual path-following interior-point algorithms for semidefinite programming, *Math. Programming*, 81, pp. 281–299.

[73] J. J. Moré (1993), Generalizations of the trust region problem, *Optim. Methods Software*, 2, pp. 189–209.

[74] J.J. Moré and D.C. Sorensen (1983), Computing a trust region step, *SIAM J. Sci. Statist. Comput.*, 4, pp. 553–572.

[75] J.J. Moré and D.C. Sorensen (1984), Newton's method, in G.H. Golub, ed., *Studies in Numerical Analysis*, vol. 24 of *MAA Studies in Mathematics*, The Mathematical Association of America.

[76] Y.E. Nesterov and A.S. Nemirovski (1988), Polynomial barrier methods in convex programming, *Èkonom. i Mat. Metody*, 24(6), pp. 1084–1091.

[77] Y.E. Nesterov and A.S. Nemirovski (1994), *Interior Point Polynomial Algorithms in Convex Programming*, SIAM Publications, SIAM, Philadelphia, USA.

[78] Y.E. Nesterov, H. Wolkowicz and Y. Ye (2000), Semidefinite programming relaxations of nonconvex quadratic optimization, in H. Wolkowicz, R. Saigal and L. Vandenberghe, eds, *Handbook of Semidefinite Programming: Theory, Algorithms and Applications*, Kluwer Academic Publishers, Boston, MA.

[79] J. W. Nieuwenhuis (1980), Another application of Guignard's generalized Kuhn–Tucker conditions, *J. Optim. Theory Appl.*, 30(1), pp. 117–125.

[80] M.L. Overton and R.S. Womersley (1993), Optimality conditions and duality theory for minimizing sums of the largest eigenvalues of symmetric matrices, *Math. Programming*, 62, pp. 321–357.

[81] P.M. Pardalos (1991), Quadratic programming with one negative eigenvalue is NP-hard, *J. Global Optim.*, 1, pp. 15–22.

[82] P. C. Parks (1992), A. M. Lyapunov's stability theory—100 years on, *IMA J. Math. Control Inform.*, 9(4), pp. 275–303.

[83] G. Pataki (1998), On the rank of extreme matrices in semidefinite programs and the multiplicity of optimal eigenvalues, *Math. Oper. Res.*, 23(2), pp. 339–358.

[84] G. Pataki (2000), Geometry of Semidefinite Programming, in H. Wolkowicz, R. Saigal and L. Vandenberghe, eds, *Handbook of Semidefinite Programming: Theory, Algorithms and Applications*, Kluwer Academic Publishers, Boston, MA.

[85] G. Pataki and L. Tuncel (1997), On the generic properties of convex optimization problems in conic form, Technical Report CORR 97-16, University of Waterloo, Waterloo, Canada.

[86] J. Peng and Y. Yuan (1997), Optimality conditions for the minimization of a quadratic with two quadratic constraints, *SIAM J. Optim.*, 7(3), pp. 579–594.

[87] A.L. Peressini (1967), *Ordered Topological Vector Spaces*, Harper & Row Publishers, New York.

[88] E.L. Peterson and J.G. Ecker (1969), Geometric programming: duality in quadratic programming and l_p-approximation, II. Canonical programs, *SIAM J. Appl. Math.*, 17, pp. 317–340.

[89] E.L. Peterson and J.G. Ecker (1970), Geometric programming: duality in quadratic programming and l_p-approximation, III. Degenerate programs, *J. Math. Anal. Appl.*, 29, pp. 365–383.

[90] E.L. Peterson and J.G. Ecker (1970), Geometric programming: duality in quadratic programming and l_p-approximation. I., in *Proceedings of the Princeton Symposium on Mathematical Programming (Princeton Univ., 1967)*, Princeton Univ. Press, pp. 445–480.

[91] S. Poljak, F. Rendl and H. Wolkowicz (1995), A recipe for semidefinite relaxation for $(0,1)$-quadratic programming, *J. Global Optim.*, 7(1), pp. 51–73.

[92] S. Poljak and H. Wolkowicz (1995), Convex relaxations of $(0,1)$-quadratic programming, *Math. Oper. Res.*, 20(3), pp. 550–561.

[93] G.D. Poole and M. Laidacker (1988), Projectionally exposed cones in $\mathbf{r}^3$, *Linear Algebra Appl.*, 111, pp. 183–190.

[94] A.J. Quist, E. De Klerk, C. Roos and T. Terlaky (1998), Copositive relaxation for general quadratic programming, *Optim. Methods Softw.*, 9, pp. 185–208. Special Issue Celebrating the 60th Birthday of Professor Naum Shor.

[95] M. Ramana, L. Tuncel and H. Wolkowicz (1997), Strong duality for semidefinite programming, *SIAM J. Optim.*, 7(3), pp. 641–662.

[96] M.V. Ramana (1993), *An Algorithmic Analysis of Multiquadratic and Semidefinite Programming Problems*, PhD thesis, Johns Hopkins University, Baltimore.

[97] M.V. Ramana (1997), An exact duality theory for semidefinite programming and its complexity implications, *Math. Programming*, 77, pp. 129–162.

[98] F. Rendl and H. Wolkowicz (1997), A semidefinite framework for trust region subproblems with applications to large scale minimization, *Math. Programming*, 77, pp. 273–299.

[99] S.A. Santos and D.C. Sorensen (1995), A new matrix-free algorithm for the large-scale trust-region subproblem, Technical Report TR95-20, Rice University, Houston, TX.

[100] A. Shapiro (1997), First and second order analysis of nonlinear semidefinite programs, *Math. Programming*, 77, pp. 301–320.

[101] A. Shapiro (2000), Duality and optimality conditions, in H. Wolkowicz, R. Saigal and L. Vandenberghe, eds, *Handbook of Semidefinite Programming: Theory, Algorithms and Applications*, Kluwer Academic Publishers, Boston, MA.

[102] H.D. Sherali and W.P. Adams (1996), Computational advances using the reformulation-linearization technique (rlt) to solve discrete and continuous nonconvex problems, *Optima*, 49, pp. 1–6.

[103] N.Z. Shor (1987), Quadratic optimization problems, *Izv. Akad. Nauk SSSR Tekhn. Kibernet.*, 222(1), pp. 128–139.

[104] R. Stern and H. Wolkowicz (1995), Indefinite trust region subproblems and nonsymmetric eigenvalue perturbations, *SIAM J. Optim.*, 5(2), pp. 286–313.

[105] J.F. Sturm (1997), *Primal-Dual Interior Point Approach to Semidefinite Programming*, PhD thesis, Erasmus University, Rotterdam, The Netherlands.

[106] J.F. Sturm (1998), Using SeDuMi 1.02, a MATLAB toolbox for optimization over symmetric cones, Technical report, Communications Research Laboratory, McMaster University, Hamilton, Canada. To appear in *Optimization Methods and Software.*

[107] C. H. Sung and B.-S. Tam (1990), A study of projectionally exposed cones, *Linear Algebra Appl.*, 139, pp. 225–252.

[108] O. Taussky (1968), Positive-definite matrices and their role in the study of the characteristic roots of general matrices, *Advances in Math.*, 2, pp. 175–186.

[109] T. Terlaky (1985), On l_p programming, *European J. Oper. Res.*, 22, pp. 70–100.

[110] M.J. Todd (1999), A study of search directions in primal-dual interior-point methods for semidefinite programming, *Optim. Methods Softw.*, 11&12, pp. 1–46.

[111] L. Tuncel and S. Xu (1999), On homogeneous convex cones, Caratheodory number and duality mapping semidefinite liftings, Technical Report CORR 99-21, University of Waterloo, Waterloo, Canada.

[112] H. Wolkowicz and Q. Zhao (1999), Semidefinite relaxations for the graph partitioning problem, *Discrete Applied Math.*, 96-97(1-3), pp. 461–479.

[113] S. Wright (2000), Recent developments in interior-point methods, in M.J.D. Powell and S. Scholtes, eds, *System Modelling and Optimization: Proceedings of 19th IFIP TC7 Conference, July, 1999, Cambridge*, Kluwer Academic Publishers, Boston, MA.

[114] V.A. Yakubovich (1967), The method of matrix inequalities in the stability theory of nonlinear control systems, *I*, *II*, *III*, *Automation and Remote Control*, 25-26(4), pp. 905–917, 577–592, 753–763.

[115] V.A. Yakubovich (1973), Minimization of quadratic functionals under the quadratic constraints and the necessity of a frequency condition in the quadratic criterion for absolute stability of nonlinear control systems, *Dokl. Akad. Nauk SSSR*, 209(14), pp. 1039–1042.

[116] V.A. Yakubovich (1973), The S-procedure and duality theorems for nonconvex problems of quadratic programming, *Vestnik Leningrad Univ.*, 1973(1), pp. 81–87.

[117] Y. Ye (1992), A new complexity result on minimization of a quadratic function with a sphere constraint, in *Recent Advances in Global Optimization*, Princeton University Press, pp. 19–31.

[118] Y. Yuan (1983), Some properties of trust region algorithms for nonsmooth optimization, Technical Report DAMTP 1983/NA14, Department of Applied Mathematics and Theoretical Physics, University of Cambridge.

[119] Y. Yuan (1990), On a subproblem of trust region algorithms for constrained optimization, *Math. Programming*, 47, pp. 53–63.

[120] Y. Yuan (1991), A dual algorithm for minimizing a quadratic function with two quadratic constraints, *Journal of Computational Mathematics*, 9, pp. 348–359.

[121] Q. Zhao, S.E. Karisch, F. Rendl and H. Wolkowicz (1998), Semidefinite programming relaxations for the quadratic assignment problem, *J. Comb. Optim.*, 2(1), pp. 71–109.

[122] S. Zlobec (1971), Optimality conditions for mathematical programming problems, *Glasnik Mat. Ser. III*, 6(26), pp. 187–192.

RECENT DEVELOPMENTS IN INTERIOR-POINT METHODS

Stephen J. Wright
Argonne National Laboratory
Argonne IL 60439, USA.
wright@mcs.anl.gov

Abstract The modern era of interior-point methods dates to 1984, when Karmarkar proposed his algorithm for linear programming. In the years since then, algorithms and software for linear programming have become quite sophisticated, while extensions to more general classes of problems, such as convex quadratic programming, semidefinite programming, and nonconvex and nonlinear problems, have reached varying levels of maturity. Interior-point methodology has been used as part of the solution strategy in many other optimization contexts as well, including analytic center methods and column-generation algorithms for large linear programs. We review some core developments in the area.

Keywords: optimization, interior-point methods

1. INTRODUCTION

Interior-point methods have been a topic of intense scrutiny by the optimization community during the past 15 years. Although methods of this type had been proposed in the 1950s, and investigated quite extensively during the 1960s [9], it was the announcement of an algorithm with intriguing complexity results and good practical performance by Karmarkar [20] that ushered in the modern era. This work placed interior-point methods at the top of the agenda for a large and diverse body of researchers and led to a series of remarkable advances in various areas of convex optimization.

Today, interior-point methods for linear programming have become quite mature both in theory and in practice, and several high-quality codes are available. For the rival algorithm, the simplex method, the sudden appearance of credible competition spurred significant improvements in the software, resulting in a quantum advance in the state of the art in computational linear programming since 1988.

M.J.D. Powell and S. Scholtes (Eds.), *System Modelling and Optimization: Methods, Theory and Applications.*

The theory of interior-point methods in other areas of convex programming and monotone complementarity also appears to have reached a fairly advanced stage. The computational picture is less clear than for general linear programming, however. In some areas, such as semidefinite programming, there is no apparent alternative algorithm whose practical efficiency is comparable to the interior-point approach, while in others, such as quadratic programming, active-set methods (which descend from the simplex method for linear programming) provide strong competition.

Investigation of the use of interior-point methods in various areas of nonconvex optimization, including discrete optimization, is in a much less advanced stage. The eventual prospects are still unclear, though early results in some areas (for example, nonlinear programming) show distinct promise. A thread common to many approaches is the use of interior-point methods to find inexact solutions of convex subproblems that arise during the course of the larger algorithm.

We start in Section 2 by outlining the state of the art of interior-point methods in linear programming, discussing the pedigree of the most important algorithms, computational issues, and customization of the approach to structured problems. In Section 3, we discuss the straightforward extensions to quadratic programming and linear complementarity, and compare the resulting algorithms with active-set methods. The extension to semidefinite programming is discussed in Section 4, along with the theoretical work on self-concordant functionals and self-scaled cones that forms the underpinning of some of this work. Finally, we present some conclusions in Section 5.

A great deal of literature is available to the reader interested in delving further into this area. A number of recent books (Ye [44], Roos, Vial, and Terlaky [35], Wright [42]) give overviews of the area, from first principles to new results and practical considerations. Theoretical background on self-concordant functionals and related developments is described by Nesterov and Nemirovskii [28] and Renegar [34]. Technical reports from the past five years can be obtained from the Interior-Point Methods Online Web site at `www.mcs.anl.gov/otc/InteriorPoint`.

For lack of space, we have omitted discussion of many interesting areas in which interior-point approaches are making an impact. Convex programming problems of the form

$$\min_x f(x) \quad \text{s.t.} \quad g_i(x) \leq 0, \quad i = 1, 2, \ldots, m,$$

where f and g_i, $i = 1, 2, \ldots, m$, are convex functions, can be solved by extensions of the primal-dual approach of Section 3; see, for example, Ralph and Wright [32]. Interestingly, it is possible to prove superlinear convergence of the resulting algorithms without assuming linear independence of the active constraints at the solution. This observation prompted recent work on improving the convergence properties of other algorithms, notably sequential quadratic programming. A number of researchers have used interior-point methods in algorithms for combinatorial and integer programming problems. (In some cases, the interior-point method is used to find an inexact solution of related problems in which the integrality constraints are relaxed.) Recent computational results are presented in Mitchell [24], and a comprehensive survey is given by Mitchell, Pardalos, and Resende [26]. In decomposition methods for large linear and convex problems, such as Dantzig-Wolfe/column generation and Benders' decomposition, interior-point methods have been used to find inexact solutions of the large master problems, or to approximately solve analytic center subproblems to generate test points. Approaches such as these are described by Gondzio and Sarkissian [16], Gondzio and Kouwenverg [15], and in the survey paper of Goffin and Vial [13]. Additionally, application of interior-point methodology to nonconvex nonlinear programming has occupied many researchers for some time now. The methods that have been proposed to date contain many ingredients, including primal-dual steps, barrier and merit functions, and scaled trust regions. Recent work in this area includes the reports of Byrd, Hribar, and Nocedal [5], Conn et al. [7], Gay, Overton, and Wright [11], and Forsgren and Gill [10].

2. LINEAR PROGRAMMING

We consider first the linear programming problem, which we state in standard form:

$$\min_x c^T x \quad \text{s.t.} \quad Ax = b, \quad x \geq 0, \tag{1}$$

where $x \in \mathbb{R}^n$ and $A \in \mathbb{R}^{m \times n}$. We assume that this problem has a strict interior, that is, the set

$$\mathcal{F}^\circ \stackrel{\text{def}}{=} \{x \,|\, Ax = b, \; x > 0\}$$

is nonempty, and that the objective function is bounded below on the set of feasible points. Under these assumptions, (1) has a (not necessarily unique) solution.

By using a logarithmic barrier function to account for the bounds $x \geq 0$, we obtain the parametrized optimization problem

$$\min_x f(x; \hat{\mu}) \stackrel{\text{def}}{=} \frac{1}{\hat{\mu}} c^T x - \sum_{i=1}^{n} \log x_i \quad \text{s.t.} \quad Ax = b, \tag{2}$$

where log denotes the natural logarithm, and $\hat{\mu} > 0$ denotes the barrier parameter. Because the logarithmic function requires its arguments to be positive, the solution $x(\hat{\mu})$ of (2) must belong to $\mathcal{F}^\circ$. It is well known (see, for example, Wright [40, Theorem 5]) that for any sequence $\{\hat{\mu}_k\}$ with $\hat{\mu}_k \downarrow 0$, all limit points of $\{x(\hat{\mu}_k)\}$ are solutions of (1).

The traditional SUMT approach of Fiacco and McCormick [9] accounts for equality constraints by including a quadratic penalty term in the objective. When the constraints are linear, as in (1), it is simpler and more appropriate to handle them explicitly. By doing so, we devise a *primal barrier algorithm* in which a projected Newton method is used to find an approximate solution of (2) for a certain value of $\hat{\mu}$, and then $\hat{\mu}$ is decreased. The projected Newton step Δx from a point x satisfies the following system:

$$\begin{bmatrix} \hat{\mu}X^{-2} & -A^T \\ A & 0 \end{bmatrix} \begin{bmatrix} \Delta x \\ \lambda^+ \end{bmatrix} = - \begin{bmatrix} c - \hat{\mu}X^{-1}e \\ Ax - b \end{bmatrix}, \tag{3}$$

where $X = \text{diag}(x_1, x_2, \ldots, x_n)$ and $e = (1, 1, \ldots, 1)^T$. Note that

$$\nabla^2_{xx} f(x; \hat{\mu}) = X^{-2}, \quad \nabla_x f(x; \hat{\mu}) = (1/\hat{\mu})c - X^{-1}e,$$

so that the equations (3) are the same as those that arise from a sequential quadratic programming algorithm applied to (2), modulo the scaling by $\hat{\mu}$ in the first line of (3). A line search can be performed along Δx to find a new iterate $x + \alpha\Delta x$, where $\alpha > 0$ is the step length.

The prototype primal barrier algorithm can be specified as follows:

primal barrier algorithm
Given $x^0 \in \mathcal{F}^\circ$ and $\hat{\mu}_0 > 0$;
Set $k \leftarrow 0$;
repeat
  Obtain x^{k+1} by performing one or more Newton steps (3),
    starting at $x = x^k$, and fixing $\hat{\mu} = \hat{\mu}_k$;
  Choose $\hat{\mu}_{k+1} \in (0, \hat{\mu}_k)$; $k \leftarrow k + 1$;
until some termination test is satisfied.

A *short-step* version of this algorithm takes a single Newton step at each iteration, with step length $\alpha = 1$, and sets

$$\hat{\mu}_{k+1} = \hat{\mu}_k \Big/ \left(1 + \frac{1}{8\sqrt{n}}\right). \tag{4}$$

It is known (see, for instance, Renegar [34, Section 2.4]) that, if the feasible region of (1) is bounded, and x^0 is sufficiently close to $x(\hat{\mu}_0)$ in a certain sense, then we obtain a point x^k whose objective value $c^T x^k$ is within ε of the optimal value after

$$O\left(\sqrt{n}\log\frac{n\hat{\mu}_0}{\varepsilon}\right) \quad \text{iterations,} \tag{5}$$

where the constant factor disguised by the $O(\cdot)$ depends on the properties of (1) but is independent of n and ε.

The rate of decrease of $\hat{\mu}$ in short-step methods is too slow to allow good practical behavior, so *long-step* variants were proposed that decreased $\hat{\mu}$ more rapidly, while possibly taking more than one Newton step for each $\hat{\mu}_k$ and also using a line search. Although long-step algorithms have better practical behavior, the complexity estimates associated with them typically are no better than the estimate (5) for the short-step approach; see Renegar [34, Section 2.4] and Gonzaga [17]. In fact, a recurring theme of worst-case complexity estimates for linear programming algorithms is that no useful relationship exists between the estimate and the practical behavior of the algorithm.

Better practical algorithms are obtained from the primal-dual framework. These methods recognize the importance of the path of solutions $x(\hat{\mu})$ to (2) in the design of algorithms, but differ from the approach above in that they treat the dual variables explicitly in the problem, rather than as adjuncts to the calculation of the primal iterates.

The dual problem for (1) is

$$\max_{(\lambda,s)} b^T\lambda \quad \text{s.t.} \quad A^T\lambda + s = c, \; s \geq 0, \tag{6}$$

where $s \in \mathbb{R}^n$ and $\lambda \in \mathbb{R}^m$. The optimality conditions for x^* to be a solution of (1) and (λ^*, s^*) to be a solution of (6) are that $(x, \lambda, s) = (x^*, \lambda^*, s^*)$ satisfies

$$
\begin{aligned}
Ax &= b, && \text{(7a)}\\
A^T\lambda + s &= c, && \text{(7b)}\\
XSe &= 0, && \text{(7c)}\\
(x, s) &\geq 0, && \text{(7d)}
\end{aligned}
$$

where $X = \operatorname{diag}(x_1, x_2, \ldots, x_n)$ and $S = \operatorname{diag}(s_1, s_2, \ldots, s_n)$, and where $(x, s) \geq 0$ indicates that all the components of x and s are nonnegative. Primal-dual methods solve (1) and (6) simultaneously by generating a sequence of iterates (x^k, λ^k, s^k) that in the limit satisfies the conditions (7). As mentioned above, the *central path* defined by the following perturbed variant of (7) plays an imporant role in algorithm design:

$$Ax = b, \tag{8a}$$

$$
\begin{aligned}
A^T\lambda + s &= c, & \text{(8b)}\\
XSe &= \hat{\mu}e, & \text{(8c)}\\
(x,s) &> 0, & \text{(8d)}
\end{aligned}
$$

where $\hat{\mu} > 0$ parametrizes the path. Note that these conditions are simply the optimality conditions for the problem (2): If $(x(\hat{\mu}), \lambda(\hat{\mu}), s(\hat{\mu}))$ satisfies (8), then $x(\hat{\mu})$ is a solution of (2). We have from (8c) that a key feature of the central path is that

$$x_i s_i = \mu, \quad \text{for all } i = 1, 2, \ldots, n, \tag{9}$$

that is, the pairwise products $x_i s_i$ are identical for all i.

In primal-dual algorithms, steps are generated by fixing $\hat{\mu}$ at some appropriate value (discussed below) and applying a perturbed Newton method to the three equalities (8a), (8b), and (8c), which form a nonlinear system in which the number of equations equals the number of unknowns. We constrain all iterates (x^k, λ^k, s^k) to have $(x^k, s^k) > 0$, so that the matrices X and S remain positive diagonal throughout, ensuring that the perturbed Newton steps are well defined. Supposing that we are at a point (x, λ, s) with $(x, s) > 0$ and the feasibility conditions $Ax = b$ and $A^T\lambda + s = c$ are satisfied, the primal-dual step $(\Delta x, \Delta\lambda, \Delta s)$ is obtained from following system:

$$
\begin{bmatrix} 0 & A & 0 \\ A^T & 0 & I \\ 0 & S & X \end{bmatrix}
\begin{bmatrix} \Delta\lambda \\ \Delta x \\ \Delta s \end{bmatrix}
= -
\begin{bmatrix} 0 \\ 0 \\ XSe - \sigma\mu e + r \end{bmatrix}, \tag{10}
$$

where $\mu = x^T s/n$, $\sigma \in [0, 1]$, and r is a perturbation term, possibly chosen to incorporate higher-order information about the system (8), or additional terms to improve proximity to the central path. If the perturbation r were not present, (10) would simply be the Newton system for (8a), (8b), and (8c), where the value of $\hat{\mu}$ is fixed at $\sigma\mu$.

Using the general step (10), we can state the basic framework for primal-dual methods as follows:

primal-dual algorithm
Given (x^0, λ^0, s^0) with $(x^0, s^0) > 0$;
Set $k \leftarrow 0$ and $\mu_0 = (x^0)^T s^0/n$;
repeat
  Choose σ_k and r^k;
  Solve (10) with $\mu = \mu_k$, $\sigma = \sigma_k$ and $r = r^k$
    to obtain $(\Delta x^k, \Delta\lambda^k, \Delta s^k)$;
  Set

$$(x^{k+1}, \lambda^{k+1}, s^{k+1}) \leftarrow (x^k, \lambda^k, s^k) + \alpha_k(\Delta x^k, \Delta\lambda^k, \Delta s^k),$$

choosing $\alpha_k \in (0, 1]$ to ensure that $(x^{k+1}, s^{k+1}) > 0$;
Set $\mu_{k+1} \leftarrow (x^{k+1})^T s^{k+1}/n$; $k \leftarrow k + 1$;
until some termination test is satisfied.

The various algorithms that use this framework differ in the way that they choose the starting point, the centering parameter σ_k, the perturbation vector r^k, and the step α_k. The simplest algorithm—a short-step path-following method similar to the primal algorithm described above—sets

$$r^k = 0, \quad \sigma_k \equiv 1 - \frac{0.4}{\sqrt{n}}, \quad \alpha_k \equiv 1,$$

and, for suitable choice of starting point, achieves convergence to a feasible point (x, λ, s) with $x^T s/n \leq \varepsilon$ for a given ε in

$$O\left(\sqrt{n} \log \frac{\mu_0}{\varepsilon}\right) \text{ iterations.} \tag{11}$$

Note the similarity of both the algorithm and its complexity estimate to the corresponding primal algorithm. As in that case, algorithms with better practical performance, but not necessarily better complexity estimates, can be obtained through more aggressive, adaptive choices of the centering parameter (that is, σ_k closer to zero). They use a line search to maintain proximity to the central path. The proximity requirement dictates, implicitly or explicitly, that while the condition (9) may be violated, the pairwise products must not be too different from each other. Many such algorithms, including path-following, potential-reduction, and predictor-corrector algorithms, are discussed in Wright [42].

Most interior-point software for linear programming is based on Mehrotra's predictor-corrector algorithm [22], often with the higher-order enhancements described by Gondzio [14]. This approach uses an adaptive choice of σ_k, selected by first solving for the pure Newton step (i.e., setting $r = 0$ and $\sigma = 0$ in (10). If this step makes good progress in reducing μ, we choose σ_k small so that the step actually taken is quite close to this pure Newton step. Otherwise, we enforce more centering and calculate a conservative direction by setting σ_k closer to 1. The perturbation vector r^k is chosen to improve the similarity between the system (10) and the original system (8) that it approximates. Gondzio's technique further enhances r^k by performing further solves of the system (10) with a variety of right-hand sides, where each solve reuses the factorization of the matrix, and is therefore not too expensive to perform.

To turn this basic algorithmic approach into a useful piece of software, we must address many issues. These include problem formulation, presolving to reduce the problem size, choice of the step length, linear algebra techniques for solving (10), and user interfaces and input formats.

Possibly the most interesting issues are associated with the linear algebra. Most codes deal with a partially eliminated form of (10), either eliminating Δs to obtain

$$\begin{bmatrix} 0 & A \\ A^T & -X^{-1}S \end{bmatrix} \begin{bmatrix} \Delta\lambda \\ \Delta x \end{bmatrix} = - \begin{bmatrix} 0 \\ -X^{-1}(XSe - \sigma\mu e + r) \end{bmatrix}, \tag{12}$$

or eliminating both Δs and Δx to obtain a system of the form

$$A(S^{-1}X)A^T\Delta\lambda = t, \tag{13}$$

to which a sparse Cholesky algorithm is applied. A modified version of the latter form is used when dense columns are present in A. These columns may be treated as a low-rank update and handled via the Sherman–Morrison–Woodbury formula or, equivalently, via a Schur complement strategy applied to a system intermediate between (12) and (13). In many problems, the matrix in (13) becomes increasingly ill-conditioned as the iterates progress, eventually causing the Cholesky process to break down as negative pivot elements are encountered. A number of simple (and in some cases counterintuitive) patches have been proposed for overcoming this difficulty while still producing useful approximate solutions of (13) efficiently; see, for example, Andersen [2] and Wright [43].

Despite many attempts, iterative solvers have not shown much promise as a means to solve (13), at least for general linear programs. A possible reason is that, besides its poor conditioning, the matrix lacks the regular spectral properties of matrices obtained from discretizations of continuous operators. Some codes do, however, use preconditioned conjugate gradient as an alternative to iterative refinement for improving the accuracy, when the direct approach for solving (13) fails to produce a solution of sufficient accuracy. The preconditioner used in this case is simply the computed factorization of the matrix $A(S^{-1}X)A^T$.

A number of interior-point linear programming codes are now available, both commercially and free of charge. Information can be obtained from the World-Wide Web via the URL mentioned earlier. It is difficult to make blanket statements about the relative efficiency of interior-point and simplex methods for linear programming, as improvements to the implementations of both techniques continue to be made. Interior-point methods tend to be faster on large problems and can better exploit multiprocessor platforms, because the expensive operations such as Cholesky factorization of (13) can be parallelized to some extent. They are not able to exploit "warm start" information—a good prior estimate of the solution, for instance—to the same extent as simplex methods. For this reason, they are not well suited for use in contexts such as branch-and-bound or branch-and-cut algorithms for integer programming, which solve many closely related linear programs.

Several researchers have devised special interior-point algorithms for special cases of (1) that exploit the special properties of these cases in solving the linear systems at each iteration. For network flow problems, Mehrotra and Wang consider preconditioned conjugate-gradient methods for solving (13), in which the preconditioner is built from a spanning tree for the underlying network (see Mehrotra and Wang [23]). For multicommodity flow problems, Castro [6] describes an algorithm for solving a version of (13) in which the block-diagonal part of the matrix is used to eliminate many of the variables, and a preconditioned conjugate-gradient method is applied to the remaining Schur complement. Techniques for stochastic programming (two-stage linear problems with recourse) are described by Birge and Qi [4] and Birge and Louveaux [3, Section 5.6].

3. SIMPLE EXTENSIONS OF THE PRIMAL-DUAL APPROACH

The primal-dual algorithms of the preceding section are readily extended to convex quadratic programming (QP) and monotone linear complementarity (LCP), both classes being generalizations of linear programming. Indeed, many of the convergence and complexity properties of primal-dual algorithms were first elucidated in the literature with regard to monotone LCP.

We state the convex QP as

$$\min_x \; c^T x + \tfrac{1}{2} x^T Q x \quad \text{s.t.} \quad Ax = b, \;\; x \geq 0, \tag{14}$$

where Q is a positive semidefinite matrix. The monotone LCP is defined by square matrices M and N and a vector q, where M and N satisfy a monotonicity property: all vectors y and z that satisfy $My + Nz = 0$ have $y^T z \geq 0$. This problem requires us to identify vectors y and z such that

$$My + Nz = q, \quad (y, z) \geq 0, \quad y^T z = 0. \tag{15}$$

With some transformations, we can express the optimality conditions (7) for linear programming, and also the optimality conditions for (14), as a monotone LCP. Other problems fit under the LCP umbrella as well, including bimatrix games and equilibrium problems. The central path for this problem is defined by the following system, parametrized as in (8) by the positive scalar μ:

$$\begin{aligned} My + Nz &= q, && \text{(16a)} \\ YZe &= \mu e, && \text{(16b)} \\ (y, z) &> 0, && \text{(16c)} \end{aligned}$$

and a search direction from a point (y, z) satisfying (16a) and (16c) is obtained by solving a system of the form

$$\begin{bmatrix} M & N \\ Z & Y \end{bmatrix} \begin{bmatrix} \Delta y \\ \Delta z \end{bmatrix} = - \begin{bmatrix} 0 \\ YZe - \sigma\mu e + r \end{bmatrix}, \qquad (17)$$

where $\mu = y^T z/n$, $\sigma \in [0, 1]$, and, as before, r is a perturbation term. The corresponding search direction system for the quadratic program (14) is identical to (10) except that the $(2, 2)$ block in the coefficient matrix is replaced by $-Q$. The primal-dual algorithmic framework and the many variations within this framework are identical to the case of linear programming with the minor difference that the step length should be the same for all variables. (In linear programming, different step lengths can be, and often are, taken for the primal variable x and the dual variables (λ, s).)

Complexity results are also similar to those obtained for the corresponding linear programming algorithm. For an appropriately chosen starting point (y^0, z^0) with $\mu_0 = (y^0)^T z^0/n$, we obtain convergence to a point with $\mu \leq \varepsilon$ in

$$O\left(n^\tau \log \frac{\mu_0}{\varepsilon}\right) \quad \text{iterations},$$

where $\tau = 1/2$, 1, or 2, depending on the algorithm. Fast local convergence results typically require an additional strict complementarity assumption that is not necessary in the case of linear programming (see Monteiro and Wright [27]), although some authors have proposed superlinear algorithms that do not require this assumption. Algorithms of the latter type require accurate identification of the set of degenerate indices before the fast convergence becomes effective. This property makes them of limited interest, since by the time the degenerate set has been identified, the problem is essentially solved.

The LCP algorithms can, in fact, be extended to a wider class of problems involving so-called sufficient matrices. Instead of requiring M and N to satisfy the monotonicity property defined above, we require that there exist a nonnegative constant κ such that

$$y^T z \geq -4\kappa \sum_{i \,|\, y_i z_i > 0} y_i z_i, \quad \text{for all } y, z \text{ with } My + Nz = 0.$$

The complexity estimate for interior-point methods applied to such problems depends on the parameter κ; that is, the complexity is not polynomial on the whole class of sufficient matrices.

Primal-dual methods have been applied to many practical applications of (14) and (15). For example, an application to Markowitz's formulation of the portfolio optimization problem is described by Takehara [36]; applications to optimal control and model predictive control are described by Wright [41] and Rao, Wright, and Rawlings [33]; an application to ℓ_1 regression is described by Portnoy and Koenker [31].

The interior-point approach has a number of advantages over the active-set approach from a computational point of view. It is difficult for an active-set algorithm to exploit any structure inherent in both Q and A, without redesigning most of the complex operations that make up this algorithm (adding a constraint to the active set, deleting a constraint, evaluating Lagrange multiplier estimates, calculating the search direction, and so on). In the interior-point approach, on the other hand, the only complex operation is the solution of the linear system (17)—and this operation is fairly straightforward by comparison with the operations in an active-set method. Since the structure and dimension of the linear system remain the same at all iterations, the routines for solving the linear systems can be designed to fully exploit the properties of the systems arising from each problem class. In fact, the algorithm can be implemented to high efficiency using an object-oriented approach, in which the programmer of each new problem class needs to supply only code for the factorization and solution of the systems (17), optimized for the structure of the new class, along with a number of simple operations such as inner-product calculations. Code that implements upper-level decisions (choice of parameter σ, vector r, steplength α) remains efficient across the gamut of applications of (15) and can simply be reused by all applications.

We note, however, that active-set methods may still require much less execution time than interior-point methods in many contexts, especially when "warm start" information is available, and when the problem is generic enough that not much benefit is gained by exploiting its structure.

The extension of primal-dual algorithms from linear programming to convex quadratic programming is so straightforward that a number of the interior-point linear programming codes have recently been extended to handle problems in the class (14) as well. In their linear algebra calculations, these codes treat both Q and A as general sparse matrices, and hence are efficient across a wide range of applications. By contrast, as noted in Gould and Toint [18, Section 4], implementations of active-set methods for (14) that are capable of handling even moderately sized problems have not been widely available.

4. SEMIDEFINITE PROGRAMMING

Here we discuss extensions of interior-point techniques to broad classes of problems that include semidefinite programming (SDP) and second-order cone programming. The SDP problem can be stated as

$$\min_X C \bullet X, \quad \text{s.t. } X \succeq 0, \quad A_i \bullet X = b_i, \; i = 1, 2, \ldots, m, \tag{18}$$

where X, C, and A_i, $i = 1, 2, \ldots, m$, are $n \times n$ symmetric matrices $\mathcal{S}\mathbb{R}^{n \times n}$, $X \succeq 0$ denotes the constraint that X be positive definite, and "$\bullet$" denotes the inner product $P \bullet Q = \sum_{i,j} P_{ij} Q_{ij}$. By further restricting X, C, and A_i all to be diagonal, we recover the linear programming problem (1). The class (18) has been studied intensively during the past seven years, in part because of its importance in applications to control systems and because many combinatorial problems have powerful SDP relaxations. The second-order cone programming problem is

$$\begin{aligned} &\min\nolimits_{x_1, t_1, \ldots, x_N, t_N} \sum\nolimits_{i=1}^N c_i^T x_i + \beta_i t_i \quad \text{s.t.} \\ &\sum\nolimits_{i=1}^N B_i x_i + d_i t_i = b, \quad \|x_i\|_2 \le t_i, \; i = 1, 2, \ldots, N, \end{aligned} \tag{19}$$

where each x_i is a vector of length $n_i \ge 1$, each B_i is an $m_0 \times n_i$ matrix, b and each d_i are vectors of length m_0, and each t_i is a scalar. Convex quadratically constrained quadratic programs can be posed in the form (19), along with sum-of-norms problems and many other applications (see Lobo et al. [21]).

The key to extending efficient interior-point algorithms to these and other convex problems was provided by Nesterov and Nemirovskii [28]. The authors explored the properties of self-concordant functions. They showed that algorithms with polynomial complexity could be constructed by using barrier functions of this type for the inequality constraint, and then applying a projected Newton's method to the resulting linearly constrained problem.

Self-concordant functions are convex functions with the special property that their third derivative can be bounded by some expression involving their second derivative at each point in their domain. This property implies that the second derivative does not fluctuate too rapidly in a relative sense, so that the function does not deviate too much from the second-order approximation on which Newton's method is based. For this reason, we can expect Newton's method to perform reasonably well on such a function.

Given a finite-dimensional real vector space $\mathcal{V}$, an open, nonempty convex set $\mathcal{S} \subset \mathcal{V}$, and a closed convex set $\mathcal{T} \subset \mathcal{V}$ with nonempty interior, we have the following formal definition.

Definition 1 *The function $F : \mathcal{S} \to \mathbb{R}$ is* self-concordant *if it is convex and if the following inequality holds for all $x \in \mathcal{S}$ and all $h \in \mathcal{V}$:*

$$\left| D^3 F(x)[h,h,h] \right| \le 2 \left(D^2 F(x)[h,h] \right)^{3/2}, \tag{20}$$

where $D^k F[h_1, h_2, \dots, h_k]$ denotes the kth differential of F along the directions $h_1, h_2, \dots, h_k$.

F is called strongly self-concordant *if $F(x_i) \to \infty$ for all sequences $x_i \in \mathcal{S}$ that converge to a point on the boundary of $\mathcal{S}$.*

*F is a ϑ-*self-concordant barrier *for $\mathcal{T}$ if it is a strongly self-concordant function for* int$\mathcal{T}$*, and the parameter*

$$\vartheta \stackrel{\text{def}}{=} \sup_{x \in \text{int}\mathcal{T}} F'(x)^T \left[F''(x) \right]^{-1} F'(x) \tag{21}$$

is finite.

Note that the exponent 3/2 on the right-hand side of (20) makes the condition independent of the scaling of h. It is shown by Nesterov and Nemirovskii [28, Corollary 2.3.3] that, if $\mathcal{T} \neq \mathcal{V}$, then the parameter ϑ is no smaller than 1.

It is easy to show that log-barrier function of Section 2 is an n-self-concordant barrier for the positive orthant $\mathbb{R}^n_+$ (that is, it satisfies (21) for $\vartheta = n$) if we take

$$\mathcal{V} = \mathbb{R}^n, \quad \mathcal{S} = \mathbb{R}^n_{++}, \quad F(x) = -\sum_{i=1}^n \log x_i,$$

where $\mathbb{R}^n_{++}$ denotes the strictly positive orthant. Another interesting case is the second-order cone (or "ice-cream cone"), for which we have

$$\mathcal{V} = \mathbb{R}^{n+1}, \quad \mathcal{S} = \{(x,t) \mid \|x\|_2 \le t\}, \quad F(x,t) = -\log \left(t^2 - \|x\|^2 \right), \tag{22}$$

where $t \in \mathbb{R}$ and $x \in \mathbb{R}^n$. In this case, F is a 2-self-concordant barrier and is appropriate for the inequality constraints in (19). A third important case is the cone of positive semidefinite matrices, for which we have

$$\begin{aligned} \mathcal{V} &= n \times n \text{ symmetric matrices} \\ \mathcal{S} &= n \times n \text{ symmetric positive semidefinite matrices} \\ F(X) &= -\log\det X, \end{aligned}$$

where F is an n-self-concordant barrier. This barrier function can be used to model the constraint $X \succeq 0$ in (18).

Self-concordant barrier functions allow us to generalize the primal barrier method of Section 2 to problems of the form

$$\min \langle c, x \rangle \quad \text{s.t.} \quad Ax = b, \;\; x \in \mathcal{T}, \tag{23}$$

where $\mathcal{T}$ is a closed convex set, $\langle c, x \rangle$ denotes a linear functional on the underlying vector space $\mathcal{V}$, and A is a linear operator. Similarly to (2), we define the barrier subproblem to be

$$\min_x \; f(x;\mu) \stackrel{\text{def}}{=} \frac{1}{\mu}\langle c, x \rangle + F(x) \quad \text{s.t.} \quad Ax = b, \tag{24}$$

where $F(x)$ is a self-concordant barrier and $\mu > 0$ is the barrier parameter. Note that, by the Definition 1, $f(x;\mu)$ is also a strongly self-concordant function. The primal barrier algorithm for (23) based on (24) is as follows:

primal barrier algorithm
Given $x^0 \in \text{int}\mathcal{T}$ and $\mu_0 > 0$;
Set $k \leftarrow 0$;
repeat
  Obtain $x^{k+1} \in \text{int}\mathcal{T}$ by performing one or more projected
    Newton steps for $f(\cdot;\mu_k)$, starting at $x = x^k$;
  Choose $\mu_{k+1} \in (0, \mu_k)$; $k \leftarrow k+1$;
until some termination test is satisfied.

Remarkably, the worst-case complexity of algorithms of this type depends on the parameter ϑ associated with F, but not on any properties of the data that defines the problem instance. For example, we can define a short-step method in which a single full Newton step is taken for each value of k, and μ is decreased according to

$$\mu_{k+1} = \mu_k \Big/ \left(1 + \frac{1}{8\sqrt{\vartheta}}\right).$$

Given a starting point with appropriate properties, we obtain an iterate x^k whose objective $\langle c, x^k \rangle$ is within ε of the optimum in

$$O\left(\sqrt{\vartheta}\log\frac{\vartheta\mu_0}{\varepsilon}\right) \quad \text{iterations.}$$

Long-step variants also are discussed by Nesterov and Nemirovskii [28]. The practical behavior of these methods does, of course, depend strongly on the properties of the particular problem instance.

The primal-dual algorithms of Section 2 can also be extended to more general problems by means of the theory of self-scaled cones developed by Nesterov and Todd [29, 30]. The basic problem considered is the conic programming problem

$$\min \langle c, x \rangle \quad \text{s.t.} \quad Ax = b, \quad x \in K, \tag{25}$$

where $K \subset \mathbb{R}^n$ is a closed convex cone, that is, a closed convex set for which $x \in K \Rightarrow tx \in K$ for all nonnegative scalars t, and A denotes a linear operator from $\mathbb{R}^n$ to $\mathbb{R}^m$. The dual cone for K is denoted by K^* and defined as

$$K^* \stackrel{\text{def}}{=} \{s \mid \langle s, x \rangle \geq 0 \text{ for all } x \in K\},$$

and we can write the dual instance of (25) as

$$\max \langle b, \lambda \rangle \quad \text{s.t.} \quad A^*\lambda + s = c, \quad s \in K^*, \tag{26}$$

where A^* denotes the adjoint of A. The duality relationships between (25) and (26) are more complex than in linear programming, but if either problem has a feasible point that lies in the interior of K or K^*, respectively, the strong duality property holds. That is, if the optimal value of either (25) or (26) is finite, then both problems have finite optimal values, and these values are the same.

K is a self-scaled cone when its interior $\text{int}K$ is the domain of a self-concordant barrier function F with certain strong properties that allow us to define algorithms in which the primal and dual variables are treated in a perfectly symmetric fashion and play interchangeable roles. In particular, we have $K^* = K$ for such cones. The full elucidation of the properties of self-scaled cones is quite complicated, but it suffices to note here that the three cones mentioned above—the positive orthant $\mathbb{R}^n_+$, the second-order cone (22), and the cone of positive semidefinite symmetric matrices—are the most interesting self-scaled cones. Their associated barrier functions are the logarithmic functions already mentioned.

To build algorithms from the properties of self-scaled cones and their barrier functions, the Nesterov–Todd theory defines a *scaling point* for a given pair $x \in \text{int}K$, $s \in \text{int}K$ to be the unique point w such that $H(w)x = s$, where $H(\cdot)$ is the Hessian of the barrier function. In the case of linear programming, it is easy to verify that w is the vector in $\mathbb{R}^n$ whose elements are $\sqrt{x_i/s_i}$. The Nesterov–Todd search directions are obtained as projected steepest descent directions for the primal and dual barrier subproblems (that is, (24) and its dual counterpart), where a weighted inner product involving the matrix $H(w)$ is used to define the projections onto the spaces defined by the linear constraints $Ax = b$ and $A^*\lambda + s = c$, respectively. The resulting directions satisfy the following linear system:

$$\begin{bmatrix} 0 & A & 0 \\ A^* & 0 & I \\ 0 & H(w) & I \end{bmatrix} \begin{bmatrix} \Delta\lambda \\ \Delta x \\ \Delta s \end{bmatrix} = - \begin{bmatrix} 0 \\ 0 \\ s + \sigma\mu\nabla F(x) \end{bmatrix}, \tag{27}$$

where $\mu = \langle x, s\rangle/\vartheta$. (The correspondence with (10) is complete if we choose the perturbation term to be $r = 0$.) By choosing the starting point appropriately, and designing schemes for choosing the parameters σ and step lengths to take along these directions, we obtain polynomial algorithms for this general setting.

Primal-dual algorithms for (25), where K is a self-scaled cone, are also studied by Faybusovich [8], who takes the viewpoint of differential geometry and, in particular, uses a Jordan algebra framework.

In the important case of semidefinite programming (18), the Nesterov–Todd framework is far from the only means for devising primal-dual methods. Many algorithms proposed before and since this framework was described do not fall under its umbrella, yet have strong theoretical properties and, in some cases, much better practical behavior. To outline a few of these methods, we write the dual of (18) as

$$\max_{y,S} \; b^T\lambda \quad \text{s.t.} \quad \sum_{i=1}^m \lambda_i A_i + S = C, \quad S \succeq 0, \tag{28}$$

where $S \in \mathcal{S}\mathbb{R}^{n\times n}$ and $\lambda \in \mathbb{R}^m$. Points on the central path for (18), (28) are defined by the following parametrized system:

$$A_i \bullet X = b_i, \quad i = 1, 2, \ldots, m, \tag{29a}$$

$$\sum_{i=1}^m \lambda_i A_i + S = C, \tag{29b}$$

$$XS = \mu I, \tag{29c}$$

$$X \succeq 0, \qquad S \succeq 0, \tag{29d}$$

where as usual μ is the positive parameter. Unlike the corresponding equations (8) for linear programming, the system (29b), (29a), (29c) is not quite "square," since the variables reside in the space $\mathcal{S}\mathbb{R}^{n\times n} \times \mathbb{R}^m \times \mathcal{S}\mathbb{R}^{n\times n}$ while the range space of the equations is $\mathcal{S}\mathbb{R}^{n\times n} \times \mathbb{R}^m \times \mathbb{R}^{n\times n}$. In particular, the product of two symmetric matrices (see (29c) is not necessarily symmetric. Before Newton's method can be applied to (29b)—the fundamental operation in primal-dual algorithms—the domain and range have to match. The different primal-dual algorithms differ in the ways that they reconcile the domain and range of these equations.

The paper of Todd [37] is witness to the intensity of research in SDP interior-point methods: It describes twenty techniques for obtaining search directions for SDP. In many of these, the equation (29c) is replaced by one whose range lies in $\mathcal{S}\mathbb{R}^{n\times n}$. That is, it is "symmetrized" and replaced with a mapping

$$\Theta(X, S) = 0. \tag{30}$$

In deriving the step $(\Delta X, \Delta\lambda, \Delta S)$, we approximate the mapping $\Theta(X + \Delta X, S + \Delta S)$ with a linear approximation of the form

$$\Theta(X, S) + \mathcal{E}\Delta X + \mathcal{F}\Delta S, \tag{31}$$

for certain operators $\mathcal{E}$ and $\mathcal{F}$. We derive primal-dual methods by using (31) along with the linear equations (29b) and (29a). The heuristics associated with linear programming algorithms—Mehrotra and Gondzio corrections, step length determination, and so on—translate in a fairly straightforward way to this setting. The implementations are much more complex, however, since the linear problem to be solved at each iteration has a much more complicated structure than that of (10). It is noted in Haeberly, Nayakkankuppam, and Overton [19] that the benefits of higher-order corrections in the SDP context are even more pronounced than in linear programming, since the cost of factoring the coefficient matrix relative to the cost of solving for a different right-hand side is much greater for SDP.

Examples of the symmetrizations (30) include the Monteiro–Zhang family, in which

$$\Theta(X, S) = \frac{1}{2}\left(P(XS)P^{-1} + P^{-T}(XS)^T P^T\right) - \mu I,$$

for some nonsingular P. The Alizadeh–Haeberly–Overton direction [1], which appears to be the most promising one from a practical point of view, is obtained by setting $P = I$, while the Nesterov–Todd direction is obtained from

$$P^2 = S^{1/2}(S^{1/2}XS^{1/2})^{-1/2}S^{1/2}.$$

A survey of the applications of SDP, ranging across eigenvalue optimization, structural optimization, control and systems theory, statistics, and combinatorial optimization, is given by Vandenberghe and Boyd [38]. The paper of Wolkowicz [39] in this volume discusses the theory and algorithms associated with applications to combinatorial problems. The main use of SDP in combinatorial optimization is in finding

SDP relaxations (that is, problems of the form (18) that contain all the feasible points of the underlying combinatorial problem in their feasible sets) that yield high quality approximate solutions to the combinatorial problem. We illustrate the technique with possibly the most famous instance to date: the technique of Goemans and Williamson [12], which yields an approximate solution whose value is within 13% of optimality for the MAX CUT problem.

In MAX CUT, we are presented with an undirected graph with N vertices whose edges have nonnegative weights w_{ij}. The problem is to choose a subset $\mathcal{S} \subset \{1, 2, \ldots, N\}$ of the vertices so that the sum of weights of the edges that cross from $\mathcal{S}$ to its complement is maximized. In other words, we aim to choose $\mathcal{S}$ to maximize the objective

$$w(\mathcal{S}) \stackrel{\text{def}}{=} \sum_{i \in \mathcal{S}, j \notin \mathcal{S}} w_{ij}.$$

This problem can be restated as an integer quadratic program by introducing variables y_i, $i = 1, 2, \ldots, N$, such that $y_i = 1$ for $i \in \mathcal{S}$ and $y_i = -1$ for $i \notin \mathcal{S}$. We then have

$$\max_y \tfrac{1}{2} \sum_{i<j} w_{ij}(1 - y_i y_j) \quad \text{s.t.} \quad y_i \in \{-1, 1\}, \;\; i = 1, 2, \ldots, N. \tag{32}$$

This problem is NP-complete. Goemans and Williamson replace the variables $y_i \in \mathbb{R}$ by vectors $v_i \in \mathbb{R}^N$ and consider instead the problem

$$\max_{v_1, v_2, \ldots, v_N} \tfrac{1}{2} \sum_{i<j} w_{ij}(1 - v_i^T v_j) \quad \text{s.t.} \quad \|v_i\| = 1, \;\; i = 1, 2, \ldots, N. \tag{33}$$

This problem is a relaxation of (32), because any feasible point y for (32) corresponds to a feasible point

$$v_i = (y_i, 0, 0, \ldots, 0)^T, \;\; i = 1, 2, \ldots, N,$$

for (33). The problem (33) can be formulated as an SDP by changing the variables $v_1, v_2, \ldots, v_N$ to a matrix $Y \in \mathbb{R}^{N \times N}$, such that

$$Y = V^T V, \;\; \text{where } V = [v_1, v_2, \ldots, v_N].$$

The constraints $\|v_i\| = 1$ can be expressed simply as $Y_{ii} = 1$, and, since $Y = V^T V$, we must have Y semidefinite. The transformed version of (33) is then

$$\max \tfrac{1}{2} \sum_{i<j} w_{ij}(1 - Y_{ij}) \quad \text{s.t.} \quad Y_{ii} = 1, \;\; i = 1, 2, \ldots, N, \;\; Y \succeq 0,$$

which has the form (18) for appropriate definitions of C and A_i, $i = 1, 2, \ldots, N$. We can recover V from Y by performing a Cholesky factorization. The final step of recovering an approximate solution to the original problem (32) is performed by choosing a random vector $r \in \mathbb{R}^N$, and setting

$$y_i = \begin{cases} 1, & \text{if } r^T v_i > 0, \\ -1, & \text{if } r^T v_i \leq 0. \end{cases}$$

A fairly simple geometric argument shows that the expected value of the solution so obtained has objective value at least .87856 of the optimal solution to (32).

Similar relaxations have been obtained for many other combinatorial problems, showing that is possible to find good approximate solutions to many NP-complete problems by using polynomial algorithms. Such relaxations are also useful if we seek *exact* solutions of the combinatorial problem by means of a branch-and-bound or branch-and-cut strategy. Relaxations can be solved at each node of the tree (in which some of the degrees of freedom are eliminated and some additional constraints are introduced) to obtain both a bound on the optimal solution and in some cases a candidate feasible solution for the original problem. Since the relaxations to be solved at adjacent nodes of the tree are similar, it is desirable to use solution information at one node to "warm start" the SDP algorithm at a child node. Mitchell [25] discusses an efficient strategy along these lines for the branch-and-cut strategy.

5. CONCLUSIONS

Interior-point methods remains an active and fruitful area of research, although the frenetic pace that has characterized the area has slowed in recent years. Linear programming codes have become mainstream and continue to undergo development, although they face continuing stiff competition from the simplex method. Semidefinite programming has proved to be an area of major impact. Applications to quadratic programming show considerable promise, because of the superior ability of the interior-point approach to exploit problem structure efficiently. The influence on nonlinear programming theory and practice has yet to be determined, even though substantial research has already been devoted to this topic. Use of the interior-point approach in decomposition methods appears promising, though no rigorous comparative studies with alternative approaches have been performed. Applications to integer programming problems have been tried by a number of researchers, but the interior-point approach is hamstrung here by competition from the simplex method with its superior warm-start capabilities.

Acknowledgments

I thank M. J. D. Powell and the other organizers of the IFIP TC7 '99 conference for arranging a most enjoyable and interesting meeting, and for a close reading of the paper which resulted in many improvements.

This work was supported by the Mathematical, Information, and Computational Sciences Division subprogram of the Office of Advanced Scientific Computing Research, U.S. Department of Energy, under Contract W-31-109-Eng-38.

References

[1] F. Alizadeh, J.-P.A. Haeberly and M.L. Overton (1998), Primal-dual interior-point methods for semidefinite programming: Convergence rates, stability, and numerical results, *SIAM Journal on Optimization*, 8, pp. 746–768.

[2] K.D. Andersen (1996), A modified Schur complement method for handling dense columns in interior-point methods for linear programming, *ACM Transaction on Mathematical Software*, 22(3), pp. 348–356.

[3] J.R. Birge and F. Louveaux (1997), *Introduction to Stochastic Programming*, Springer Series in Operations Research, Springer.

[4] J.R. Birge and L. Qi (1988), Computing block-angular Karmarkar projections with applications to stochastic programming, *Management Science*, 34, pp. 1472–1479.

[5] R.H. Byrd, M. Hribar and J. Nocedal (1997), An interior point algorithm for large scale nonlinear programming, OTC Technical Report 97/05, Optimization Technology Center.

[6] J. Castro (1998), A specialized interior-point algorithms for multicommodity network flows, Technical report, Statistics and Operations Research, Universitat Rovira i Virgili, Tarragona, Spain.

[7] A.R. Conn, N.I.M. Gould, D. Orban and P. Toint (1999), A primal-dual trust-region algorithm for minimizing a non-convex function subject to general inequality and linear equality constraints, Technical Report RAL-TR-1999-054, Atlas Centre, Rutherford Appleton Laboratory.

[8] L. Faybusovich (1997), Linear systems in Jordan algebras and primal-dual interior-point algorithms, *Journal of Computational and Applied Mathematics*, 86, pp. 149–175.

[9] A.V. Fiacco and G.P. McCormick (1968), *Nonlinear Programming: Sequential Unconstrained Minimization Techniques*, Wiley, New York. Reprinted by SIAM Publications, 1990.

[10] A. Forsgren and P.E. Gill (1998), Primal-dual interior-point methods for nonconvex nonlinear programming, *SIAM Journal on Optimization*, 8(4), pp. 1132–1152.

[11] D.M. Gay, M.L. Overton and M.H. Wright (1997), A primal-dual interior method for nonconvex nonlinear programming, Technical Report 97-4-08, Computing Sciences Research, Bell Laboratories, Murray Hill, NJ.

[12] M.X. Goemans and D.P. Williamson (1995), Improved approximation algorithms for maximum cut and satisfiability problems using semidefinite programming, *Journal of the Association for Computing Machinery*, 42(6), pp. 1115–1145.

[13] J. Goffin and J. Vial (1999), Convex nondifferentiable optimization: A survey based on the analytic center cutting plane method, Technical Report 99.02, Logilab, HEC, Section of Management Studies, University of Geneva.

[14] J. Gondzio (1996), Multiple centrality corrections in a primal-dual method for linear programming, *Computational Optimization and Applications*, 6, pp. 137–156.

[15] J. Gondzio and R. Kouwenberg (1999), High performance computing for asset liability management, Technical Report MS-99-004, Department of Mathematics and Statistics, The University of Edinburgh.

[16] J. Gondzio and R. Sarkissian (1996), Column generation with a primal-dual method, Technical Report 96.9, Logilab, HEC, Section of Management Studies, University of Geneva. Revised October, 1997.

[17] C. Gonzaga (1991), Large-step path-following methods for linear programming, *SIAM Journal on Optimization*, 1, pp. 268–279.

[18] N.I.M. Gould and P.L. Toint (1999), SQP methods for large-scale nonlinear programming, Technical Report RAL-TR-1999-055, Atlas Centre, Rutherford Appleton Laboratory.

[19] J. Haeberly, M.V. Nayakkankuppam and M.L. Overton (1999), Extending Mehrotra and Gondzio higher order methods to mixed semidefinite-quadratic-linear programming, to appear in Optimization Methods and Software.

[20] N. Karmarkar (1984), A new polynomial-time algorithm for linear programming, *Combinatorica*, 4, pp. 373–395.

[21] M.S. Lobo, L. Vandenberghe, S. Boyd and H. Lebret (1998), Applications of second-order cone programming, *Linear Algebra and Its Applications*, 248, pp. 193–228.

[22] S. Mehrotra (1992), Asymptotic convergence in a generalized predictor-corrector method, Technical Report, Dept. of Industrial Engineering and Management Science, Northwestern University, Evanston, Ill.

[23] S. Mehrotra and J.-S. Wang (1995), Conjugate gradient based implementation of interior point methods for network flow problems, Technical Report 95–70, Department of Industrial Engineering and Management Sciences, Northwestern University, Evanston, Ill.

[24] J.E. Mitchell (1997), Computational experience with an interior-point cutting plane algorithm, Technical report, Mathematical Sciences Department, Rensselaer Polytechnic Institute. Revised March 1999.

[25] J.E. Mitchell (1999), Restarting after branching in the SDP approach to MAX-CUT and similar combinatorial optimization problems, Technical report, Mathematical Sciences Department, Rensselaer Polytechnic Institute, Troy, NY.

[26] J.E. Mitchell, P.M. Pardalos and M. Resende (1998), Interior-point methods for combinatorial optimization, in D. Du and P.M. Pardalos, editors, *Handbook of Combinatorial Optimization*, volume 1, Kluwer Academic Publishers.

[27] R.D.C. Monteiro and S.J. Wright (1994), Local convergence of interior-point algorithms for degenerate monotone LCP, *Computational Optimization and Applications*, 3, pp. 131–155.

[28] Y.E. Nesterov and A.S. Nemirovskii (1994), *Interior Point Polynomial Methods in Convex Programming*, SIAM Publications, Philadelphia.

[29] Y.E. Nesterov and M.J. Todd (1997), Self-scaled barriers and interior-point methods for convex programming, *Mathematics of Operations Research*, 22, pp. 1–42.

[30] Y.E. Nesterov and M.J. Todd (1998), Primal-dual interior-point methods for self-scaled cones, *SIAM Journal on Optimization*, 8, pp. 324–362.

[31] S. Portnoy and R. Koenker (1997), The Gaussian hare and the Laplacian tortoise: Computability of squared-error vs. absolute-error estimators, *Statistical Science*, 12, pp. 279–300.

[32] D. Ralph and S.J. Wright (1996), Superlinear convergence of an interior-point method despite dependent constraints, Preprint ANL.MCS-P622-1196, Mathematics and Computer Science Division, Argonne National Laboratory, Argonne, Ill.

[33] C.V. Rao, S.J. Wright and J.B. Rawlings (1998), Application of interior-point methods to model predictive control, *Journal of Optimization Theory and Applications*, 99, pp. 723–757.

[34] J. Renegar (1999), A mathematical view of interior-point methods in convex optimization, Unpublished notes.

[35] C. Roos, J.-P. Vial and T. Terlaky (1997), *Theory and Algorithms for Linear Optimization : An Interior Point Approach*, Wiley-Interscience Series in Discrete Mathematics and Optimization, John Wiley and Sons.

[36] H. Takehara (1993), An interior-point algorithm for large-scale portfolio optimization, *Annals of Operations Research*, 45, pp. 373–386.

[37] M.J. Todd (1999), A study of search directions in primal-dual interior-point methods for semidefinite programming, Technical report, School of Operations Research and Industrial Engineering, Cornell University, Ithaca, NY.

[38] L. Vandenberghe and S. Boyd (1996), Semidefinite programming, *SIAM Review*, 38, pp. 49–95.

[39] H. Wolkowicz (2000), Semidefinite and Lagrangian relaxation algorithms for hard combinatorial problems, *Proceedings of the 19th IFIP TC7 Conference on System Modeling and Optimization, July, 1999, Cambridge*, Kluwer Academic Publishers (to appear).

[40] M.H. Wright (1992), Interior methods for constrained optimization, in *Acta Numerica 1*, pp. 341–407.

[41] S.J. Wright (1993), Interior point methods for optimal control of discrete-time systems, *Journal of Optimization Theory and Applications*, 77, pp. 161–187.

[42] S.J. Wright (1997), *Primal-Dual Interior-Point Methods*, SIAM Publications, Philadelphia.

[43] S.J. Wright (1999), Modified Cholesky factorizations in interior-point algorithms for linear programming, *SIAM Journal on Optimization*, 9, pp. 1159–1191.

[44] Y. Ye (1997), *Interior Point Algorithms : Theory and Analysis*, Wiley-Interscience Series in Discrete Mathematics and Optimization, John Wiley and Sons.

OTHER PAPERS THAT WERE PRESENTED AT THE CONFERENCE

Algorithms for Optimization Calculations

A Two-Dimensional Search in a Newton Method
Michael Bartholomew-Biggs (University of Hertfordshire, UK), Simge Kahvecioglu and Robert Keil.

Numerical Performance of the Optimization Algorithms in TOMLAB
Kenneth Holmström and Mattias Björkman (Mälardalen University, Sweden).

Quadratic Programming Problems Arising from the Simulation of Micromagnetism
P.T. Boggs (Sandia National Laboratories, USA), A.J. Kearsley and J.W. Tolle.

A Class of Globally Convergent Conjugate Gradient Methods
Yu-Hong Dai (Chinese Academy of Sciences) and Ya-Xiang Yuan.

The Adventures of a Simple Algorithm
Achiya Dax (Hydrological Service, Israel).

Implementation of an Infeasible Primal Dual Interior Point Method for Large-Scale Nonlinear Convex Programming
Olivier Epelly (University of Geneva, Switzerland), Jacek Gondzio and Jean-Philippe Vial.

Conveying Problem Structure to Optimization Algorithms
Robert Fourer and David M. Gay (Bell Labs, USA).

SQP Methods for Large-Scale Optimization
Philip E. Gill (University of California, San Diego, USA).

Quadratic and Multiple Cuts in the Analytic Center Cutting Plane Method
Jean-Louis Goffin (McGill University, Montreal, Canada).

M.J.D. Powell and S. Scholtes (Eds.), *System Modelling and Optimization: Methods, Theory and Applications.*

Solving Second Order Cone Programs in an Efficient and Numerically Stable Way
Donald Goldfarb (Columbia University, USA) and Katya Scheinberg.

Interior–Point Solver for Structured Linear Programs
Jacek Gondzio (University of Edinburgh, UK) and Robert Sarkissian.

Convergence Conditions and Linesearch Algorithms for the Polak-Ribiere Conjugate Gradient Method
Luigi Grippo (Università di Roma "La Sapienza", Italy) and Stefano Lucidi.

A Radial Basis Function Method for Global Optimization
Hans-Martin Gutmann (University of Cambridge, UK).

Computational Study of NOXCB: An Optimization Code for the Nonlinear Network Flow Problem with Linear Side Constraints
F. Javier Heredia (Universitat Politecnica de Catalunya, Spain).

FAIPA – Feasible Arc Interior Point Algorithm for Large Scale Nonlinear Constrained Optimization
Jose Herskovits (Federal University of Rio de Janeiro, Brazil).

A Class of Faster and Robust Gradient Methods for Unconstrained Minimisation Problems
Wenbin Liu (University of Kent, UK) and Yu-Hong Dai.

Globally Convergent Derivative Free Methods for Linearly Constrained Minimization
Stefano Lucidi (Università di Roma "La Sapienza", Italy), Marco Sciandrone and Paul Tseng.

A New Hybrid Method for Large-Scale Unconstrained Optimization
José Luis Morales (Instituto Technológico Autónomo de México, Mexico) and Jorge Nocedal.

Solving Large Bound-Constrained Optimization Problems
Chih-Jen Lin and Jorge J. Moré (Argonne National Laboratory).

Handling Ill-Conditioned Constraint Jacobians in Large Scale Optimization
M. Marazzi, J. Nocedal (Northwestern University, USA) and R. Waltz.

A Truncated Newton Method for Large Scale Constrained Optimization
Gianni Di Pillo, Stefano Lucidi and Laura Palagi (Università di Roma "La Sapienza", Italy).

Parallelization of Continuous Verified Global Optimization
N. Revol (University of Lille, France), Y. Denneulin, J.-F. Mehaut and B. Planquelle.

A Truncated Newton Method for Large Scale Unconstrained Optimization Based on a Planar Conjugate Gradient Scheme
Giovanni Fasano, Stefano Lucidi and Massimo Roma (Università di Roma "La Sapienza", Italy).

A Polynomial-Time Algorithm for Semidefinite Optimization Based on a New Search Direction
E. de Klerk, J. Peng, C. Roos (Delft University of Technology, Netherlands) and T. Terlaky.

Convergence Properties of a Regularization Scheme for Complementarity Constrained NLPs
Stefan Scholtes (University of Cambridge, UK).

Combining Search Directions in Nonlinear Optimization
H.D. Scolnik (University of Buenos Aires, Argentina), M. Marazzi and C. Ruscitti.

Parallel Global Optimization Algorithms Using Local Tuning to the Shape of the Objective Function
Yaroslav D. Sergeyev (University of Calabria, Rende, Italy).

Interior-Point Methods for Nonconvex Nonlinear Programming: Orderings and Higher Order Methods
David F. Shanno (Rutgers University, USA) and Robert J. Vanderbei.

A Computationally Efficient Feasible SQP Algorithm
Craig T. Lawrence and André Tits (University of Maryland, USA).

Interior-Point Methods for Semidefinite Programming
Michael J. Todd (Cornell University, USA).

Global Convergence of Trust-Region SQP–Filter Algorithms for General Nonlinear Programming
Roger Fletcher, Nicholas I.M. Gould, Sven Leyffer and Philippe Toint (FUNDP, Namur, Belgium).

Algorithms for Identification of Redundancies in Linear Programming Problems
V.R. Uthariaraj (Anna University, Chennai, India), T.R. Natesan and V. Sankaranarayan.

A Predictor–Corrector Algorithm for QSQP Combining Dinkin-type and Newton Centering Steps
Jia-Wang Nie and Ya-Xiang Yuan (Chinese Academy of Sciences).

Some Dogleg Path Trust Region Methods for Indefinite Working Matrices
Jianzhong Zhang (City University of Hong Kong) and Chengxian Xu.

Select and Clone: A Statistical Approach to Global Optimization
Antanas Žilinskas (Vytautas Magnus University, Vilnius, Lithuania).

Applications

Augmented Lagrangean Relaxation and Decomposition Applied to the Short-Term Hydrothermal Coordination Problem
Cesar Beltran (UPC–Barcelona, Spain) and F. Javier Heredia.

Optimization of Inventory Consolidation at Multiple Locations
Rosa H. Birjandi (i2 Technologies Advanced Solutions Center, White Plains, USA) and Dmitry V. Golovashkin.

Exploiting Dual Optimality Conditions in Solution of the Unit Commitment Problem
Erik Dotzauer (Mälardalen University, Sweden) and Hans F. Ravn.

A Parallel Optimization Approach to ATM Network Survivability
F.J. Gonzalez-Castano (Universidad de Vigo, Spain) and U.M. Garcia-Palomares.

Congestion Toll Pricing Models
Donald W. Hearn (University of Florida, USA) and Mehmet Bayram Yildirim.

Robust Algorithms for Chemical Equilibrium Analysis
Kenneth Holmström (Mälardalen University, Sweden) and Yulia Gel'.

Dynamic Multiobjective Heating Optimization
R.P. Hamalainen and J. Mantysaari (Helsinki University of Technology, Finland).

Probabilistic Long-Term Coordination of Hydrothermal Electric Power Generation using a Multi-Interval Bloom and Gallant's Model and Expected Hydrogeneration Contribution Functions
Narcís Nabona (Universitat Politecnica de Catalunya, Spain).

Optimal Policies for Batch Processes
Dulce C.M. Silva and Nuno M.C. Oliveira (University of Coimbra, Portugal).

Global Survivability in Telecommunication Networks
Raja Rébaï (INRIA, Rocqencourt, France), J. Frédéric Bonnans, Mounir Haddou and Abdel Lisser.

Optimization of Vehicle Fleet Compatibility
A.J.G. Schoofs (Eindhoven University of Technology, Netherlands), T. Nastic and H.G. Mooi.

Discrete Optimization

A Time–Space Network Model for the Berth Allocation Problem
Chuen-Yih Chen (National Cheng Kung University, Taiwan) and Tung-Wei Hsieh.

Approximate Dynamic Programming Approach for Integer Programming
Dimitris Bertsimas and Ramazan Demir (MIT Operations Research Center, USA).

A $O(n \log n)$ Procedure for Identifying a Class of Facets of the Knapsack Polytope from Strong Minimal Covers
Laureano F. Escudero, Araceli Garin (Universidad del Pais Vasco, Bilbao, Spain) and Gloria Perez.

Dual Ascent Methods for Some Classes of Integer Linear Programs
Manlio Gaudioso and Giovanni Giallombardo (Università della Calabria, Rende, Italy).

Proper Circular Arc Graph Models for Periodic Allocation Problems
Giuseppe Confessore, Paolo Dell'Olmo and Stefano Giordani (Università di Roma "Tor Vergata", Italy).

The Maximization of Submodular Functions: Old and New Proofs for the Correctness of Dichotomy Algorithms
Boris Goldengorin (University of Groningen, Netherlands) and Gert A. Tijssen.

Metaheuristic Approaches for Target-Radar Allocation
Magnus Hindsberger (Technical University of Denmark) and Rene Victor Valqui Vidal.

A Column Generation Approach to the Direct Flight Problem
Tore Grünert, Stefan Irnich (Rheinisch-Westfälische Technische Hochschule Aachen, Germany) and Hans-Jürgen Sebastian.

Quality of Solutions for Perturbed Combinatorial Optimization Problems
Marek Libura (Polish Academy of Sciences).

Minimizing the Makespan in a Two-Machine Flowshop with Fabrication and Assembly Operations
B.M.T. Lin (Ming Chuan University, Taiwan) and T.C.E. Cheng.

Train Scheduling in Public Rail Transport
Thomas Lindner (Technical University of Braunschweig, Germany) and Uwe T. Zimmermann.

Project Risk Management: Dealing with Random Influences in Scheduling and Project Control
Rolf H. Möhring (Technical University of Berlin, Germany).

Solving Hard Generalized Assignment Problems: Optimizing Methods versus Heuristic Approaches
Robert M. Nauss (University of Missouri, St. Louis, USA).

Improved Dynamic Programming for Knapsack Problems
Ulrich Pferschy (University of Graz, Austria).

Perfect Distribution Phenomenon in Combinatorics and Its Application to System Optimization
V. Riznyk (University of Technology and Agriculture, Bydgoszcz, Poland).

A Genetic Algorithm Approach for Large Scale Resource Constrained Production Scheduling Problems
Olli Kamarainen, Vesa Ek, Sampo Ruuth (Helsinki University of Technology, Finland) and Kimmo Nieminen.

Variations of Column Generation and Airline Crew Scheduling Problems
Shu-Hui Liou, Ruey-Lin Sheu (National Cheng Kung University, Taiwan) and Weichung Wang.

Mixed Linear and Semidefinite Programming for Combinatorial and Quadratic Optimization
Steven J. Benson, Yinyu Ye (University of Iowa, USA) and Xiong Zhang.

A Column Generation Approach to Engine Scheduling
Marco E. Lübbecke and Uwe T. Zimmermann (Technical University of Braunschweig, Germany).

Worst–Case Performance of Priority Algorithms for Scheduling on Parallel Processors
Gaurav Singh and Yakov Zinder (University of Technology, Sydney, Australia).

Distributed Parameter Systems

Influence of Boundary Variations in the Stationary Navier-Stokes Equations with Small Viscosity
Sebastien Boisgerault (Ecole des Mines de Paris, France) and Jean-Paul Zolésio.

Boundary Control Problems with Indefinite Quadratic Cost Functionals
Francesca Bucci (Università degli Studi di Firenze, Italy).

Differentiation of the Solution of the Wave Equation, with Respect to a Non-Smooth Domain
John Cagnol (Ecole des Mines de Paris, France) and Jean-Paul Zolésio.

Shape-Sensitivity Analysis for Nonlinear Heat Convection
(presented by J. Cagnol)
R. Dziri (Université de Tunis, Tunisia).

Mapping Method in the Optimal Shape Design Problems Governed by Hemivariational Inequalities
Lezek Gasiński (Jagiellonian University, Krakow, Poland).

Existence of Free-Boundary for a Two Non-Newtonian Fluids Problem
Nicolas Gomez (Ecole des Mines de Paris, France) and Jean-Paul Zolésio.

Proper Orthogonal Decomposition in Control of Fluid Flow
Michael Hinze (Technische Universität Berlin, Germany).

Boundary Stabilization of Full von Karman Models with Thermoelasticity
Irena Lasiecka (University of Virginia, USA).

Optimization Methods for Solving Elliptic Control Problems with Control and State Constraints: Boundary and Distributed Control
Helmut Maurer (Universität Münster, Germany) and Hans D. Mittelmann.

Control Problems for Semilinear Parabolic Equations with Controls Localized on Manifolds
Phuong Anh Nguyen (Université Paul Sabatier, Toulouse, France) and Jean-Pierre Raymond.

Feedback Laws for the Optimal Control of Parabolic Variational Inequalities
Catalin Popa (Universitatea "Al.I.Cuza", Iasi, Romania).

Application of Special Smoothing Procedure to Numerical Solutions of Inverse Problems for Real 2-D Systems
Edward Rydygier (Warsaw University of Technology, Poland) and Zdzislaw Trzaska.

Spectral Operators Generated by Damped Hyperbolic Equations and their Applications to Control Problems
Marianna A. Shubov (Texas Tech University, USA).

Shape Sensitivity Analysis in Domains with Cracks
Gilles Fremiot and Jan Sokolowski (Université Henri Poincaré Nancy I, France).

Inverse-Type Inequalities for Multi-Dimensional Hyperbolic and Schrödinger Equations with Variable Coefficients
Roberto Triggiani (University of Virginia, USA).

Some New Problems Occuring in Modelling of Oxygen Sensors
J.P. Yvon (INSA, Rennes, France), J. Henry and A. Viel.

The Topological Derivative Method and Artificial Neural Networks for Numerical Solution of Shape Inverse Problems
L. Jackowska-Strumillo, J. Sokolowski and A. Zochowski (Polish Academy of Sciences).

Modelling

Maximizing a Class of Separable Non-Concave Functions under a Linear Constraint
Djangir A. Babayev (US West Advanced Technologies, Boulder, USA).

Identification of Hammerstein/Wiener Dynamic Models
Aldo Balestrino and Andrea Caiti (University of Siena, Italy).

Dynamics of Genetic Algorithms: Theory and Applications
Alonzo Jarman (Queen Mary and Westfield College, London, UK).

Efficient Training of Neural Networks for Pattern Recognition
Francesco Lampariello (CNR, Italy) and Marco Sciandrone.

An Implementation of Newton-Like Multiplier Methods on Nonlinear Networks with Nonlinear Inequality Side Constraints
Eugenio Mijangos (Basque Country University, Spain).

Modeling and Computational Considerations for Design Optimization of Large Engineering Systems
Panos Papalambros (University of Michigan, USA), Nestor Michelena and Sigurd Nelson III.

New Challenges in Ecological Modelling: Toward the Explanation of "Patchy" Spatio-Temporal Dynamics
Sergei Petrovskii (Shirshov Institute of Oceanology, Moscow, Russia) and Horst Malchow.

Neural Networks and their Application in Finance - A Numerical View
Ekkehard Sachs and Michaela Schulze (University of Trier, Germany).

Globally Convergent Decomposition Algorithms for Training RBF Neural Networks
Celina Buzzi, Luigi Grippo and Marco Sciandrone (CNR, Italy).

A Kind of Intelligent Multi-Layer Frequency Planning Method
Li Xu, Song Junde (Beijing University of Post & Telecommunications, China) and Hou Jindong.

Using Genetic Algorithms to Solve Difficult Goal Programming Problems
Mehrdad Tamiz (University of Portsmouth, UK).

Optimal Control

Convergence of SQP Methods for Discretizations of Nonlinear Optimal Control Problems with Mixed Control-State Constraints
Walter Alt (University of Jena, Germany).

Sensitivity Analysis and Real-Time Control of State Constrained Control Problems
Dirk Augustin (WWU Münster, Germany) and Helmut Maurer.

Dynamic Optimization of a Class of Multistage System using a Hybrid Stochastic-Deterministic Method
Eva Balsa Canto (CSIC, Vigo, Spain), Vassilios S. Vassiliadis and Julio R. Banga.

Discrete Relaxed Method for Nonconvex Optimal Control and Variational Problems
I. Chryssoverghi (National Technical University, Athens, Greece), J. Coletsos and B. Kokkinis.

On the Use of Nonlinear Programming for Modelling and Control of Natural Resources Exploitation
Joao Lauro Dorneles Facó (Universidade Federal do Rio de Janeiro, Brazil).

Upper Lipschitz Stability of Stationary Solutions in Nonlinear Optimization
Diethard Klatte (Universität Zürich, Switzerland).

A LMI-Based Algorithm for Designing Suboptimal Static $\mathcal{H}_2/\mathcal{H}_\infty$ Output Feedback Controllers
Friedemann Leibfritz (University of Trier, Germany).

Convergence of the Euler Approximation to State and Control Constrained Optimal Control Problems
A.L. Dontchev, W.W. Hager and K. Malanowski (Polish Academy of Sciences).

An Algorithm for Computing Optimal Controls for Systems Described by Mixed Equality-Inequality Differential-Algebraic Systems
M. Bell and R.W.H. Sargent (Imperial College, London, UK).

An Optimal Control Model for Collision Avoidance in Air Navigation
Dick Sutherland (Dalhousie University, Canada).

An Optimal Control Problem Related to Wastewater Treatment
Aurea Martinez, Carmen Rodriguez and Miguel E. Vazquez-Mendez (University of Santiago de Compostela, Spain).

Wellposedness by Perturbations of Optimal Control Problems
T. Zolezzi (Università di Genova, Italy).

Optimization Methods in Engineering Design

Maximal Slope Stability of Soils by Use of Reinforcing Material: Numerical Aspects
Jean Deteix (Centre de Recherches Mathematiques, Canada).

Parallel Solution of Contact Problems
Zdeněk Dostál, Francisco Gomes (University of Campinas, Brazil) and Sandra Santos.

Cascading – An Approach to Robust Structural Design
Michal Kočvara (University of Erlangen, Germany) and Jochem Zowe.

Space Mappings for Optimal Engineering Design
John W. Bandler and Kaj Madsen (Technical University of Denmark).

Domain Optimization for Unilateral Problems by an Embedding Domain Method
A. Myslinski (Warsaw University of Technology, Poland).

A New Approach for the Optimization of Dynamically Loaded Structures with respect to the Reliability
H. Weber (Ruhr-University Bochum, Germany), D. Hartmann and A. Lenzen.

Stochastic Optimization

Multiserver Retrial Queues: Optimization of the Retrial Rate
J.R. Artalejo (Universidad Complutense de Madrid, Spain).

Modeling and Estimation of Stochastic Volatility and Application to Option Pricing
Shin-Ichi Aihara and Arunabha Bagchi (University of Twente, Netherlands).

Symbolic Computation of Stochastic Sensitivities in Engineering Design
H.P. Wynn and R.A. Bates (University of Warwick, UK).

Scenario-Based Stochastic Programs: Changes of the Probability Distribution
Jitka Dupačová (Charles University, Prague, Czech Republic) and Werner Römisch.

Stochastic Optimization of Catastrophic Risk Portfolios
Y.M. Ermoliev (IIASA, Laxenburg, Austria), T.Y. Ermolieva, G.J. MacDonald and V.I. Norkin.

Optimization of Systems with Discrete Design Variables and Stochastic Responses by Sequential Linear Integer Programming
S.J. Abspoel, L.F.P. Etman (Eindhoven University of Technology, Netherlands), A.J.G. Schoofs and J.E. Rooda.

Solving a Sequence of Successively Discretized Multistage Stochastic Linear Programs
Karl Frauendorfer and Gido Haarbrücker (University of St. Gallen, Switzerland).

S–Estimation in Regression and Stochastic Optimization
Keith Knight (University of Toronto, Canada).

Approximate Solution of Stochastic Programs with Probability Functionals
Riho Lepp (Tallinn Technical University, Estonia).

Busy Period of an M/G/1 Retrial Queue: Optimization of Entropy Functionals
M.J. Lopez-Herrero (Universidad Complutense de Madrid, Spain).

Adaptive Control of Robots by Means of Stochastic Programming Techniques
Kurt Marti (Universität der Bundeswehr, Munich, Germany).

On the Numerical Solution of Jointly Chance Constrained Problems
János Mayer (University of Zürich, Switzerland).

A Stochastic Control Problem for Renewable Resource Exploitation
Sara Pasquali (Università degli Studi di Padova, Italy).

Concentrator Location Problem with Stochastic Demands
Takayuki Shiina (Central Research Institute of Electric Power Industry, Japan).

Constrained Global Optimization Using Stochastic Differential Equations on Manifolds
Annelie Stöhr (TU Munich, Germany).

Approximation of a Stochastic Integer Program with Applications in Telecommunications
Shane Dye, Leen Stougie and Asgeir Tomasgard (SINTEF Industrial Management, Norway).

On Newton's Methods for Multi-Stage Stochastic Nonlinear Programming
Gongyun Zhao (National University of Singapore).

Theory

An Interior Linearization Method of Computing Fixed Points of Equilibrium Programming Problems
Anatoly Antipin (Russian Academy of Sciences).

Method of Chebyshev Points of Simplices in Convex Programming
Valerian P. Bulatov (Russian Academy of Sciences) and Tatjana I. Belykh.

On the Resolution of the Generalized Nonlinear Complementarity Problem
R. Andreani, Ana Friedlander (State University of Campinas, Brazil) and Sandra A. Santos.

Similarity Transformation Approach to Identifiability Results: An Algorithm and Some Comments
Lilianne Denis-Vidal, Ghislaine Joly-Blanchard (University of Technology of Compiegne, France), Celine Noiret and Michel Petitot.

On Some Relations Between Generalized Partial Derivatives and Convex Functions
Peter Recht (University of Dortmund, Germany).

A Rate Independent Evolution Variational Inequality with a Nonlinear Elliptic Part
A.H. Siddiqi (King Fahd University of Petroleum and Minerals, Dhahran, Saudi Arabia) and M. Brokate.

An Efficient Optimization Algorithm for Estimating the Norm of Inverse Lyapunov Operators
Vasile Sima (Katholieke Universiteit Leuven, Belgium), Petko Petkov and Sabine Van Huffel.

New Characterizations of Weak Sharp and Strict Local Minimizers in Nonlinear Programming
Marcin Studniarski (University of Łódź, Poland) and Monika Studniarska.

Proximal Methods for Variational Inequalities with Composed Monotone Operators
A. Kaplan and R. Tichatschke (University of Trier, Germany).

Application of Wavelet Transform to the Determination of Resonance Frequencies
Mariusz Ziólko (University of Mining and Metallurgy, Krakow, Poland), Wojciech Batko and Tomasz Korbiel.